Advances in Industrial Control

More information about this series at http://www.springer.com/series/1412

Ioan Doré Landau
Tudor-Bogdan Airimitoaie
Abraham Castellanos-Silva
Aurelian Constantinescu

Adaptive and Robust Active Vibration Control

Methodology and Tests

Springer

Ioan Doré Landau
Department of Automatic Control
GIPSA-LAB (Grenoble INP/CNRS/UJF/Stendhal)
Saint Martin d'Heres
France

Tudor-Bogdan Airimitoaie
IMS-lab (University of Bordeaux, Bordeaux INP, CNRS) UMR 5218
Talence
France

Abraham Castellanos-Silva
Grenoble
France

Aurelian Constantinescu
Saint-Lazare, QC
Canada

ISSN 1430-9491 ISSN 2193-1577 (electronic)
Advances in Industrial Control
ISBN 978-3-319-41449-2 ISBN 978-3-319-41450-8 (eBook)
DOI 10.1007/978-3-319-41450-8

Library of Congress Control Number: 2016945847

Printed on acid-free paper

This Springer imprint is published by Springer Nature
The registered company is Springer International Publishing AG Switzerland

Ce qui est simple est toujours faux
Ce qui ne l'est pas est inutilisable

Paul Valéry
Mauvaises Pensées

Series Editors' Foreword

The series *Advances in Industrial Control* aims to report and encourage technology transfer in control engineering. The rapid development of control technology has an impact on all areas of the control discipline. New theory, new controllers, actuators, sensors, new industrial processes, computer methods, new applications, new design philosophies…, new challenges. Much of this development work resides in industrial reports, feasibility study papers and the reports of advanced collaborative projects. The series offers an opportunity for researchers to present an extended exposition of such new work in all aspects of industrial control for wider and rapid dissemination.

It is easy to find examples in everyday life where one experiences the effects of vibration and noise. In fact, there are so many applications where vibration control and/or noise suppression is required that it is difficult to make some form of classification. Virtually every application, domestic and industrial, involving a motor, engine or turbine, for its drive system will have a problem with vibration and/or noise and will invariably require mechanisms or techniques to suppress them. Over a number of years this diversity of applications has been investigated from a control system's point of view by Prof. Ioan Landau and his team of researchers and engineering consultants. The comprehensive results of this study and research are presented in this *Advances in Industrial Control* monograph *Adaptive and Robust Active Vibration Control: Methodology and Tests* by Ioan Doré Landau, Tudor-Bogdan Airimitoaie, Abraham Castellanos-Silva and Aurelian Constantinescu.

The advantage of a systems-engineering approach to any diverse set of similar industrial problems, in a wide range of applications, is the emergence of a generic study framework that rises above the specific applications. This also allows the application of solution methods from the now well-developed and extensive toolbox of control design methods. At a later stage, such generic solutions can then be tailored for particular application studies.

Professor Landau and his team followed this approach with a focus on active vibration control and active noise control, identifying and classifying along the way the type of system disturbances for which rejection is required. In many ways the focus on the signal structure of the disturbance represents a way of classifying each type of application. This in-depth understanding of the disturbances enables Prof. Landau and his team to construct solutions from the robustness and adaptive control design paradigms.

An important contribution reported in the monograph is the three physical benchmark laboratory units at the GIPSA-LAB Grenoble. These benchmark installations are described in the monograph and are used to test different aspects of vibration control. The monograph also archives the models and data so that other researchers in the field may use them for further study.

What can the reader expect from this excellent monograph? Professor Ioan Doré Landau is an eminent control engineer who has had a long and fruitful career in the control community, notably acting as the Research Director at the National Centre for Scientific Research at Grenoble. His books, monographs and papers are known for their clarity, and scholarship. He and his team of researchers have produced a monograph that continues this trademark characteristic in the presentation of control concepts and results. The monograph is complete with laboratory experimental results and the theory is supported by instructive appendices. It is an exemplar for the *Advances in Industrial Control* monograph series.

Michael J. Grimble
Michael A. Johnson
Industrial Control Centre
University of Strathclyde
Glasgow, Scotland, UK

Preface

Attenuation of vibration and noise constitutes a growing concern in today's human activities. For more than 45 years, it was realized that passive attenuation of vibration and noise via dedicated absorbers has limits and the concepts of active vibration and noise control have emerged. Active vibration and noise control are strongly related to control methodology even if in the past the control community was not the driving force in this field. Almost from the beginning, the uncertainties and changes in the characteristics of the environment (vibrations, noise, system dynamics) have prompted the idea of using an adaptive approach in active vibration or noise control. Addressing some of these issues from a robustness point of view is a much more recent tendency in the field. Practical experience has shown also the limitations of using only physical models for designing active vibration or noise control systems bringing to light the need of dynamic model identification directly from input/output data.

The aim of this book is to approach the design of active vibration control systems from the perspective of today's control methodology. In that sense the first objective is to formulate from the beginning the various design problems encountered in active vibration control as control problems and search for the most appropriate control tools to solve them. Experimental validation of the proposed solutions on relevant test benches is another issue addressed in this book. To make these techniques widely accepted, an appropriate presentation should be given, eliminating theoretical developments unnecessary for the users (which can be found elsewhere) and focusing on algorithms' presentation and their use. Nevertheless, the proposed solutions cannot be fully understood and creatively exploited without a clear understanding of the basic concepts and methods and so these are given in-depth coverage. The book is mainly based on the work done in a number of Ph.D. theses prepared at Gipsa-lab (INPG/UJF/CNRS), Grenoble, France:

- A. Constantinescu "Robust and adaptive control of an active suspension" [1];
- M. Alma "Adaptive rejection of disturbances in active vibration control" [2];

- T.B. Airimitoaie “Robust control and tuning of active vibration control systems” [3]; and
- A. Castellanos-Silva “Feedback adaptive compensation for active vibration control in the presence of plant parameter uncertainties” [4];

as well as on the results of an international experimental benchmark on adaptive feedback vibration attenuation [5].[1]

All the methods and algorithms proposed in the book have been thoroughly validated experimentally on three test benches (designed by Mathieu Noé from Paulstra—Vibrachoc, Paris) and located at the Gipsa-lab (INPG/UJF/CNRS) in Grenoble, France.

The idea of writing this book arose when I was asked to present a tutorial on control tools for active vibration control at the 4ème Colloque francophone “Analyse Vibratoire Expérimentale”, Blois, France, November 2014 (Chairman: Roger Serra, INSA Centre Val de Loire). On that occasion, I listed the concepts, methods and algorithms that have been used to provide solutions for active damping, feedback and feedforward attenuation of vibration. All these concepts and methods, which form the basis of the solutions proposed, are taught separately in various control courses or can be found in various books, so it appeared reasonable to try to bring them together and present them accessibly for those interested in using modern control concepts in active vibration control. With this knowledge to hand, the various solutions proposed for active vibration control can be easily understood and used. The need for including experimental results in order to allow readers to assess the potential of the various solutions is obvious.

Three major problems are addressed in the book:

- Active damping (for improving the performance of passive absorbers);
- Adaptive feedback attenuation of single and multiple tonal vibrations; and
- Feedforward and feedback attenuation of broad-band vibrations.

With few exceptions, the analytical details have been skipped and the reference to the appropriate journal papers has been made. The focus is on enhancing motivations, algorithms presentation and experimental evaluations.

Once I had a clear view of how this book should be, I solicited the collaboration of Tudor-Bogdan Airimitoaie, Abraham Castellanos-Silva and Aurelian Constantinescu in order to realize it.

Website

Complementary information and material for teaching (simulators, algorithms and data files) can be found on the book website: http://www.landau-adaptivecontrol.org/

---

[1]http://www.gipsa-lab.grenoble-inp.fr/~ioandore.landau/benchmark_adaptive_regulation

Expected Audience

The book may be considered as the basis of a course for graduate students in mechanical, mechatronic, industrial electronic, aerospace and naval engineering.

Part of the book may be used to illustrate the applicability of various graduate control courses (system identification, adaptive control, robust control).

The book is of interest for practising engineers in the field of active vibration control wishing to acquire new concepts and techniques well validated in practice.

The book is also of interest for the people concerned with active noise control, since the techniques presented can, to a large extent, be used for active noise control too. Researchers in the field of active vibration control may also find inspiring material that opens paths toward new developments.

About the Content

The book is divided into six parts. The introductory part (Chaps. 1 and 2) presents the problems addressed in the book and the test benches used for experimental validation.

The second part is dedicated to the presentation of the control techniques used effectively in active vibration control. Chapter 3 discusses the discrete-time model representation used throughout the book. Chapter 4 is dedicated to the presentation of the parameter adaptation algorithms that will be used throughout the book. Chapter 5 gives a compact presentation of system identification techniques focusing on the specific algorithms used in active vibration control. Chapter 6 illustrates the use of these identification techniques for identifying the dynamic models of the three test benches already presented in Chap. 2. Chapter 7 reviews basic methods for the design of digital controllers that have been used in active vibration control. Chapter 8 provides effective solutions for identification in closed-loop operation allowing the improvement of the dynamic models identified in open-loop operation or re-tuning of the controller. Chapter 9 addresses the problem of controller order reduction because the result of the design is often a high-order controller; since on the one hand the models of the system are of high dimension and on the other the robustness constraints contribute to the increase of the order of the controller.

The third part is dedicated to the problem of active damping (Chap. 10). The design aspects and the experimental evaluation are discussed in detail.

The fourth part is concerned with the robust and adaptive attenuation of vibrations by feedback. Chapter 11 treats the problem of robust feedback attenuation of narrow-band (tonal) disturbances subject to limited frequency variations. Chapter 12 introduces the basic algorithm for adaptive attenuation of narrow-band disturbances. Experimental evaluations on two test benches are presented. Performance comparison of robust and adaptive solutions is provided. Chapter 13 is specifically dedicated to the problem of attenuating multiple unknown and time-varying vibrations. Two algorithms specifically developed for this problem will be presented and their performance and complexity will be compared with those of the basic algorithm presented in Chap. 12.

In the fifth part of the book we consider feedforward compensation of disturbances, which has to be used when the bandwidth of disturbances (vibrations) is such that the performance/robustness compromise cannot be conveniently satisfied by feedback alone. Chapter 14 examines the linear design, which has to be done from data (since the model of the disturbance is unknown and it should be identified from data). Chapter 15 provides adaptive solutions for infinite impulse response (IIR) feedforward compensation as well as experimental results illustrating the performance of such systems in various situations. Chapter 16 provides adaptive solutions for Youla–Kučera feedforward compensator configuration. Experimental comparison between the two configurations concludes the chapter.

Part six of the book contains five appendices. Appendix A is dedicated to the *generalized stability margin* and the *Vinnicombe distance* between two transfer functions: two very useful concepts in system identification in closed-loop operation and controller reduction. Appendix B details the numerically safe implementation of parameter adaptation algorithms in real time. Appendix C details the derivation of an adaptation algorithm used in Chap. 13 for rejection of narrow-band disturbances. Appendix D details the derivation of explicit equations for the residual force or acceleration in the context of adaptive feedforward compensation. These equations allow the straightforward definition of the appropriate parameter adaptation algorithm. Finally, Appendix E gives the details and experimental evaluation of an *integral plus proportional* parameter adaptation algorithm (IP-PAA adaptation), which adds a "proportional" component to the classical "integral" parameter adaptation algorithms.

There are 271 references disseminated at the end of each chapter.

Pathways Through the Book

For a course on the subject, the Chaps. 1–9 have to be covered first followed, in no particular order, by Parts III, IV or V.

For experts in digital, robust and adaptive control, Chaps. 3, 4, 5, 7, 8, and 9 can be skipped and again Parts III, IV and V can be read in any order.

An image of the applicability of the results can be easily obtained by reading Chap. 2 and the sections concerning experimental results in Chaps. 10–16.

Figure 1 gives a view of the interdependence of the various chapters.

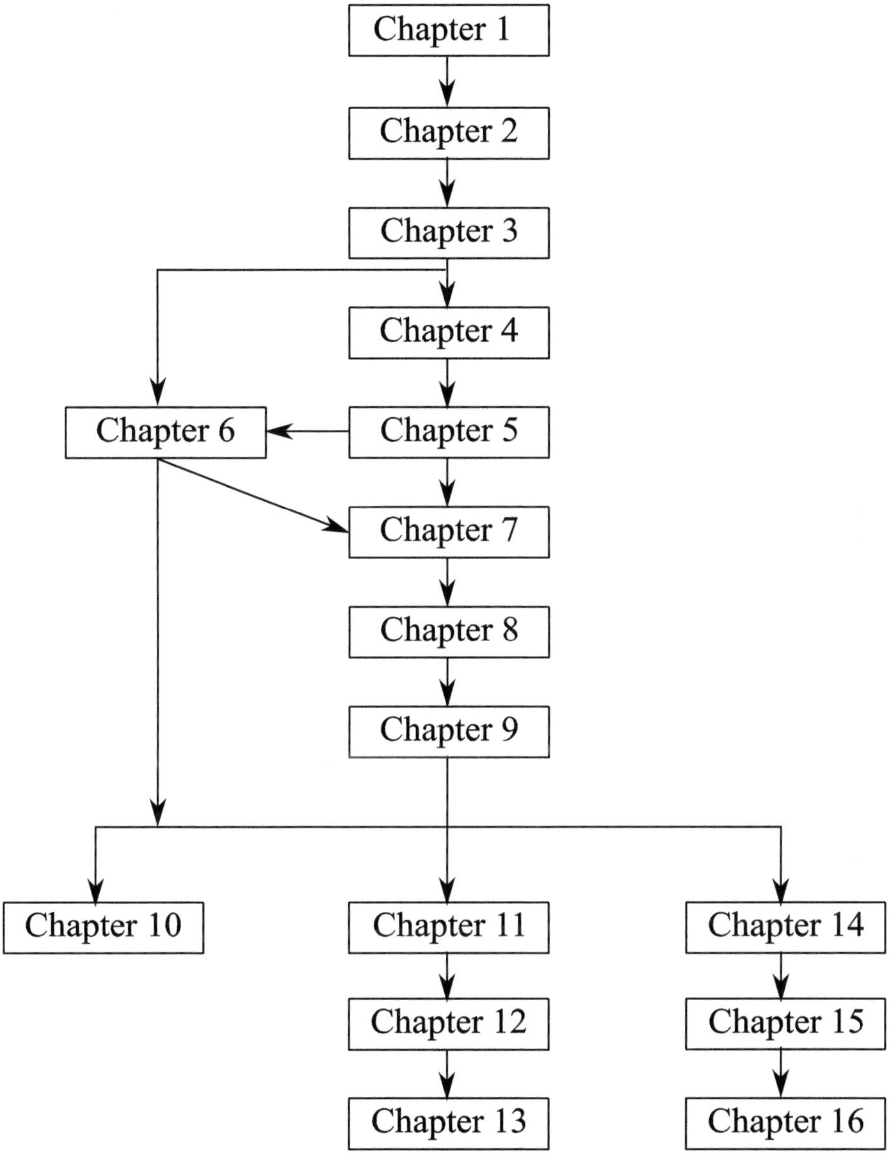

Fig. 1 Pathways through the book

Acknowledgements

I would like first to thank M. Noé, who on the one hand designed the bench tests and on the other hand pointed out the pertinent problems to be solved in active vibration control. The long, steady interaction between us was a major driving factor in our research on active vibration control.

I would like to thank M. Alma, whose contributions to the field of active vibration control are reflected in the book.

I would like to thank D. Rey, G. Buche and A. Franco for their involvement in the research project and the technical support in operating the test benches.

Over the years, working in the area of active vibration control, I have had the privilege of interacting with a number of colleagues among whom I would like to mention: B.D.O. Anderson, S. Aranovski, F. Ben Amara, R.B. Bitmead, D. Bonvin, M. Bodson, R.A. de Callafon, X. Chen, L. Dugard, T. Hélie, P. Ioannou, C.R. Johnson, A. Karimi, J. Langer, F.L. Lewis, J.J. Martinez, G. Ruget, R. Serra, M. Tomizuka, S. Valentinotti, Z. Wu. I would like to express my appreciation for their contributions.

The long-term support of the Centre National de la Recherche Scientifique (CNRS) and of the GIPSA-LAB Grenoble (Institut National Polytechnique de Grenoble, Université Joseph Fourier, CNRS) is gratefully acknowledged.

I would also like to thank Oliver Jackson from Springer, whose enthusiasm and professionalism has helped us to finalize this book.

Grenoble, France Ioan Doré Landau
April 2016

References

[1] Constantinescu, A.: Commande robuste et adaptative d'une suspension active. Thèse de doctorat, Institut National Polytechnique de Grenoble (2001)

[2] Alma, M.: Rejet adaptatif de perturbations en contrôle actif de vibrations. Ph.D. thesis, Université de Grenoble (2011)

[3] Airimitoaie, T.B.: Robust design and tuning of active vibration control systems. Ph.D. thesis, University of Grenoble, France, and University "Politehnica" of Bucharest, Romania (2012)

[4] Castellanos-Silva, A.: Compensation adaptative par feedback pour le contrôle actif de vibrations en présence d'incertitudes sur les paramétres du procédé. Ph.D. thesis, Université de Grenoble (2014)

[5] Landau, I.D., Silva, A.C., Airimitoaie, T.B., Buche, G., Noé, M.: Benchmark on adaptive regulation—rejection of unknown/time-varying multiple narrow band disturbances. European Journal of Control **19**(4), 237—252 (2013). http://dx.doi.org/10.1016/j.ejcon.2013.05.007#1

Contents

Acronyms

ADC	Analog to digital converter
AF-CLOE	Adaptive filtered closed-loop output error algorithm
AFOLOE	Open-loop output error with adaptive filtered observations
ANC	Active noise control
ANVC	Active noise and vibration control
ARMA	Autoregressive moving average
ARMAX	Autoregressive moving average with exogenous input
ARX	Autoregressive with exogenous input
a.s.	Asymptotically stable
AVC	Active vibration control
b	Vinnicombe stability margin
CLIM	Closed-loop input matching algorithm
CLOE	Closed-loop output error recursive algorithm
CLOM	Closed-loop output matching algorithm
DAC	Digital to Analog Converter
ELS	Extended least squares
F-CLOE	Filtered closed-loop output error algorithm
FIR	Finite impulse response
FOLOE	Open-loop output error with filtered observations
FULMS	Filtered-U least mean squares algorithm
FUPLR	Filtered-U pseudo linear regression algorithm
FUSBA	Filtered-U stability based algorithm
FXLMS	Filtered-X least mean squares algorithm
GPC	Generalized predictive control
H_∞	H infinity control
IIR	Infinite impulse response
IMP	Internal model principle
I-PAA	"Integral" parameter adaptation algorithm
IP-PAA	"Integral + proportional" parameter adaptation algorithm
ITAE	Integral over time of absolute value of error

LMS	Least mean squares
LQC	Linear-quadratic control
LQR	Linear-quadratic regulator
LS	Least squares
LTI	Linear time-invariant
MBC	Model based control
OE	Output error
OEFC	Output error with fixed compensator
OLOE	Open-loop output error
PAA	Parameter adaptation algorithm
PRBS	Pseudo-random binary sequence
PSD	Power spectral density
Q	Youla–Kučera filter
RELS	Recursive extended least squares
RLS	Recursive least squares
RML	Recursive maximum likelihood algorithm
RS	Polynomial digital controller
t	Normalized sampling time (except in Sect. 3.1.2 where it is denoted by k)
T_s	Sampling period
ν-gap	Vinnicombe gap
X-CLOE	Extended closed-loop output error algorithm
XOLOE	Output error with extended prediction model algorithm
YK	Youla–Kučera
ZOH	Zero order hold

Part I
Introduction to Adaptive and Robust Active Vibration Control

Chapter 1
Introduction to Adaptive and Robust Active Vibration Control

1.1 Active Vibration Control: Why and How

Vibrations are present almost everywhere and their presence often causes problems for the operation of the various systems. Vibrations are disturbances that affect a system (see also [1, 2]). They have a variety of origins, some examples of which are: geological disturbances, traffic, mechanical engines, motor operation, and electrical drives.

High-precision measurements, high-precision drives (like memory disc drives, Blu-ray, DVD and CD drives), photo and video cameras, and stabilized platforms require solutions for keeping the effect of these disturbances at a very low level. In transport systems (ground, water, or air) the vibrations can have a destructive effect and they also affect the comfort of the passengers. Reduction of vibration effects is mandatory in manufacturing systems and stabilized platforms in which the tolerances imposed on their various tasks are of a similar magnitude to the effect of the vibration.

The use of passive vibration absorbers (fluid dampers, elastomers, and so on) is a well-known solution but unfortunately not sufficient in many cases for assuring the desired level of attenuation over the desired frequency range. Semi-active (semi-passive) dampers, for which the properties of the absorbing material can be changed allow an improvement of the expected performance in some cases; however, when high-performance attenuation is needed an active control solution should be considered.

From a mechanical point of view, one makes a distinction between *active vibration isolation* and *active vibration control* (AVC). In active vibration isolation, an active damper (suspension) is inserted between the source of vibration and the mass that has to be isolated. In active vibration control a force, which will counteract the effect of the incoming vibrations, is generated through an actuator driven from the available measurements (force or acceleration) of disturbance. In short, the compensating force should be of the same magnitude but in opposite phase.

An active hydraulic isolation system is presented in Fig. 1.1. The size of the main chamber of the elastomer cone located between the source of vibrations and the

I.D. Landau et al., *Adaptive and Robust Active Vibration Control*,
Advances in Industrial Control, DOI 10.1007/978-3-319-41450-8_1

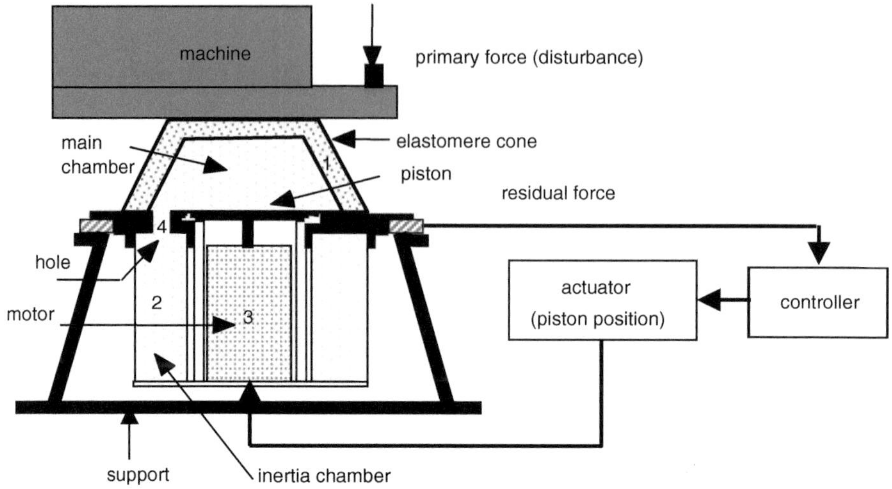

Fig. 1.1 Active isolation system used to reduce the effect of vibrations on the chassis

chassis is modified by the effect of a piston controlled through a linear motor (which develops a force). An AVC system is shown in Fig. 1.2. In this example, the objective is to reduce the vibrations created by the motor at the level of the chassis. By means of actuators, an opposite vibration force is introduced[1] to the chassis with a shift phase of 180°.

Vibrations are usually measured by accelerometers or force sensors. Actuators are typically active dampers, inertial motors (working on the same principle as loudspeakers), piezoelectric actuators.

From a control point of view, active vibration control and active vibration isolation are almost identical problems that can be solved using *feedback control* or *feedforward disturbance compensation* if information on the disturbance is available.

Another problem, related to active isolation, is the *active damping*. Despite the fact that they provide good attenuation over a wide band of frequencies, passive dampers have a significant resonant peak at a certain frequency in the range of operation. Adding active damping by feedback will correct this feature. Figure 1.3 illustrates this behaviour by showing the power spectral density (PSD) of the residual force without and with active damping. One can see that the resonance effect around 30 Hz has been attenuated with negligible deterioration of the damping performances at other frequencies. Active damping consists of damping the corresponding vibration mode without changing its frequency.[2]

In active vibration (isolation) control one distinguishes between two "paths":

---

[1]In these two examples the actuators are driven by a feedback controller, but in other cases the actuator can be driven by a feedforward compensator.

[2]Light mechanical structures are characterized by multiple low damped vibration modes. These modes have to be damped since on the one hand they can become a source of vibration and on the other environmental disturbances can lead to inadmissible movements of the structure.

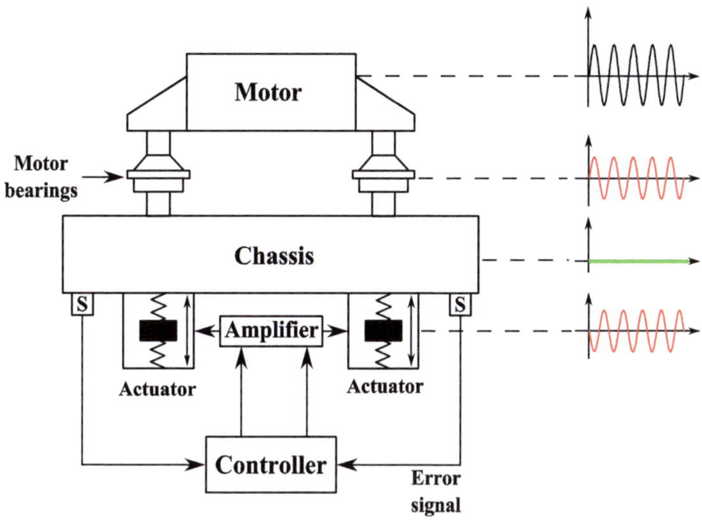

Fig. 1.2 Active vibration control used to reduce the effect of vibrations on the chassis

Fig. 1.3 Power spectral density of the residual force without and with active damping on an active suspension system

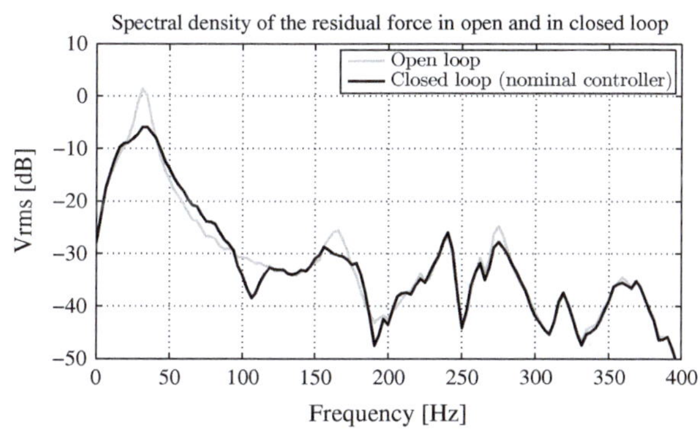

- the "primary path" through which the disturbances propagate towards the system; and
- the "secondary path" through which the compensation effect is applied.

Since from a control point of view there are no differences between active isolation and active vibration control we will use the generic term "Active Vibration Control" (AVC).

The principles of AVC and active noise control (ANC) are similar. Of course the range of frequencies and the type of instrumentation involved are different but the same control techniques can be used; however, the present book will focus on adaptive and robust active vibration control and the applications will concern this field.

The books [3, 4] give a compact and clear presentation of the origin and evolution of active vibration control techniques. It should be mentioned that these techniques have often been invented by researchers in the areas of vibration isolation and signal processing. The book [5] focuses on the dynamic modelling of active structures from physics equations and develops continuous-time feedback strategies based on these models.

The interest of the automatic control community in AVC is much more recent (it started essentially in the 1990s). The objective of the present book is to look at the problem of AVC from the perspective provided by automatic control methodology. From this perspective, the vibrations that we would like to attenuate strongly (or eliminate) are generically termed "disturbances".

Two of the major objectives of automatic control are:

- attenuation (or total rejection) of disturbances by feedback and feedforward actions; and
- damping of vibration modes.

These two problems are different. Adding damping is related to the assignment by feedback of desired closed-loop poles while strong attenuation (or total rejection) of disturbances are related to the introduction of the disturbance model in the controller (the internal model principle).

In AVC and ANC, disturbances can be characterized by their frequency content and their location in a specific region in the frequency domain. The disturbances can be of *narrow-band* type (simple or multiple) or of *broad-band* type. Of course, a combination of both is possible and, in certain cases, what we call broad-band may be several finite-band disturbances over separated small regions in the frequency domain; however, the distinction between these two types of disturbances is convenient in order to examine the techniques used for their compensation.

Fundamentally, in active control a compensator system is introduced, which will generate a "secondary" source. This compensator (acting through the "secondary path") will, when conveniently driven, interfere destructively with the disturbance coming from the "original" (in general non-accessible) primary source through what is called the "primary path". In the control terminology the "secondary path" is the *plant to be controlled* in order to reduce, as much as possible, the effect of the disturbance on the controlled output, which in the case of AVC is the measured residual acceleration or force. To achieve this, generically a feedback controller will be used (see Fig. 1.4).

An important concept, which allows one to assess the disturbance attenuation properties, damping of the vibration modes, stability of the feedback control loop, and robustness, is the so-called "output sensitivity function" (the transfer function between the disturbance and the measured output, i.e. between $p(t)$ and $y(t)$ in Fig. 1.4). There are some fundamental issues when approaching the problem of attenuating the disturbances by feedback. The first is related to the properties of the famous "Bode integral" on the modulus of the output sensitivity function expressed in dB,

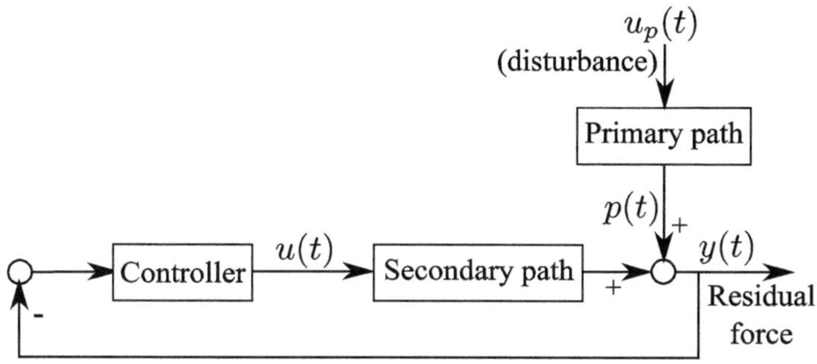

Fig. 1.4 Block diagram of active vibration compensation by feedback

which has value zero if the system is open-loop stable[3] (i.e. the sum of the areas above and under the 0 dB axis taken with their sign is zero). Since the objective is to strongly attenuate (or even totally reject asymptotically) the disturbance, this may require significant holes (low values) in the magnitude of the sensitivity function, which in turn (even with a very careful design) may lead to an unacceptable "water bed" effect, both in terms of performance (one amplifies at certain frequencies where some disturbance can still be present) as well as in terms of robustness (the modulus margin may become unacceptable[4]). Figure 1.5 illustrates the Bode Integral. As the attenuation is augmented in a certain frequency range, the maximum of the modulus of the output sensitivity function increases. Therefore there are inherent limitations in using feedback for active vibration control.[5]

A fundamental result in feedback control, which is of great interest for the problem of vibration attenuation, is the "internal model principle" (IMP), which stipulates that the disturbance will be asymptotically cancelled if, and only if, the controller contains the "model of the disturbance".

This brings in view the concepts of "plant model" and "disturbance model". In order to design the feedback controller properly, knowledge of the "plant model" and the "disturbance model" is necessary. The control methodology is a "model-based design" known as MBC (model based control).

One should distinguish between a "knowledge plant model" and a "dynamic plant model". The "knowledge plant model" is obtained from the law of physics and mechanics describing the operation of the compensator system. Unfortunately, these models are often not precise enough, since on one hand their precision depends on the perfect knowledge of some physical parameters (which is hard to get) and on the other hand it is difficult to model all the elements constituting the compensator

[3]Both the controller and the plant to be controlled are stable.

[4]The modulus margin is the minimum distance between the open-loop transfer function hodograph and the Nyquist point and is equal to the inverse of the maximum of the modulus of the sensitivity function [6].

[5]For example, narrow-band disturbances can be rejected by feedback up to a certain number while keeping an acceptable profile for the output sensitivity function (at least 3 or 4—see [7] and Chap. 13). Sufficiently "narrow" finite-band disturbances can also be handled by feedback alone.

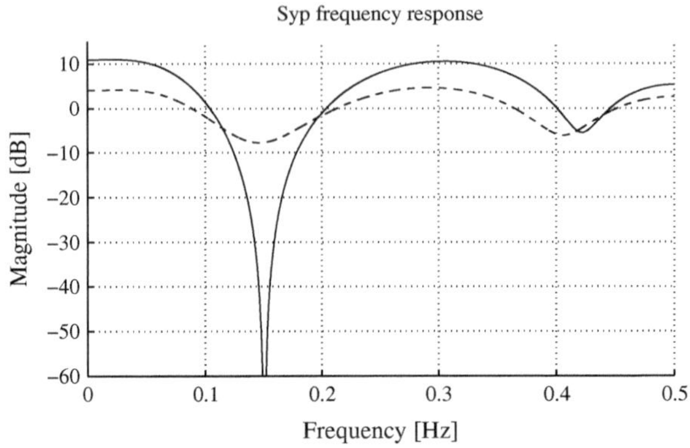

Fig. 1.5 Modulus of the output sensitivity functions for various attenuations

system. For this reason one uses what is called the "control dynamic plant model", i.e. a kind of filter (parametric model) that describes the dynamical relationship between the variations of the control input and the variations of the output of the system. This kind of model, necessary for design, can be obtained directly from an experimental test using the techniques of "System Identification" (this will be discussed in Chaps. 5 and 6).

In most AVC systems the characteristics of the compensator systems remain almost unchanged during operation. This means that the associated dynamic control model remains almost unchanged and therefore the parameters of the identified model are almost constant.

Nevertheless, for controller design we need the "model of the disturbance" in addition. A common framework is the assumption that the disturbance is the result of white noise or a Dirac impulse passed through the model of the disturbance. The knowledge of this model together with the knowledge of the model of the secondary path (compensator) allows the design of an appropriate control strategy. In practice, in most of the cases the characteristics of these disturbances are unknown or time-varying. While in some particular cases (with a limited range of variations in the frequency of the vibrations) a robust design can be considered (see the example given in Sect. 11.3 as well as [8–10]), in most situations, as a consequence of the high level of attenuation required, an adaptive approach is necessary to obtain a good tuning with respect to the disturbance characteristics (note that the adaptive loop can be added on top of a robust controller—see Sect. 12.2).

When the limitations induced by the Bode integral do not allow the achievement of the desired performances by feedback (in particular for the case of broad-band disturbances), one has to consider adding a feedforward compensation, which requires a "source" correlated with the disturbance to be attenuated.[6]

---

[6]The source is located upstream with respect to the location where the residual force (acceleration) or noise is measured.

In a number of applications of AVC and ANC, an image of the disturbances (a correlated measurement) acting upon the system can be made available. This information is very useful for attenuating the disturbances using a feedforward compensation scheme; however, the feedforward compensator filter will depend not only upon the dynamics of the compensator system (the plant) but also upon the characteristics of the disturbances and of the primary path (the transfer function between the source and the residual acceleration or force).

1.2 A Conceptual Feedback Framework

Figure 1.6 represents an active noise and vibration control (ANVC) system using both feedforward and feedback compensators. The system has two inputs and two outputs. The first input is the *disturbance* $w(t)$, which is generated by the unknown disturbance source $s(t)$ passed through a filter with unknown characteristics. The second input is the *control signal*, $u(t)$. The first output is the measurement of the *residual acceleration* (force, noise) $e(t)$ (also called the *performance variable*) and the second output is a *signal correlated with the unknown disturbance*, $y_1(t)$ in Fig. 1.6. This correlation is a result of the physical characteristics of the system.

As shown in Fig. 1.6, the path that transmits the filtered disturbance, $w(t)$, to the residual acceleration is called the *primary path*. The control signal, on the other hand, is transmitted to the residual acceleration through the *secondary path*. The residual acceleration (the performance variable) is formed by addition between the output of the primary path, denoted $x(t)$, and the output of the secondary path, denoted $z(t)$.

In general, ANVC systems also present a *positive coupling path* (also called the *reverse path*) between the control signal $u(t)$ and the measured signal $y_1(t)$, which is shown in Fig. 1.6. This results in an internal positive feedback that can destabilize the ANVC system if not taken into account. The objective is that of minimizing the performance variable, $e(t)$, and stabilizing the system, by computing an appropriate control, $u(t)$, based on the measurements $e(t)$ and $y_1(t)$.

One can see that, in the control system architecture presented in Fig. 1.6, the control signal $u(t)$ is obtained by the subtraction of the *feedback control*, $u_2(t)$, from the *feedforward control*, $u_1(t)$. The measurements obtained from the system can be put into a vector form as $y(t) = [y_1(t),\ y_2(t)]^T = [y_1(t),\ e(t)]^T$. As a consequence, the controller also has a vector representation $\kappa = [N,\ \ -K]^T$, where N and K denote, respectively, the *feedforward* and the *feedback* compensators.

With this notation, the equation relating the measurements to the control signal is given by

$$u(t) = u_1(t) - u_2(t) = N \cdot y_1(t) - K \cdot y_2(t) = \kappa^T \cdot y(t). \tag{1.1}$$

The feedforward controller denomination attributed to N is motivated by the fact that $y_1(t)$, called the *correlated image of the disturbance*, is measured upstream of the performance variable (see the bench test described in Sect. 2.3). This also

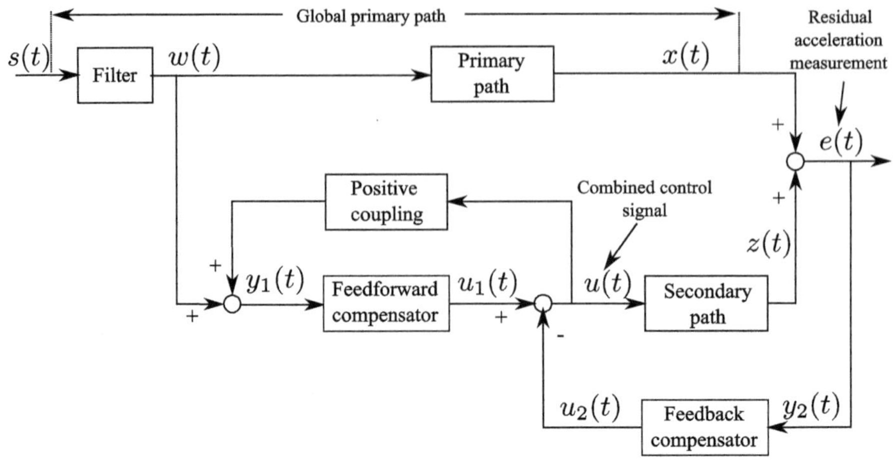

Fig. 1.6 Block diagram representation of the combined feedforward–feedback control problem

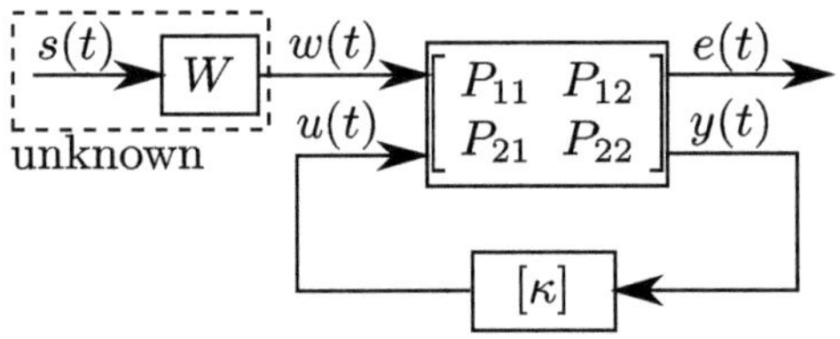

Fig. 1.7 Generalized ANVC system representation

assumes that it is physically possible to obtain such a measurement. The situations where this is not possible constitute feedback control problems, while the others are more generally referred in the literature as hybrid control. A standard feedback representation in the form of a 2-inputs–2-outputs system as shown in Fig. 1.7 can also be considered. This representation is very well known in robust and optimal control (see also [11]).

The equations associated with the feedback system representation are

$$\begin{bmatrix} e(t) \\ y(t) \end{bmatrix} = \begin{bmatrix} P_{11} & P_{12} \\ P_{21} & P_{22} \end{bmatrix} \begin{bmatrix} w(t) \\ u(t) \end{bmatrix} = \left[\begin{array}{c|c} D & G \\ \hline 1 & M \\ D & G \end{array}\right] \begin{bmatrix} w(t) \\ u(t) \end{bmatrix}, \tag{1.2}$$

where D, G, and M correspond to the models of the primary, secondary, and reverse paths. The control is given by (1.1).

1.3 Active Damping

As indicated previously, active damping concerns augmentation of the damping of some vibration modes characterizing a mechanical structure (the frequencies of these modes are not changed). Nevertheless, damping of these low-frequency vibration modes will influence the "output sensitivity function" as a consequence of the Bode integral property. Adding strong damping at a resonance will induce a deterioration of the attenuation performances in the nearby frequency region. In fact active damping requires careful shaping of the output sensitivity function in order to bound the loss of performance at other frequencies by distributing the "water bed" effect across a wide frequency band (see Fig. 1.3).[7] The design of active damping will be discussed and illustrated in Chap. 10.

1.4 The Robust Regulation Paradigm

In the context of AVC (as well as for ANC) the primary sense of *robustness* is the capacity of attenuating disturbances located in a given range of frequencies but whose frequency characteristics are not exactly known. The characteristics (the model) of the disturbances are generally unknown and may be time-varying. As a consequence, their location in the frequency domain will change. It is not possible to design a robust linear controller that introduces a strong attenuation over a wide frequency range (as a consequence of the Bode integral). Therefore, a compromise is required between the width of the frequency region where the disturbance may be located and the attenuation that can be achieved. If one absolutely needs a strong attenuation, the tolerated range of frequency variation will be small. Conversely, if there is great uncertainty in the location of the disturbances in the frequency domain, then the attenuation that can be achieved will be small.

The secondary sense of *robustness* is the capacity of the linear controller to handle small uncertainties on the system model parameters in the vicinity of their nominal values. The system parameter uncertainties will be handled by respecting constraints on the modulus of the output and input sensitivity functions.[8]

The situations where a robust controller solution for AVC provides satisfactory results in practice, depend upon the compromise between the level of attenuation required and the range of frequency variations of the disturbances (see Chap. 11 for applications of linear robust control design).

---

[7]The resulting controller may be of high order and this raises the problem of controller order reduction, which will be discussed in Chap. 9.

[8]The input sensitivity function is the transfer function between the disturbance $p(t)$ and the control input $u(t)$ (see Fig. 1.4).

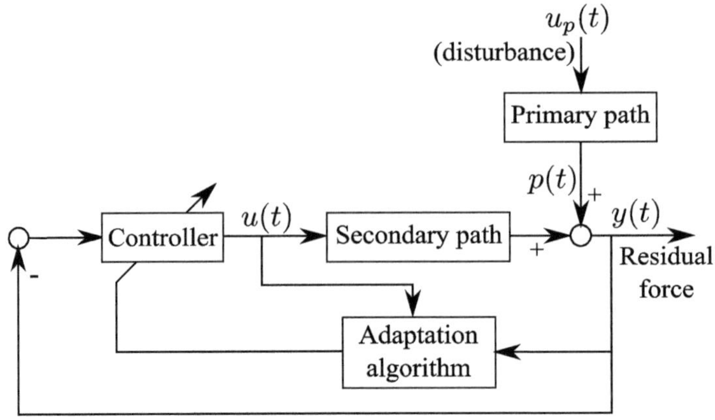

Fig. 1.8 Adaptive feedback attenuation of unknown disturbances

1.5 The Adaptive Regulation Paradigm

Since the characteristics (the models) of the disturbances are generally unknown and may be time-varying over a wide frequency range, often a single robust linear controller that achieves the desired attenuation cannot be designed. In such situations adaptive feedback or feedforward compensation has to be used.

Figure 1.8 illustrates the adaptive attenuation of disturbances by feedback. In addition to the classical feedback loop an *adaptation loop* tunes the parameters of the controller in real time. In order to do so, it uses, as primary information, both the residual acceleration (force), which is a *performance variable*, and the control input.

Figure 1.9 illustrates the adaptive rejection of unknown disturbances by feedforward compensation. A "well-located" transducer can provide a measurement highly correlated with the unknown disturbance (a good image of the disturbance). This information is applied to the control input of the secondary path through an adaptive filter whose parameters are adapted so as to minimize the effect of the disturbance on the output.

Adaptive feedforward vibration (or noise) compensation is currently used in AVC and ANC when an image of the disturbance is available [12].

Nevertheless, at the end of the 1990s it was pointed out that in most of these systems there is a physical "positive" feedback coupling between the compensator system and the measurement of the image of the disturbance (vibration or noise) [13–16] (see also Sect. 1.2).[9] The inherent internal physical positive feedback may cause instability of the AVC or ANC systems. As a consequence, the development of adaptive algorithms for feedforward compensation should take into account the internal positive feedback.

So at this point one can say that one has two types of disturbances:

[9]This will be illustrated on the experimental platform that will be presented in Sect. 2.3.

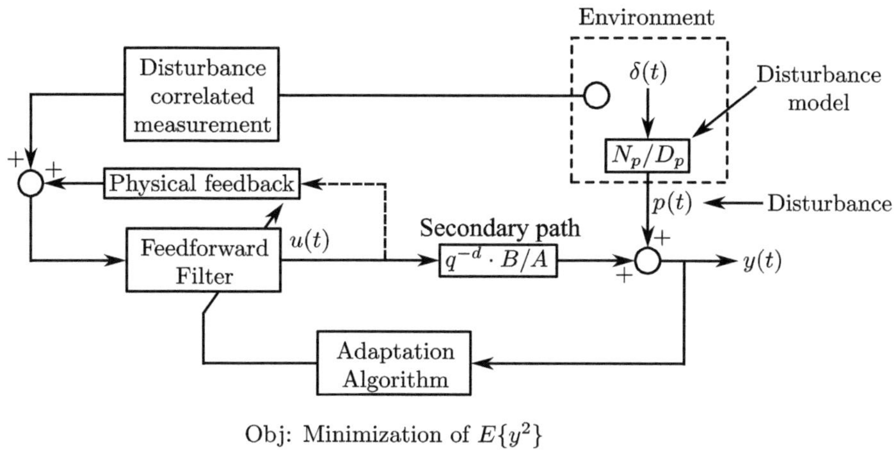

Fig. 1.9 Adaptive feedforward compensation of unknown disturbances

- single or multiple narrow-band disturbances; and
- broad (finite)-band disturbances;

and two approaches for doing adaptive disturbance attenuation:

- the adaptive feedback approach (which only requires a measurement of the residual force or acceleration); and
- the adaptive feedforward compensation (requiring an additional transducer for getting a correlated measurement with the disturbance).

In addition, there are two possible modes of operation:

- *self-tuning* operation (in which the adaptation procedure starts either on demand or when the performance is unsatisfactory and ends when the new controller is estimated); and
- *adaptive* operation (in which the adaptation procedure is performed continuously and the controller is updated at each sampling).

As indicated earlier, a common framework is the assumption that the disturbance is the result of white noise or a Dirac impulse passed through the *model of the disturbance*. Knowledge of this model allows the design of an appropriate controller. In general, the structure for such a *model of disturbance* does not change and can be assessed from data (using spectral analysis or order estimation techniques); however, the parameters of the model are unknown and may be time-varying. Therefore adaptation has to deal with the change in the parameters of the model of the disturbance.

The classical adaptive control paradigm deals with the construction of a control law when the parameters of the plant dynamic model are unknown and time-varying [17]. Nevertheless, in the present context, the plant dynamic model is almost invariant and it can be identified. The objective then is the rejection of disturbances characterized by unknown and time-varying disturbance models. It seems reasonable to call this paradigm *adaptive regulation*. Classical adaptive control focuses on adaptation of the controller parameters with respect to plant model parameters while adaptive regulation focuses on adaptation of the controller parameters with respect to variations in the disturbance model parameters.

In *adaptive regulation* the plant model is assumed to be known (obtained, for example by *system identification*). It is also assumed that the possible small variations or uncertainties of the plant model can be handled by a robust control design.

Adaptive regulation covers both adaptive feedback compensation and adaptive feedforward compensation since, on the one hand, adaptation has to deal with the change in the characteristics of the disturbances and, on the other hand, adaptive feedforward compensation is still a feedback structure as a consequence both of the internal positive coupling and of the presence of the adaptation loop, which is driven by the residual error.

The problem of adaptive regulation as defined above has been previously addressed in a number of papers ([18–28]) among others. Landau [29] presents a survey of the various techniques (up to 2010) used in adaptive feedback regulation as well as a review of a number of applications.

An international benchmark on the attenuation of multiple and unknown time-varying, narrow-band disturbances by feedback has been organized. The test bench was the AVC system that will be presented in Sect. 2.2. The results are summarized in [7] and allow the comparative evaluation of various designs.

1.6 Concluding Remarks

In order to reduce the impact of vibrations, one has several solutions related to the demanded performance:

- Passive: use materials with vibration attenuation properties.
- Semi-active: change the attenuation properties of the materials used for attenuation.
- Active: use compensation force to counteract vibrations.
- Robust AVC: when the characteristics of the vibrations are almost known and their domain of variation is small.
- Adaptive AVC: when the characteristics of vibrations are unknown and/or time-varying over a significant frequency range and high attenuation is required.

Design of robust AVC requires the model of the disturbances (and their domain of variation) as well as the models of the secondary path (for the feedback and feedforward approach) and of the primary path (for the feedforward approach).

Design of adaptive active vibration control does not require either the model of the disturbance or the model of the primary path.

1.7 Notes and Reference

The books [3–5] offer complementary perspectives to AVC and provide many practical examples. In particular the modelling aspects starting from basic laws of physics are enhanced.

Comparative evaluation of various techniques proposed is important. *The European Journal of Control*, no. 4, 2013 [7] is dedicated to a benchmark on adaptive attenuation of unknown and time-varying multiple narrow-band disturbances. The reference [30] should also be considered.

The references [29, 31–36] survey various aspects of AVC. Specific references related to the various topics will be provided at the end of the corresponding chapters.

It is not the objective of this book to provide an exhaustive reference list presenting applications of adaptive and robust AVC but a limited list of references covering applications in a number of fields include [3, 4, 26, 29, 37–44].

The special issue of *International Journal of Adaptive Control and Signal Processing* on adaptive frequency estimation with applications [45] gives a view of some recent research results in the field. This special issue includes [46–53].

References

1. Li S, Qiu J, Li J, Ji H, Zhu K (2012) Multi-modal vibration control using amended disturbance observer compensation. IET Control Theory Appl 6(1):72–83. doi:10.1049/iet-cta.2010.0573
2. Li S, Li J, Mo Y, Zhao R (2014) Composite multi-modal vibration control for a stiffened plate using non-collocated acceleration sensor and piezoelectric actuator. Smart Mater Struct 23(1):1–13
3. Fuller C, Elliott S, Nelson P (1999) Active control of vibration. Academic Press, Cambridge
4. Elliott S (2001) Signal processing for active control. Academic Press, San Diego, California
5. Preumont A (2011) Vibration control of active structures—an introduction. Springer, Heidelberg
6. Landau I, Zito G (2005) Digital control systems—design, identification and implementation. Springer, London
7. Landau ID, Silva AC, Airimitoaie TB, Buche G, Noé M (2013) Benchmark on adaptive regulation—rejection of unknown/time-varying multiple narrow band disturbances. Eur J Control 19(4):237–252. doi:10.1016/j.ejcon.2013.05.007
8. Alma M, Martinez J, Landau I, Buche G (2012) Design and tuning of reduced order H_∞ feedforward compensators for active vibration control. IEEE Trans Control Syst Technol 20(2):554–561. doi:10.1109/TCST.2011.2119485
9. Rotunno M, de Callafon R (2003) Design of model-based feedforward compensators for vibration compensation in a flexible structure. Internal report, Dept. of Mechanical and Aerospace Engineering. University of California, San Diego
10. Carmona J, Alvarado V (2000) Active noise control of a duct using robust control theory. IEEE Trans. Control Syst Technol 8(6):930–938
11. Tay TT, Mareels IMY, Moore JB (1997) High performance control. Birkhäuser Boston
12. Elliott S, Sutton T (1996) Performance of feedforward and feedback systems for active control. IEEE Trans Speech Audio Process 4(3):214–223. doi:10.1109/89.496217
13. Kuo S, Morgan D (1999) Active noise control: a tutorial review. Proc IEEE 87(6):943–973. doi:10.1109/5.763310

14. Hu J, Linn J (2000) Feedforward active noise controller design in ducts without independent noise source measurements. IEEE Trans Control Syst Technol 8(3):443–455
15. Jacobson C, Johnson CR, Jr, McCormick D, Sethares W (2001) Stability of active noise control algorithms. IEEE Signal Process Lett 8(3):74–76. doi:10.1109/97.905944
16. Zeng J, de Callafon R (2006) Recursive filter estimation for feedforward noise cancellation with acoustic coupling. J Sound Vib 291(3–5):1061–1079. doi:10.1016/j.jsv.2005.07.016
17. Landau ID, Lozano R, M'Saad M, Karimi A (2011) Adaptive control, 2nd edn. Springer, London
18. Bodson M, Douglas S (1997) Adaptive algorithms for the rejection of sinusosidal disturbances with unknown frequency. Automatica 33:2213–2221
19. Benamara F, Kabamba P, Ulsoy A (1999) Adaptive sinusoidal disturbance rejection in linear discrete-time systems—part I: Theory. J Dyn Syst Meas Control 121:648–654
20. Valentinotti S (2001) Adaptive rejection of unstable disturbances: Application to a fed-batch fermentation. Thèse de doctorat, École Polytechnique Fédérale de Lausanne
21. Marino R, Santosuosso G, Tomei P (2003) Robust adaptive compensation of biased sinusoidal disturbances with unknown frequency. Automatica 39:1755–1761
22. Ding Z (2003) Global stabilization and disturbance suppression of a class of nonlinear systems with uncertain internal model. Automatica 39(3):471–479. doi:10.1016/S0005-1098(02)00251-0
23. Landau I, Constantinescu A, Rey D (2005) Adaptive narrow band disturbance rejection applied to an active suspension—an internal model principle approach. Automatica 41(4):563–574
24. Kinney C, Fang H, de Callafon R, Alma M (2011) Robust estimation and automatic controller tuning in vibration control of time varying harmonic disturbances. In: 18th IFAC World Congress, Milano, Italy, pp 5401–5406
25. Aranovskiy S, Freidovich LB (2013) Adaptive compensation of disturbances formed as sums of sinusoidal signals with application to an active vibration control benchmark. Eur J Control 19(4), 253–265. doi:10.1016/j.ejcon.2013.05.008. (Benchmark on adaptive regulation: rejection of unknown/time-varying multiple narrow band disturbances)
26. Chen X, Tomizuka M (2012) A minimum parameter adaptive approach for rejecting multiple narrow-band disturbances with application to hard disk drives. IEEE Trans Control Syst Technol 20(2):408–415. doi:10.1109/TCST.2011.2178025
27. Emedi Z, Karimi A (2012) Fixed-order LPV controller design for rejection of a sinusoidal disturbance with time-varying frequency. In: 2012 IEEE multi-conference on systems and control, Dubrovnik
28. Marino R, Santosuosso G, Tomei P (2008) Output feedback stabilization of linear systems with unknown additive output sinusoidal disturbances. Eur J Control 14(2):131–148
29. Landau ID, Alma M, Constantinescu A, Martinez JJ, Noë M (2011) Adaptive regulation—rejection of unknown multiple narrow band disturbances (a review on algorithms and applications). Control Eng Pract 19(10):1168–1181. doi:10.1016/j.conengprac.2011.06.005
30. Castellanos-Silva A, Landau ID, Dugard L, Chen X (2016) Modified direct adaptive regulation scheme applied to a benchmark problem. Eur J Control 28:69–78. doi:10.1016/j.ejcon.2015.12.006
31. Landau ID, Airimitoaie TB, Castellanos SA (2015) Adaptive attenuation of unknown and time-varying narrow band and broadband disturbances. Int J Adapt Control Signal Process 29(11):1367–1390
32. Landau ID, Airimitoaie TB, Castellanos SA, Alma M (2015) Adaptative active vibration isolation—a control perspective. MATEC web of conferences 20, 04,001. doi:10.1051/matecconf/20152004001
33. Alkhatib R, Golnaraghi M (2003) Active structural vibration control: a review. Shock Vib Dig 35(5):367
34. Fuller C, Von Flotow A (1995) Active control of sound and vibration. IEEE Control Syst 15(6):9–19
35. Zhou S, Shi J (2001) Active balancing and vibration control of rotating machinery: a survey. Shock Vib Dig 33(5):361–371

36. Preumont A, François A, Bossens F, Abu-Hanieh A (2002) Force feedback versus acceleration feedback in active vibration isolation. J Sound Vib 257(4):605–613
37. Martinez JJ, Alma M (2012) Improving playability of blu-ray disc drives by using adaptive suppression of repetitive disturbances. Automatica 48(4):638–644
38. Taheri B (2013) Real-time pathological tremor identification and suppression. Phd thesis, Southern Methodist University
39. Taheri B, Case D, Richer E (2014) Robust controller for tremor suppression at musculoskeletal level in human wrist. IEEE Trans Neural Syst Rehabil Eng 22(2):379–388. doi:10.1109/TNSRE.2013.2295034
40. Taheri B, Case D, Richer E (2015) Adaptive suppression of severe pathological tremor by torque estimation method. IEEE/ASME Trans Mechatron 20(2):717–727. doi:10.1109/TMECH.2014.2317948
41. Bohn C, Cortabarria A, Härtel V, Kowalczyk K (2004) Active control of engine-induced vibrations in automotive vehicles using disturbance observer gain scheduling. Control Eng Pract 12(8):1029–1039
42. Karkosch H, Svaricek F, Shoureshi R, Vance J (1999) Automotive applications of active vibration control. In: Proceedings of the European control conference
43. Li Y, Horowitz R (2001) Active suspension vibration control with dual stage actuators in hard disk drives. In: Proceedings of the American control conference, 2001, vol 4, pp 2786–2791. IEEE, New York
44. Hong J, Bernstein DS (1998) Bode integral constraints, collocation, and spillover in active noise and vibration control. IEEE Trans Control Syst Technol 6(1):111–120
45. Bodson: Call for papers: Recent advances in adaptive methods for frequency estimation with applications. Int J Adapt Control Signal Process 28(6), 562–562 (2014). doi:10.1002/acs.2486
46. Chen X, Tomizuka M (2015) Overview and new results in disturbance observer based adaptive vibration rejection with application to advanced manufacturing. Int J Adapt Control Signal Process 29(11):1459–1474. doi:10.1002/acs.2546
47. Chen B, Pin G, Ng WM, Hui SYR, Parisini T (2015) A parallel prefiltering approach for the identification of a biased sinusoidal signal: Theory and experiments. Int J Adapt Control Signal Process 29(12):1591–1608. doi:10.1002/acs.2576
48. Khan NA, Boashash B (2016) Multi-component instantaneous frequency estimation using locally adaptive directional time frequency distributions. Int J Adapt Control Signal Process 30(3):429–442. doi:10.1002/acs.2583
49. Marino R, Tomei P (2016) Adaptive notch filters are local adaptive observers. Int J Adapt Control Signal Process 30(1):128–146. doi:10.1002/acs.2582
50. Carnevale D, Galeani S, Sassano M, Astolfi A (2016) Robust hybrid estimation and rejection of multi-frequency signals. Int J Adapt Control Signal Process. doi:10.1002/acs.2679
51. Jafari S, Ioannou PA (2016) Rejection of unknown periodic disturbances for continuous-time MIMO systems with dynamic uncertainties. Int J Adapt Control Signal Process. doi:10.1002/acs.2683
52. Menini L, Possieri C, Tornambè A (2015) Sinusoidal disturbance rejection in chaotic planar oscillators. Int J Adapt Control Signal Process 29(12):1578–1590. doi:10.1002/acs.2564
53. Ushirobira R, Perruquetti W, Mboup M (2016) An algebraic continuous time parameter estimation for a sum of sinusoidal waveform signals. Int J Adapt Control Signal Process. To appear

Chapter 2
The Test Benches

2.1 An Active Hydraulic Suspension System Using Feedback Compensation

The structure of the active hydraulic suspension (active isolation configuration) is presented in Fig. 2.1. Two photos of the system are presented in Fig. 2.2 (Courtesy of Hutchinson Research Center, Vibrachoc and GIPSA-LAB, Grenoble.) It consists of the active suspension, a load, a shaker and the components of the control scheme. The mechanical construction of the load is such that the vibrations produced by the shaker, fixed to the ground, are transmitted to the upper side of the active suspension. The active suspension is based on a hydraulic system allowing to reduce the overpressure at the frequencies of the vibration modes of the suspension. Its components are (see Fig. 2.1):

- an elastomer cone (1) which marks the main chamber filled up with silicon oil;
- a secondary chamber (2) marked by a flexible membrane;
- a piston (3) attached to a motor (when the piston is fixed, the suspension is passive);
- an orifice (4) allowing the oil to pass between the two chambers; and
- a force sensor located between the support and the active suspension.

The size of the main chamber of the elastomer cone is modified by the effect of the piston driven by a linear motor. The controller will act upon the piston (through a power amplifier) in order to reduce the residual force. The equivalent control scheme is shown in Fig. 2.3. The system input, $u(t)$ is the position of the piston (see Fig. 2.1), the output $y(t)$ is the residual force measured by a force sensor. The transfer function between the disturbance force, u_p, and the residual force $y(t)$ is called *primary path*. In our case (for testing purposes), the primary force is generated by a shaker controlled by a signal given by the computer. The transfer function between the input of the system, $u(t)$, and the residual force is called *secondary path*. The input of the system being a position and the output a force, the secondary path transfer function has a double differentiator behaviour. The sampling frequency used is $f_s = 800\,\text{Hz}$.

I.D. Landau et al., *Adaptive and Robust Active Vibration Control*,
Advances in Industrial Control, DOI 10.1007/978-3-319-41450-8_2

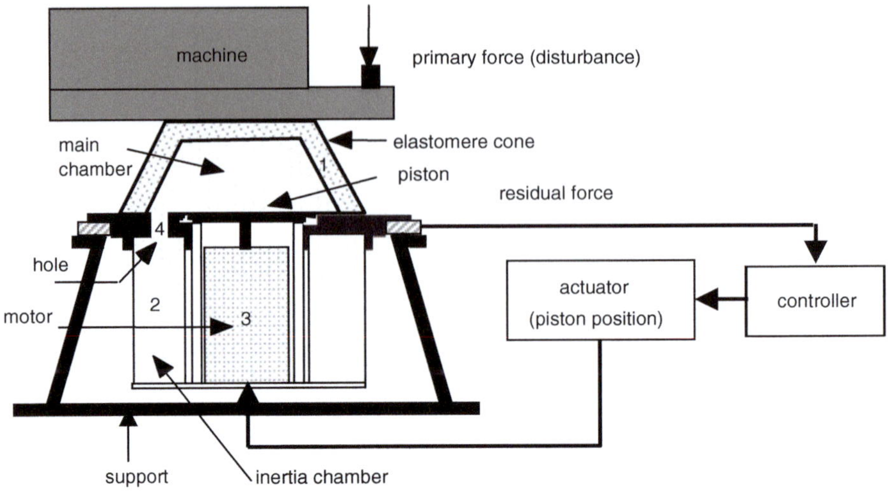

Fig. 2.1 Active suspension system (scheme)

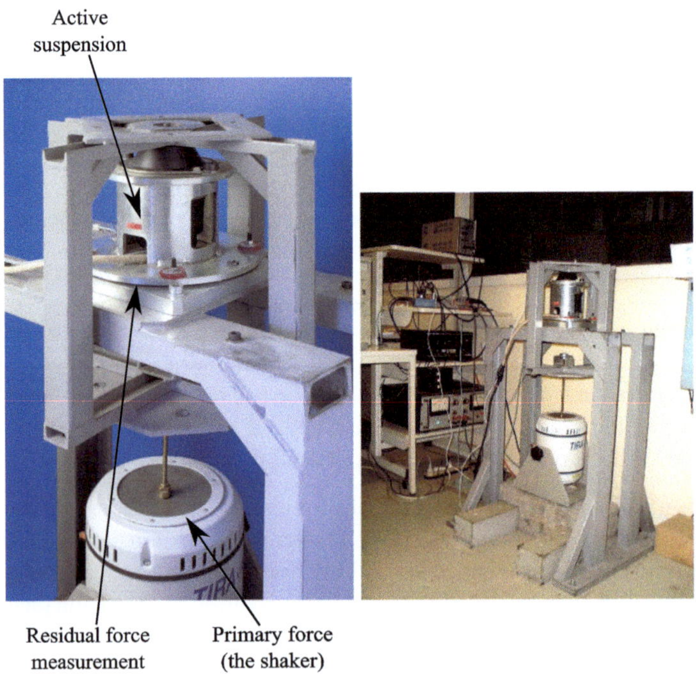

Fig. 2.2 Active suspension system (*left*). View of the experimental setting (*right*) (Courtesy of Hutchinson Research Center, Vibrachoc and Gipsa-lab, Grenoble, France.)

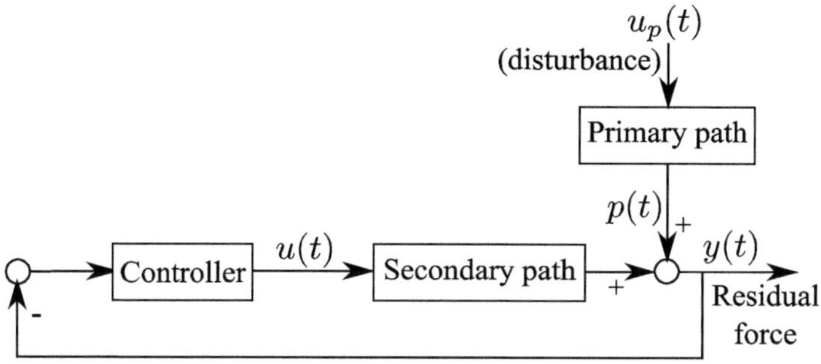

Fig. 2.3 Block diagram of active vibration control systems

The control objective is to strongly attenuate (cancel) the effect of unknown narrow-band disturbances on the output of the system (the residual force).

The system has been considered as a "black box". A system identification procedure has been used in order to obtain the dynamic model of the system (called also *control model*) to be used for control design. The identification procedure will be described in Sect. 6.1.

The frequency characteristic of the identified primary path model (open-loop identification), between the signal u_p sent to the shaker in order to generate the disturbance and the residual force $y(t)$, is presented in Fig. 2.4. The first vibration mode of the primary path model is near 32 Hz. The primary path model has been only used for simulation purposes.

The frequency characteristic of the identified secondary path model (open-loop operation) is presented also in Fig. 2.4. There exist several very low damped vibration modes on the secondary path, the first one being at 31.8 Hz with a damping factor 0.07. The identified model of the secondary path has been used for the design of the controller.

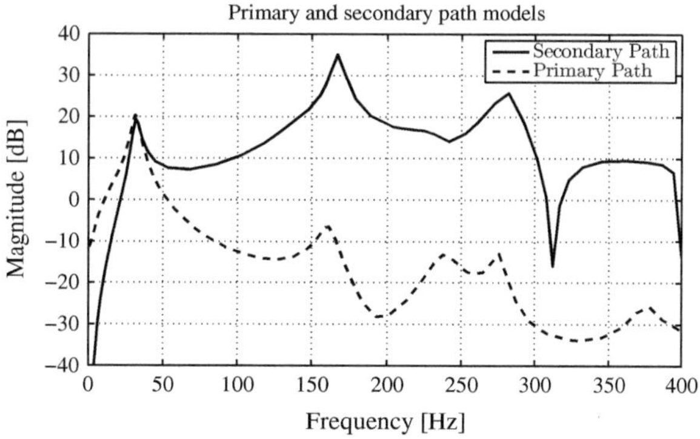

Fig. 2.4 Frequency characteristics of the primary and secondary paths

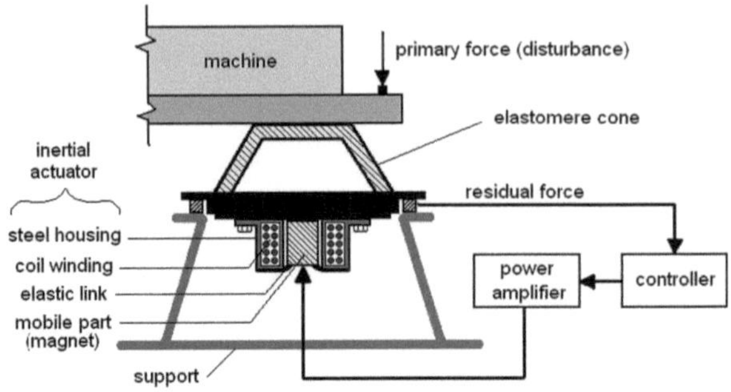

Fig. 2.5 Active vibration control using an inertial actuator (scheme)

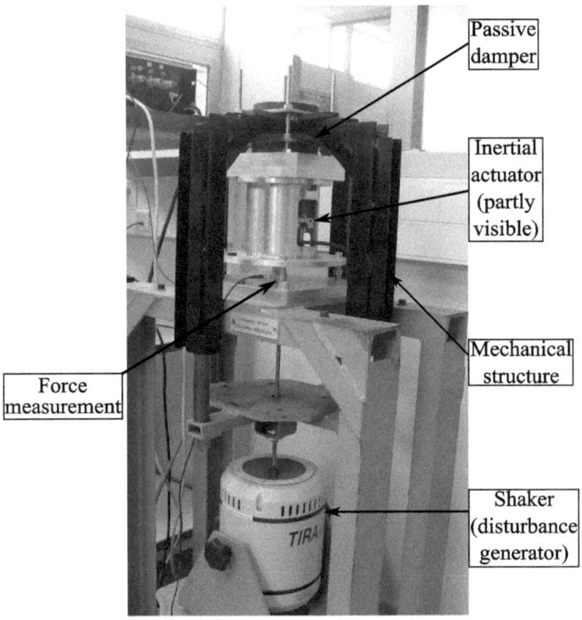

Fig. 2.6 The active vibration control system using an inertial actuator (photo)

2.2 An Active Vibration Control System Using Feedback Compensation Through an Inertial Actuator

The structure of the system is presented in Fig. 2.5. A general view (photo) of the system including the testing equipment is shown Fig. 2.6. It consists of a passive damper, an inertial actuator,[1] a chassis, a transducer for the residual force, a controller, a power amplifier, a shaker and a load which also transmits the vibration from the shaker to the chassis. The mechanical construction of the load is such that the vibrations produced by the shaker, fixed to the ground, are transmitted to the upper

[1]Inertial actuators use a similar principle as loudspeakers (see [1]).

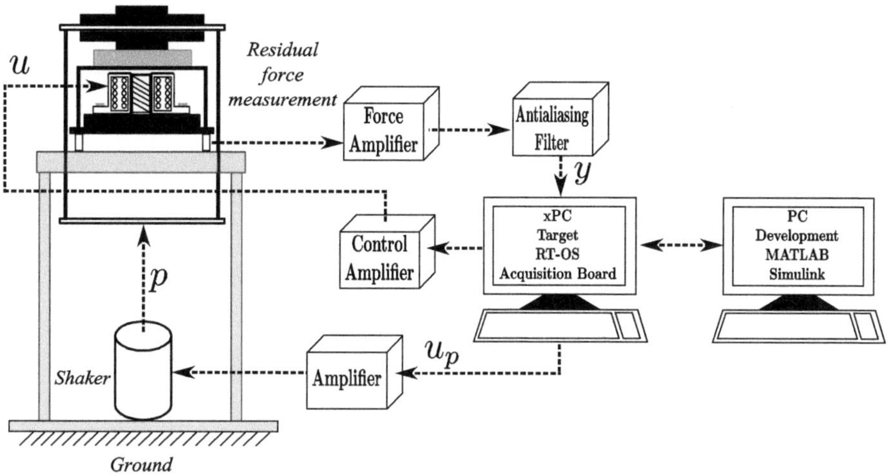

Fig. 2.7 The active vibration control system using an inertial actuator—hardware configuration

side, on top of the passive damper. The inertial actuator will create vibrational forces which can counteract the effect of vibrational disturbances.

The equivalent control scheme is shown in Fig. 2.3. The system input, $u(t)$ is the position of the mobile part (magnet) of the inertial actuator (see Fig. 2.5), the output $y(t)$ is the residual force measured by a force sensor. The transfer function between the disturbance force u_p, and the residual force $y(t)$ is called *primary path*. In our case (for testing purposes), the primary force is generated by a shaker driven by a signal delivered by the computer. The plant transfer function between the input of the inertial actuator, $u(t)$, and the residual force is called *secondary path*.

The complete hardware configuration of the system is shown in Fig. 2.7. The control objective is to cancel (or at least strongly attenuate) the effect of unknown narrow-band disturbances on the output of the system (residual force), i.e., to attenuate the vibrations transmitted from the machine to the chassis. The physical parameters of the system are not available. The system has been considered as a *black box* and the corresponding models for control have been identified from data. The details of the identification procedure will be given in Sect. 6.2. The sampling frequency is $f_s = 800\,\text{Hz}$.

Figure 2.8 gives the frequency characteristics of the identified models for the primary and secondary paths.[2] The system itself in the absence of the disturbances features a number of low damped resonance modes and low damped complex zeros (anti-resonance).

More details can be found at: http://www.gipsa-lab.grenoble-inp.fr/~ioandore.landau/benchmark_adaptive_regulation/.

[2]The primary path model is used only for simulation purposes.

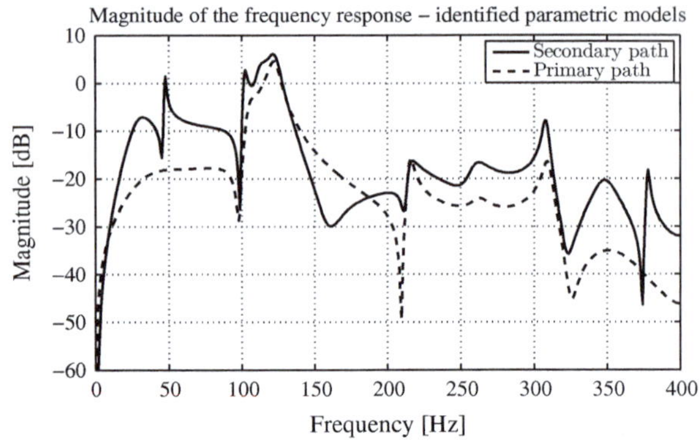

Fig. 2.8 Frequency response (magnitude) for the primary and the secondary path models

2.3 An Active Distributed Flexible Mechanical Structure with Feedforward–Feedback Compensation

Figure 2.9 shows a distributed flexible mechanical structure equipped for implementing an AVC using feedforward and feedback compensation. Figure 2.10 shows the details of the complete system including the AVC control scheme. The corresponding control block diagram is shown in Fig. 2.11.

The mechanical structure consists of five metal plates connected by springs. The uppermost and lowermost ones are rigidly jointed together by four screws. The middle three plates will be labeled for easier referencing M1, M2 and M3 (see Fig. 2.10). M1 and M3 are equipped with inertial actuators. The one on M1 serves as disturbance generator (inertial actuator I in Fig. 2.10), the one at the bottom serves for disturbance compensation (inertial actuator II in Fig. 2.10). The correlated measurement with the disturbance (image of the disturbance) is obtained from an accelerometer which is positioned on plate M1. Another sensor of the same type is positioned on plate M3

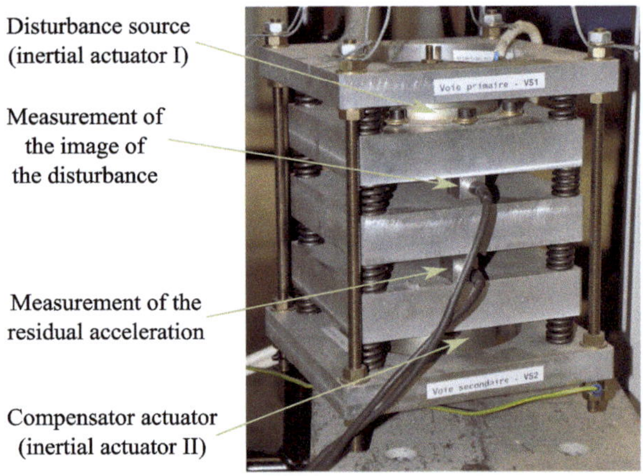

Fig. 2.9 An active distributed flexible mechanical structure equipped with sensors and actuators for feedforward–feedback disturbance compensation (photo)

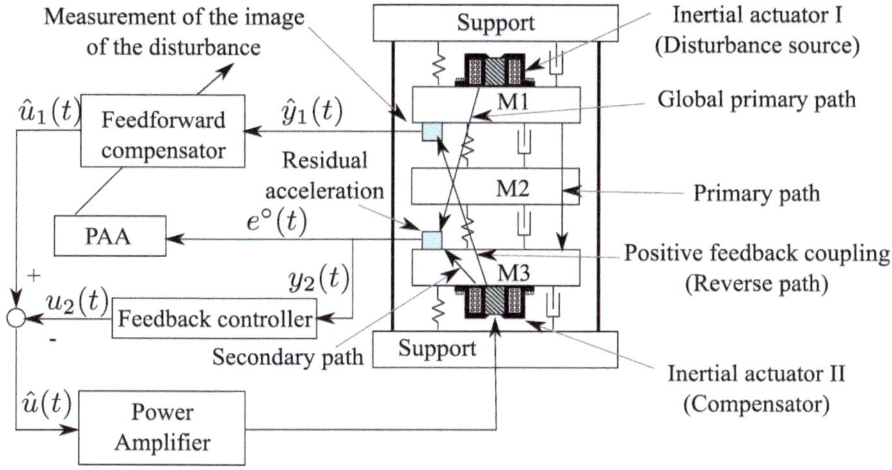

Fig. 2.10 An AVC system using feedforward and feedback compensation for the distributed flexible mechanical structure (scheme)

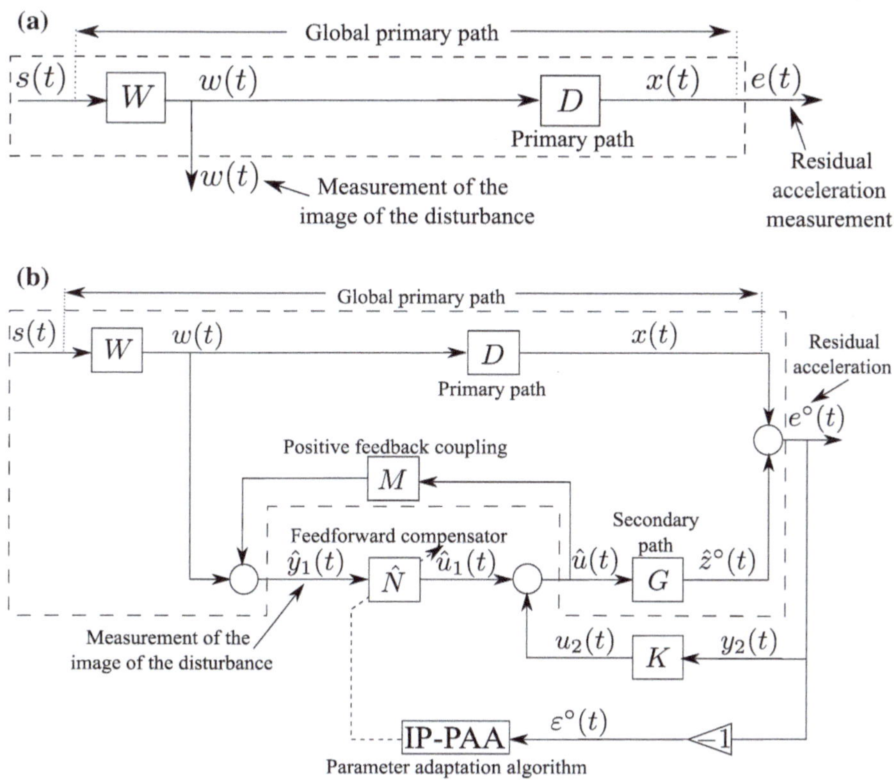

Fig. 2.11 Feedforward–feedback AVC—the control scheme: **a** in open loop and **b** with adaptive feedforward + fixed feedback compensator

and serves for measuring the residual acceleration (see Fig. 2.10). The objective is to minimize the residual acceleration measured on plate M3. This experimental setting allows to experiment both adaptive feedforward compensation (with or without additional feedback) as well as adaptive feedback disturbance compensation alone (without using the additional measurement upstream).

The disturbance is the position of the mobile part of the inertial actuator (see Figs. 2.9 and 2.10) located on top of the structure. The input to the compensator system is the position of the mobile part of the inertial actuator located on the bottom of the structure. When the compensator system is active, the actuator acts upon the residual acceleration, but also upon the measurement of the image of the disturbance through the *reverse path* (a positive feedback coupling). The measured quantity $\hat{y}_1(t)$ will be the sum of the correlated disturbance measurement $w(t)$ obtained in the absence of the feedforward compensation (see Fig. 2.11a) and of the effect of the actuator used for compensation (positive internal mechanical feedback). This is illustrated in Fig. 2.12 by the spectral densities of $\hat{y}_1(t)$ in open-loop ($w(t)$) and when feedforward compensation is active (the effect of the mechanical feedback is significant).

As from the previous experimental settings, the system is considered as a *black box* and the models for control design have been obtained by system identification from input/output data. The details of the identification procedure are given in Sect. 6.3. The sampling frequency is $f_s = 800\,\text{Hz}$. The frequency characteristics of the identified models of the primary, secondary and reverse paths are shown in Fig. 2.13.

This mechanical structure is representative for a number of situations encountered in practice and will be used to illustrate the performance of the various algorithms which will be presented in this book.

At this stage it is important to make the following remarks when the feedforward filter is absent (open-loop operation):

- very reliable models for the secondary path and the "positive" feedback path can be identified;

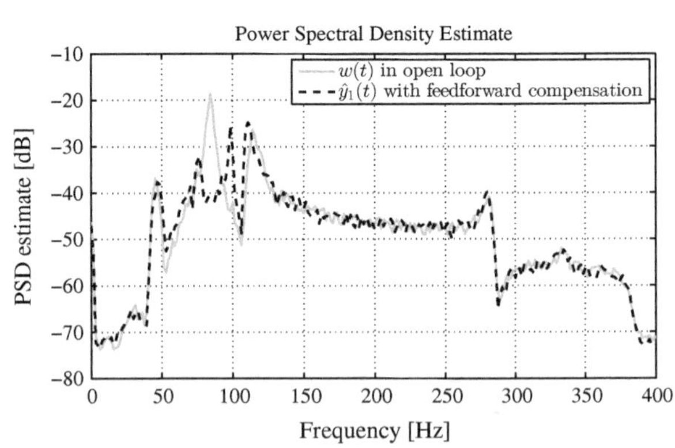

Fig. 2.12 Spectral densities of the image of the disturbance $\hat{y}_1(t)$ in open loop and when feedforward compensation scheme is active (experimental)

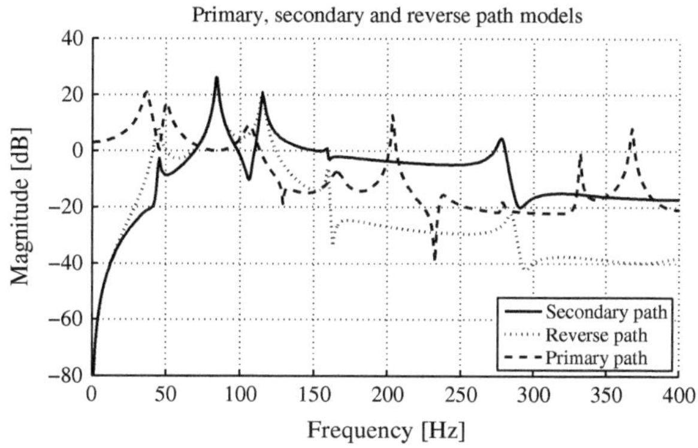

Fig. 2.13 Frequency response (magnitude) for the primary, secondary and reverse paths

- an estimation of the primary path transfer function can be obtained using the measured $w(t)$ as input and $e(t)$ as output (the compensator actuator being at rest);
- the quality of the primary path identified model will depend on the frequency characteristics of the signal $w(t)$ coming from the environment;
- design of a fixed model-based stabilizing feedforward compensator requires the knowledge of the reverse path model only;
- knowledge of the disturbance characteristics and of the primary, secondary and reverse paths models is mandatory for the design of an optimal fixed model-based feedforward compensator [2–4];
- **adaptation algorithms do not use information neither upon the primary path whose characteristics may be unknown nor upon the disturbance characteristics which may be unknown and time-varying.**

2.4 Concluding Remarks

- The test benches considered allow to evaluate different solutions for active vibration control and active damping.
- Their structure emphasizes the difficulties which may be encountered in practice.
- To obtain the complete dynamical models of these systems necessary for control design, identification of discrete time models from input/output data has been used (see Chap. 6).

2.5 Notes and References

For further details on these test benches see [5–8]. All the data for simulating the test bench presented in Sect. 2.2 is available at: http://www.gipsa-lab.grenoble-inp.fr/~ioandore.landau/benchmark_adaptive_regulation.

The book website provides input/output data and models for all the three test benches.

References

1. Landau ID, Alma M, Constantinescu A, Martinez JJ, Noë M (2011) Adaptive regulation - rejection of unknown multiple narrow band disturbances (a review on algorithms and applications). Control Eng Pract 19(10):1168–1181. doi:10.1016/j.conengprac.2011.06.005
2. Alma M, Martinez J, Landau I, Buche G (2012) Design and tuning of reduced order H_∞ feedforward compensators for active vibration control. IEEE Trans Control Syst Technol 20(2):554–561. doi:10.1109/TCST.2011.2119485
3. Rotunno M, de Callafon R (2003) Design of model-based feedforward compensators for vibration compensation in a flexible structure. Internal report, Department of Mechanical and Aerospace Engineering. University of California, San Diego
4. Carmona J, Alvarado V (2000) Active noise control of a duct using robust control theory. IEEE Trans Control Syst Technol 8(6):930–938
5. Landau I, Constantinescu A, Rey D (2005) Adaptive narrow band disturbance rejection applied to an active suspension - an internal model principle approach. Automatica 41(4):563–574
6. Landau I, Constantinescu A, Loubat P, Rey D, Franco A (2001) A methodology for the design of feedback active vibration control systems. In: European Control Conference (ECC), pp 1571–1576
7. Landau ID, Silva AC, Airimitoaie TB, Buche G, Noé M (2013) Benchmark on adaptive regulation - rejection of unknown/time-varying multiple narrow band disturbances. Eur J Control 19(4):237–252. doi:10.1016/j.ejcon.2013.05.007
8. Landau I, Alma M, Airimitoaie T (2011) Adaptive feedforward compensation algorithms for active vibration control with mechanical coupling. Automatica 47(10):2185–2196. doi:10.1016/j.automatica.2011.08.015

Part II
Techniques for Active Vibration Control

Chapter 3
Active Vibration Control Systems—Model Representation

3.1 System Description

3.1.1 Continuous-Time Versus Discrete-Time Dynamical Models

Before discussing the system description aspects, one has to take in account that the control law will be implemented on a digital computer. To do this there are two basic options:

- Represent the system in continuous time, compute the control law in continuous time and then discretize the continuous-time control law for implementation.
- Select the sampling frequency, represent the system in discrete time, compute the control law in discrete time and implement it directly.

Since one deals with mechanical systems, differential equations can be written to describe the dynamical behaviour of the various parts of the system allowing to build a "dynamical model" to be used for the design of the active vibration control system [1, 2].[1] There are however several obstacles in using a continuous-time representation of the system.

First of all, since the physical parameters are not precisely known, the model which can be obtained from the fundamental principles will not be very reliable. In addition there are parts of the systems for which it is difficult to give a perfect representation and to associate the corresponding parameters. For a high performance control design one needs an accurate dynamical model of the specific system to be controlled and therefore one has to consider identifying dynamical models from experimental input/output data, obtained by what is called an "identification protocol" (a "black box model" will be obtained).

It turns out that identification techniques are more efficient and much easier to implement if one considers the identification of discrete time dynamic models.

[1] Modern control design techniques use "model based control design".

I.D. Landau et al., *Adaptive and Robust Active Vibration Control*,
Advances in Industrial Control, DOI 10.1007/978-3-319-41450-8_3

It is also important to point out that using a continuous-time representation, the objective of the discretization of the designed control law will be to copy as much as possible the continuous-time control and this will require in general a very high sampling frequency. Further analysis is required in order to be sure that the discretized control law guarantees the robustness and performance of the system (since discretization introduces an approximation).

It turns out that if one considers the alternative situation, i.e., to discretize the input and output of the system at a sampling frequency which is related to its band-pass, one obtains through system identification a discrete-time dynamical model which can be used to design a discrete-time control algorithm.

Using a discrete-time representation of the system will require a lower sampling frequency[2] (directly related to the higher frequencies to be controlled) and the control algorithm to be implemented results directly from the design (no additional analysis is necessary since the control algorithm has been designed at the sampling frequency used).

Therefore because discrete time dynamical models allow:

- using a lower sampling frequency;
- using simpler and more efficient identification algorithms;
- getting directly the control algorithm to be implemented on a digital computer,

they will be used subsequently for representing active vibration control systems. The design of the control algorithm will be based on identified discrete-time dynamical models of the system.

3.1.2 *Digital Control Systems*

In this section, one reviews the basic requirements for the implementation of a digital control system. For a more detailed discussion of the various aspects see [3–6]. Figure 3.1 represents the structure of a digital control system. In Fig. 3.1, the set[3]: Digital to Analog Converter (D/A), Zero order hold (ZOH),[4] Continuous-time plant, Analog to Digital Converter (A/D) constitutes the discrete-time system to be controlled by a digital controller implemented on the computer used for control.

3.1.2.1 Selection of the Sampling Frequency

A good rule for the selection of the sampling frequency is [3]:

---

[2]Numerous examples show that by using this approach, the sampling frequency can be reduced with respect to the previous approach.

[3]Temporarily in this section t designates the continuous time and k the normalized sampling time ($k = \frac{time}{T_s}$). Starting from Sect. 3.1.3 the normalized discrete time will be denoted by t.

[4]ZOH keeps constant the signal delivered by the D/A converter between two sampling instants.

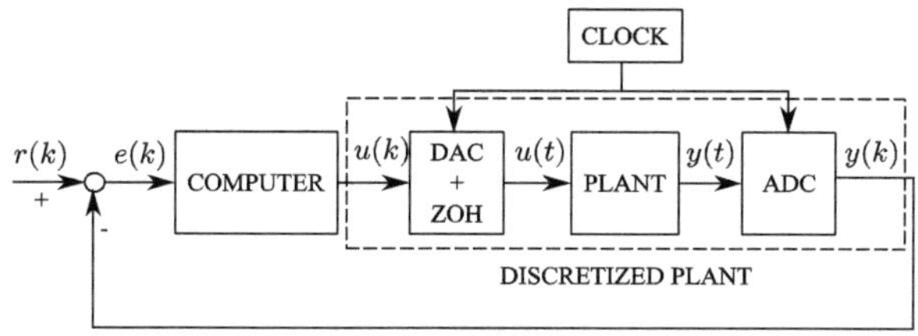

Fig. 3.1 A digital control system

$$f_s = (6 \rightarrow 25) f_B^{CL}, \tag{3.1}$$

where

- f_s = sampling frequency (in Hz);
- f_B^{CL} = desired bandwidth of the closed-loop system (in Hz).

Of course, the desired closed-loop bandwidth is related to the bandwidth of the system to be controlled. The formula (3.1) gives enough freedom for the selection of the sampling frequency.

As a general rule, one tries to select the lower sampling frequency compatible with the desired performances.

Except in very particular cases, all the discrete-time models will feature a fractional delay. Fractional delays are reflected as zeros in the transfer function of the discrete-time models (these zeros will be unstable if the fractional delay is larger that half of the sampling period [3]). For continuous-time systems with a relative degree higher or equal to 2, high frequency sampling will induce unstable zeros [5]. The consequence of the presence of unstable zeros in the discrete-time models used for control design is that control strategies based on cancelling the zeros cannot be used.

3.1.2.2 Anti-aliasing Filters

The theory of discrete-time systems [5, 6] indicates that the maximum frequency ($f_{\max}$) of a signal sent to the analog to digital converter should satisfy:

$$f_{\max} < f_s/2, \tag{3.2}$$

where f_s is the sampling frequency and $f_s/2$ is called the Nyquist or Shannon frequency.

Sending frequencies over $f_s/2$ produces distortion of the recovered discrete-time spectrum called *aliasing*. Therefore, *anti-aliasing* filters should always be introduced in order to remove the undesirable components of the signal. Anti-aliasing filters are constituted in general as several second order filters in cascade (Bessel, ITAE, Butterworth type). They should introduce a consequent attenuation of the signal beyond $f_s/2$ but their bandwidth should be larger than the desired closed-loop bandwidth.

Their design will also depend on the level of undesirable signals at frequencies beyond $f_s/2$.

The anti-aliasing filters introduce a high frequency dynamics which can in general be approximated by an additional small time delay. Since one directly estimates a discrete-time model from data, their effect is captured by the estimated model.

3.1.3 Discrete-Time System Models for Control

The discrete-time models will represent the behaviour of the controlled system from the discrete-time control applied to the system through a D/A converter and a ZOH to the output of the A/D converter which will discretize the measured output. Single-input single-output time invariant systems will be considered. They will be described by input–output discrete-time models of the form:

$$y(t) = -\sum_{i=1}^{n_A} a_i y(t-i) + \sum_{i=1}^{n_B} b_i u(t-d-i), \tag{3.3}$$

where t denotes the normalized sampling time (i.e., $t = \frac{time}{T_s}$, T_s = sampling period), $u(t)$ is the input, $y(t)$ is the output, d is the integer number of sampling periods contained in the time delay of the systems, a_i and b_i are the parameters (coefficients) of the models.

As such the output of the system at instant t is a weighted average of the past output over an horizon of n_A samples plus a weighted average of past inputs over an horizon of n_B samples (delayed by d samples).

This input–output model (3.3) can be more conveniently represented using a coding in terms of forward or backward shift operators defined as:

$$qy(t) = y(t+1); \quad q^{-1}y(t) = y(t-1) \tag{3.4}$$

Using the notations

$$1 + \sum_{i=1}^{n_A} a_i q^{-i} = A(q^{-1}) = 1 + q^{-1}A^*(q^{-1}), \tag{3.5}$$

where

$$A(q^{-1}) = 1 + a_1 q^{-1} + \cdots + a_{n_A} q^{-n_A} \tag{3.6}$$

$$A^*(q^{-1}) = a_1 + a_2 q^{-1} + \cdots + a_{n_A} q^{-n_A+1} \tag{3.7}$$

and

$$\sum_{i=1}^{n_B} b_i q^{-i} = B(q^{-1}) = q^{-1} B^*(q^{-1}), \tag{3.8}$$

where

$$B(q^{-1}) = b_1 q^{-1} + b_2 q^{-2} + \cdots + b_{n_B} q^{-n_B} \tag{3.9}$$
$$B^*(q^{-1}) = b_1 + b_2 q^{-1} + \cdots + b_{n_B} q^{-n_B+1}. \tag{3.10}$$

Equation (3.3) can be rewritten as

$$A(q^{-1})y(t) = q^{-d} B(q^{-1})u(t) = q^{-d-1} B^*(q^{-1})u(t) \tag{3.11}$$

or forward in time

$$A(q^{-1})y(t+d) = B(q^{-1})u(t) \tag{3.12}$$

as well as

$$y(t+1) = -A^* y(t) + q^{-d} B^* u(t) = -A^* y(t) + B^* u(t-d). \tag{3.13}$$

It can be observed that (3.13) can also be expressed as

$$y(t+1) = \theta^T \phi(t), \tag{3.14}$$

where θ defines the vector of parameters

$$\theta^T = [a_1, \ldots a_{n_A}, b_1, \ldots, b_{n_B}] \tag{3.15}$$

and $\phi(t)$ defines the vector of measurements (or the regressor)

$$\phi^T(t) = [-y(t) \cdots - y(t - n_A + 1), u(t-d) \cdots u(t - d - n_B + 1)] \tag{3.16}$$

The form of (3.14) will be used in order to estimate the parameters of a system model from input–output data. Filtering both left and right sides of (3.11) through a filter $1/A(q^{-1})$ one gets

$$y(t) = G(q^{-1})u(t), \tag{3.17}$$

where

$$G(q^{-1}) = \frac{q^{-d} B(q^{-1})}{A(q^{-1})} \tag{3.18}$$

is termed the *transfer operator*.[5]

---

[5]In many cases, the argument q^{-1} will be dropped out, to simplify the notations.

Computing the z-transform of (3.3), one gets the *pulse transfer function* characterizing the input–output model of (3.3)

$$G(z^{-1}) = \frac{z^{-d}B(z^{-1})}{A(z^{-1})} \tag{3.19}$$

It can be observed that the transfer function of the input–output model of (3.3) can be formally obtained from the *transfer operator* by replacing the time operator q by the complex variable z. Nevertheless, one should be careful since the domain of these variables is different. However, in the linear case with constant parameters one can use either one and their appropriate signification will result from the context.

Note also that the transfer operator $G(q^{-1})$ can be defined even if the parameters of the model (3.3) are time-varying, while the concept of pulse transfer function does not simply exist in this case.

The order n of the system model[6] (3.3), is the dimension of the minimal state space representation associated to the input–output model (3.3) and in the case of irreducible transfer function it is equal to

$$n = \max(n_A, n_B + d), \tag{3.20}$$

which corresponds also to the number of the poles of the irreducible transfer function of the system.

The order of the system is immediately obtained by expressing the transfer operator (3.18) or the transfer function (3.19) in the forward operator q and respectively the complex variable z. The passage from $H(z^{-1})$ to $H(z)$ is obtained multiplying by z^n:

$$G(z) = \frac{\bar{B}(z)}{\bar{A}(z)} = \frac{z^{r-d}B(z^{-1})}{z^r A(z^{-1})} \tag{3.21}$$

Example

$$G(z^{-1}) = \frac{z^{-3}(b_1 z^{-1} + b_2 z^{-2})}{1 + a_1 z^{-1}}$$
$$r = \max(1, 5) = 5$$
$$G(z) = \frac{b_1 z + b_2}{z^5 + a_1 z^4}$$

[6]The order of the system will be in general estimated from input/output data.

3.2 Concluding Remarks

- Recursive (differences) equations are used to describe discrete-time dynamic models.
- The delay operator q^{-1} ($q^{-1}y(t) = y(t-1)$) is a simple tool to handle recursive discrete-time equations.
- The input–output relation for a discrete-time model is conveniently described in the time domain by the pulse transfer operator $G(q^{-1})$: $y(t) = G(q^{-1})u(t)$.
- The pulse transfer function of a discrete-time linear system is expressed as function of the complex variable $z = e^{sT_s}$ (T_s = sampling period). The pulse transfer function can be derived from the pulse transfer operator $G(q^{-1})$ by replacing q^{-1} with z^{-1}.
- The asymptotic stability of a discrete-time model is ensured if, and only if, all pulse transfer function poles (in z) lie inside the unit circle.
- The order of a pulse transfer function of the form

$$G(z^{-1}) = \frac{z^{-d}B(z^{-1})}{A(z^{-1})} \tag{3.22}$$

is $n = \max(n_A, n_B + d)$, where n_A and n_B are the orders of the polynomials A and B, respectively, and d is the integer time delay in terms of sampling periods.

3.3 Notes and References

There are many excellent books on digital control systems. The books [3, 5, 6] are probably the most suited for the topics of the present book. The book [7] provides many discrete-time models obtained from the discretization of continuous-time models for various physical systems.

References

1. Preumont A (2011) Vibration control of active structures – an introduction. Springer, New York
2. Fuller C, Elliott S, Nelson P (1997) Active control of vibration. Academic Press, New York
3. Landau I, Zito G (2005) Digital control systems - design, identification and implementation. Springer, London
4. Landau I (1993) Identification et commande des systèmes, 2nd edn. Série Automatique. Hermès, Paris
5. Astrom KJ, Hagander P, Sternby J (1984) Zeros of sampled systems. Automatica 20(1):31–38. doi:10.1109/CDC.1980.271968
6. Franklin GF, Powell JD, Workman ML (1998) Digital control of dynamic systems, vol 3. Addison-Wesley, Menlo Park
7. Ogata K (1987) Discrete-time control systems, 1st edn. Prentice Hall, New Jersey

Chapter 4
Parameter Adaptation Algorithms

4.1 Introduction

Parameter adaptation algorithms (PAA) will play a fundamental role in the implementation of the various adaptive active vibration control systems. They can be introduced in various ways. We will consider two problems:

- Recursive parameter estimation in system identification.
- Parameter adaptation of the feedforward filter in adaptive feedforward vibration compensation.

These two problems will allow to introduce two basic configurations: the series parallel (equation error) parameter estimator and the parallel (output error) estimator.

The on-line parameter estimation principle for sampled models is illustrated in Fig. 4.1.

A discrete time model with adjustable parameters is implemented on the computer. The error $\varepsilon(t)$ between the system output at instant t, $y(t)$ and the output predicted by the model $\hat{y}(t)$ (called plant-model error or prediction error) is used by the *parameter adaptation algorithm*, which, at each sampling instant, will modify the model parameters in order to minimize this error (in the sense of a certain criterion).

The key element for implementing the on-line estimation of the plant model parameters is the *Parameter Adaptation Algorithm* (PAA) which drives the parameters of the adjustable prediction model from the data acquired on the system at each sampling instant. This algorithm has a *recursive* structure, i.e., the new value of the estimated parameters is equal to the previous value plus a correcting term which will depend on the most recent measurements.

In general a *parameter vector* is defined. Its components are the different parameters that should be estimated.

The recursive *parameter adaptation algorithms* have the following structure: New estimated parameter vector = previous estimated parameter vector + correcting term.

I.D. Landau et al., *Adaptive and Robust Active Vibration Control*,
Advances in Industrial Control, DOI 10.1007/978-3-319-41450-8_4

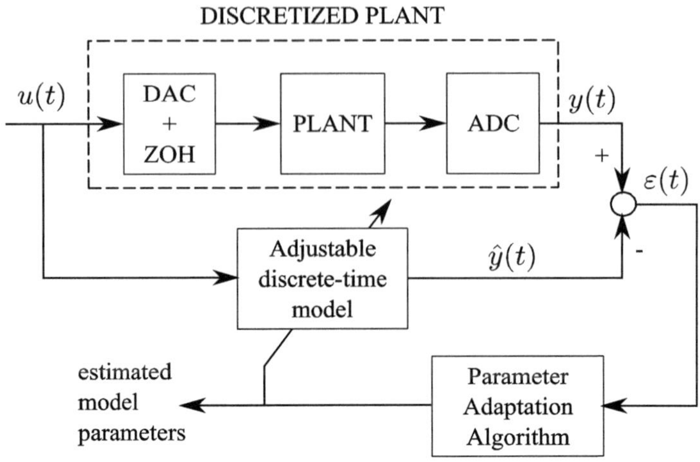

Fig. 4.1 Parameter estimation principle

The correcting term has in general the structure of a product: *Adaptation gain* × *Measurement function* × *Prediction error function*. The resulting structure is

$$\begin{bmatrix} New\ estimated \\ parameters \\ (vector) \end{bmatrix} = \begin{bmatrix} Previous\ estimated \\ parameters \\ (vector) \end{bmatrix} + \\ + \begin{bmatrix} Adaptation \\ gain \\ (matrix) \end{bmatrix} \times \begin{bmatrix} Measurement \\ function \\ (vector) \end{bmatrix} \times \begin{bmatrix} Prediction\ error \\ function \\ (scalar) \end{bmatrix}$$

This structure corresponds to the so-called *integral type adaptation algorithms* (the algorithm has memory and therefore maintains the estimated value of the parameters when the correcting terms become null). The algorithm can be viewed as a discrete-time integrator fed at each instant by the correcting term. The measurement function vector is generally called the *observation vector*. The prediction error function is generally called the *adaptation error*.

The adaptation gain plays an important role in the performance of the parameter adaptation algorithm and it may be constant or time-varying.

4.2 Structure of the Adjustable Model

4.2.1 Case (a): Recursive Configuration for System Identification—Equation Error

Consider the discrete-time model of a plant described by:

$$y(t+1) = -a_1 y(t) + b_1 u(t) = \theta^T \phi(t) \tag{4.1}$$

where $u(t)$ is the input, $y(t)$ is the output, both measurable. The unknown parameters a_1 and b_1 are the unknown parameters of the model. One defines the unknown *parameter vector* θ:

$$\theta^T = [a_1, b_1] \tag{4.2}$$

and the *measurement vector*:

$$\phi^T(t) = [-y(t), u(t)] \tag{4.3}$$

The adjustable prediction model will be described in this case by:

$$\hat{y}^\circ(t+1) = \hat{y}[(t+1)|\hat{\theta}(t)] = -\hat{a}_1(t)y(t) + \hat{b}_1(t)u(t) = \hat{\theta}^T(t)\phi(t) \tag{4.4}$$

where $\hat{y}^\circ(t+1)$ is termed the *a priori* predicted output depending on the values of the estimated parameters at instant t.

$$\hat{\theta}^T(t) = [\hat{a}_1(t), \hat{b}_1(t)] \tag{4.5}$$

is the *estimated parameter vector* at instant t.

As it will be shown later, it is very useful to consider also the *a posteriori* predicted output computed on the basis of the new estimated parameter vector at $t+1$, $\hat{\theta}(t+1)$, which will be available somewhere between $t+1$ and $t+2$. The *a posteriori* predicted output will be given by:

$$\begin{aligned} \hat{y}(t+1) &= \hat{y}[(t+1)|\hat{\theta}(t+1)] = -\hat{a}_1(t+1)y(t) + \hat{b}_1(t+1)u(t) \\ &= \hat{\theta}^T(t+1)\phi(t) \end{aligned} \tag{4.6}$$

One defines an *a priori* prediction error as:

$$\varepsilon^\circ(t+1) = y(t+1) - \hat{y}^\circ(t+1) \tag{4.7}$$

and an *a posteriori* prediction error as:

$$\varepsilon(t+1) = y(t+1) - \hat{y}(t+1) \tag{4.8}$$

The objective is to find a *recursive parameter adaptation algorithm* with memory which will minimize a certain criterion. The structure of such an algorithm is:

$$\hat{\theta}(t+1) = \hat{\theta}(t) + \Delta\hat{\theta}(t+1) = \hat{\theta}(t) + f[\hat{\theta}(t), \phi(t), \varepsilon^\circ(t+1)] \tag{4.9}$$

where the correcting term $f[.]$ is a function of $\hat{\theta}(t)$, $\phi(t)$, $\varepsilon^\circ(t+1)$.

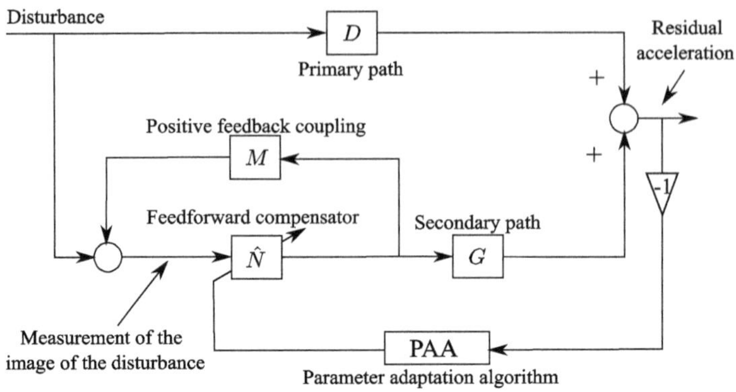

Fig. 4.2 Adaptive feedforward disturbance compensation

4.2.2 *Case (b): Adaptive Feedforward Compensation—Output Error*

Consider the basic scheme of adaptive feedforward disturbance compensation shown in Fig. 4.2. Further assume that the secondary path has a transfer function $G = 1$ and that there is no internal positive feedback, i.e., $M = 0$. Figure 4.3a represents this simplified configuration and its equivalent representations are shown in Fig. 4.3b, c. The equivalent representation shown in Fig. 4.3c is known as "output error" ($N = -D$ is unknown).

Assume that the equivalent primary path in Fig. 4.3c can be represented by the simple model (considered also in the previous section, (4.1)):

$$y(t+1) = -a_1 y(t) + b_1 u(t) = \theta^T(t)\phi(t) \tag{4.10}$$

where:

$$\theta^T = [a_1, b_1]; \ \phi^T(t) = [-y(t), u(t)] \tag{4.11}$$

Nevertheless, in this case, we do not have access to the output $y(t)$ of the primary path when the feedforward compensator acts. One has access only to the residual error (residual acceleration or force) denoted here $\varepsilon(t)$. So the predictor considered in the previous section cannot be used. Therefore, since the objective is to drive the residual error to zero, one can consider as an approximation of $y(t)$ the output of the predictor itself (which is measurable). The *output error adjustable predictor* (i.e., the feedforward compensator) is described by:

$$\hat{y}^\circ(t+1) = -\hat{a}_1(t)\hat{y}(t) + \hat{b}_1(t)u(t) = \hat{\theta}^T(t)\psi(t) \tag{4.12}$$

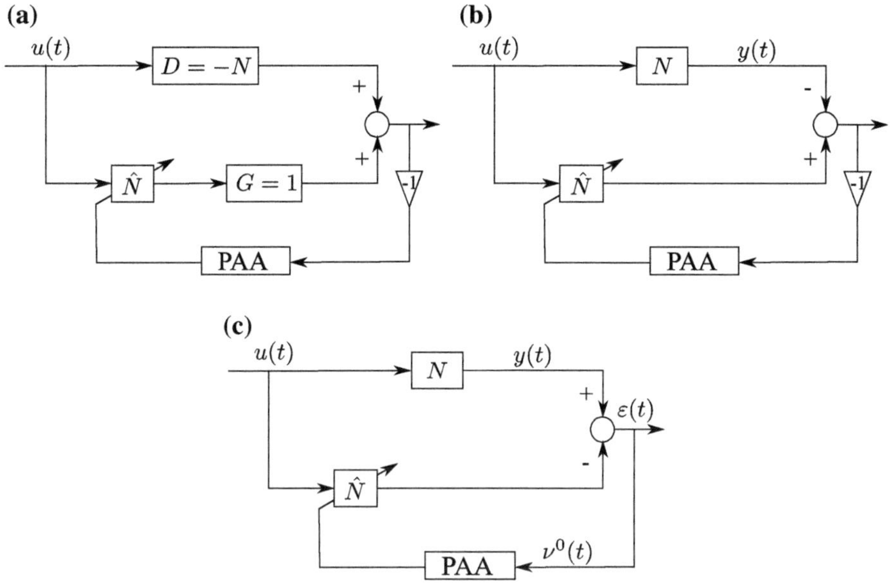

Fig. 4.3 Equivalent representation of the adaptive feedforward compensation for the case $G = 1$ and $M = 0$ (**a** $\Longrightarrow$ **b** $\Longrightarrow$ **c**)

where $\hat{y}^{\circ}(t+1)$ is the *a priori* output of the predictor, and:

$$\hat{y}(t+1) = -\hat{a}_1(t+1)\hat{y}(t) + \hat{b}_1(t+1)u(t) = \hat{\theta}^T(t+1)\psi(t) \tag{4.13}$$

is the *a posteriori* output of the predictor. One defines the vectors

$$\hat{\theta}^T(t) = [\hat{a}_1(t), \hat{b}_1(t)], \quad \psi^T(t) = [-\hat{y}(t), u(t)] \tag{4.14}$$

where $\hat{\theta}^T(t)$ is the vector of adjustable parameters and $\psi^T(t)$ is the observation vector. Since $\hat{y}(t)$ should converge asymptotically to $y(t)$, $\hat{y}(t)$ is an approximation of the output $y(t)$ which will improve over time.

The *a priori* prediction error will have the expression:

$$\varepsilon^{\circ}(t+1) = y(t+1) - \hat{y}^{\circ}(t+1) = \theta^T\phi(t) - \hat{\theta}^T(t)\psi(t) \tag{4.15}$$

and the *a posteriori* prediction error will be given by:

$$\varepsilon(t+1) = y(t+1) - \hat{y}(t+1) = \theta^T\phi(t) - \hat{\theta}^T(t+1)\psi(t) \tag{4.16}$$

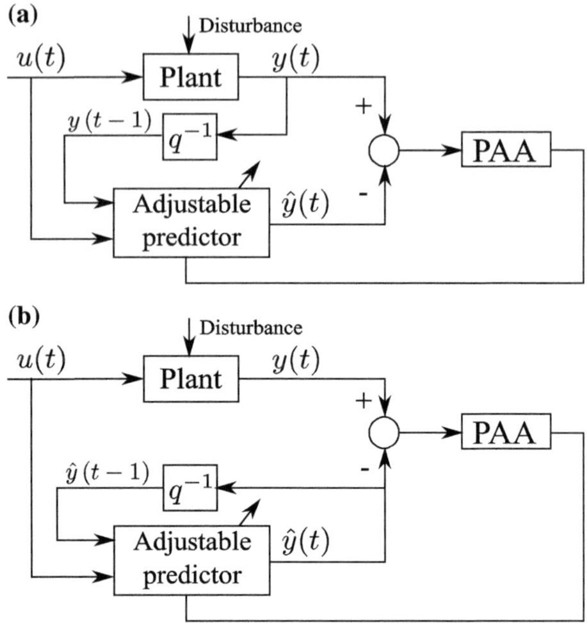

Fig. 4.4 Comparison between two adjustable predictor structures: **a** Equation error (recursive least squares predictor) and **b** Output error

The difference between the two types of predictors is illustrated in Fig. 4.4. Equation error predictors and output error predictors are also called *series-parallel predictor* and *parallel predictor*, respectively, in relation with the configurations shown in Fig. 4.4.

4.3 Basic Parameter Adaptation Algorithms

Several approaches can be considered for deriving parameter adaptation algorithms. We will consider first for pedagogical reasons the gradient technique followed by the least squares.

Nevertheless, it is the stability approach which will be used later for both synthesis and analysis of PAA.

4.3.1 Basic Gradient Algorithm

The aim of the basic gradient parameter adaptation algorithm is to minimize a quadratic criterion in terms of the *a priori* prediction error.

We will consider this approach first for the case of the equation error predictor (Eqs. (4.1)–(4.9)).

The objective is to find a recursive parameter adaptation algorithm with memory. The structure of such an algorithm is:

$$\hat{\theta}(t+1) = \hat{\theta}(t) + \Delta\hat{\theta}(t+1) = \hat{\theta}(t) + f[\hat{\theta}(t), \phi(t), \varepsilon^{\circ}(t+1)] \tag{4.17}$$

The correction term $f[\hat{\theta}(t), \phi(t), \varepsilon^{\circ}(t+1)]$ must depend solely on the information available at the instant $(t+1)$ when $y(t+1)$ is acquired (last measurement $y(t+1)$, $\hat{\theta}(t)$, and a finite amount of information at times $t, t-1, t-2, \ldots, t-n$). The correction term must enable the following criterion to be minimized at each step:

$$\min_{\hat{\theta}(t)} J(t+1) = [\varepsilon^{\circ}(t+1)]^2 \tag{4.18}$$

A solution can be provided by the gradient technique.

If the iso-criterion curves ($J = constant$) are represented in the plane of the parameters a_1 and b_1, concentric closed curves are obtained around the minimum value of the criterion, which is reduced to the point (a_1, b_1) corresponding to the parameters of the plant model. As the value of $J = const$ increases, the iso-criterion curves move further and further away from the minimum. This is illustrated in Fig. 4.5.

In order to minimize the value of the criterion, one moves in the opposite direction of the gradient to the corresponding iso-criterion curve. This will lead to a curve

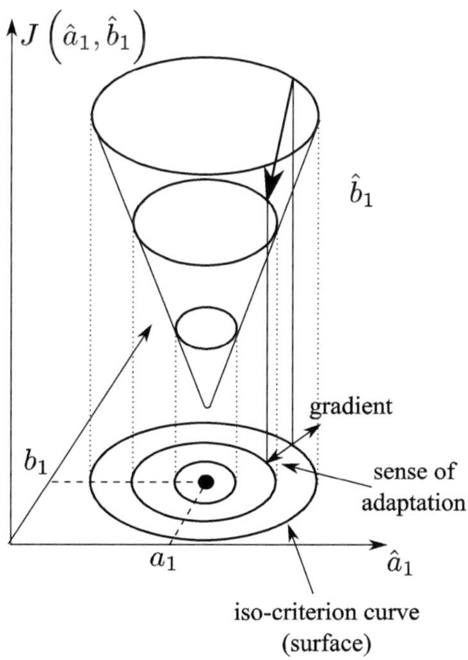

Fig. 4.5 Principle of the gradient method

corresponding to $J = const$, of a lesser value, as is shown in Fig. 4.5. The corresponding parametric adaptation algorithm will have the form:

$$\hat{\theta}(t+1) = \hat{\theta}(t) - F\frac{\partial J(t+1)}{\partial \hat{\theta}(t)} \tag{4.19}$$

where $F = \alpha I(\alpha > 0)$ is the matrix adaptation gain (I—unitary diagonal matrix) and $\partial J(t+1)/\partial \hat{\theta}(t)$ is the gradient of the criterion given in (4.18) with respect to $\hat{\theta}(t)$. From (4.18) one obtains:

$$\frac{1}{2}\frac{\partial J(t+1)}{\partial \hat{\theta}(t)} = \frac{\partial \varepsilon^{\circ}(t+1)}{\partial \hat{\theta}(t)}\varepsilon^{\circ}(t+1) \tag{4.20}$$

But:

$$\varepsilon^{\circ}(t+1) = y(t+1) - \hat{y}^{\circ}(t+1) = y(t+1) - \hat{\theta}^{T}(t)\phi(t) \tag{4.21}$$

and

$$\frac{\partial \varepsilon^{\circ}(t+1)}{\partial \hat{\theta}(t)} = -\phi(t) \tag{4.22}$$

Introducing (4.22) in (4.20), the parameter adaptation algorithm of (4.19) becomes:

$$\hat{\theta}(t+1) = \hat{\theta}(t) + F\phi(t)\varepsilon^{\circ}(t+1) \tag{4.23}$$

where F is the matrix adaptation gain.

There are two main choices for the adaptation gain:

1. $F = \alpha I; \alpha > 0$
2. $F > 0$ (positive definite matrix)[1]

The resulting algorithm has an integral structure. Therefore it has memory (for $\varepsilon^{0}(t+1) = 0$, $\hat{\theta}(t+1) = \hat{\theta}(t)$).

The geometrical interpretation of the PAA of (4.23) is given in Fig. 4.6 for the two choices of the adaptation gain.

For $F = \alpha I$, $\alpha > 0$, the correction is done in the direction of the observation vector (which in this case is the measurement vector) or within $\pm 90°$ around this direction when $F > 0$ (a positive definite matrix may cause a rotation of a vector for less than 90°).

The parameter adaptation algorithm given in (4.23) presents instability risks if a large adaptation gain (respectively a large α) is used. This can be understood by referring to Fig. 4.5. If the adaptation gain is large near the optimum, one can move

[1]A symmetric square matrix F is termed positive definite if $x^{T}Fx > 0$, $\forall\, x > 0$, $x \in \mathbb{R}^{n}$. In addition: (i) all the terms of the main diagonal are positive, (ii) the determinants of all the principal minors are positive.

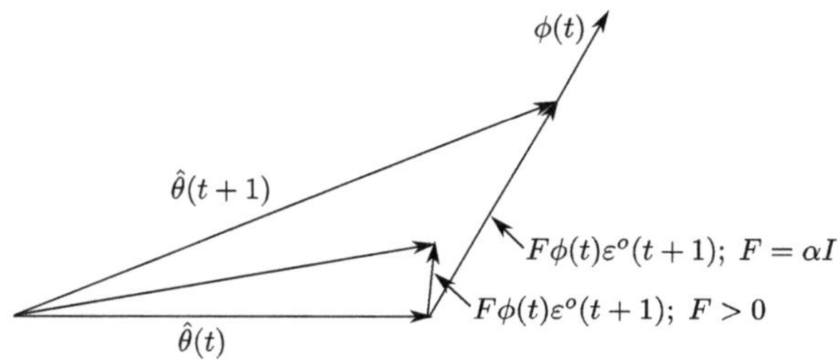

Fig. 4.6 Geometrical interpretation of the gradient adaptation algorithm

away from this minimum instead of getting closer. A necessary condition for stability (but not sufficient) is that for an adaptation gain $F = \alpha I$, α should satisfy:

$$\alpha < \frac{1}{\phi^T(t)\phi(t)}. \tag{4.24}$$

See [1] for details.

4.3.2 *Improved Gradient Algorithm*

4.3.2.1 Equation Error Predictor

In order to assure the stability of the PAA for any value of the adaptation gain α (or of the eigenvalues of the gain matrix F) in the case of the equation error predictor, the same gradient approach is used but a different criterion is considered:

$$\min_{\hat{\theta}(t+1)} J(t+1) = [\varepsilon(t+1)]^2. \tag{4.25}$$

The equation:

$$\frac{1}{2}\frac{\partial J(t+1)}{\partial \hat{\theta}(t+1)} = \frac{\partial \varepsilon(t+1)}{\partial \hat{\theta}(t+1)}\varepsilon(t+1) \tag{4.26}$$

is then obtained. From (4.6) and (4.8) it results that:

$$\varepsilon(t+1) = y(t+1) - \hat{y}(t+1) = y(t+1) - \hat{\theta}^T(t+1)\phi(t) \tag{4.27}$$

and, respectively:

$$\frac{\partial \varepsilon(t+1)}{\partial \hat{\theta}(t+1)} = -\phi(t) \tag{4.28}$$

Introducing (4.28) in (4.26), the parameter adaptation algorithm of (4.19) becomes:

$$\hat{\theta}(t+1) = \hat{\theta}(t) + F\phi(t)\varepsilon(t+1) \tag{4.29}$$

This algorithm depends on $\varepsilon(t+1)$, which is a function of $\hat{\theta}(t+1)$. For implementing this algorithm, $\varepsilon(t+1)$ must be expressed as a function of $\varepsilon^{\circ}(t+1)$, i.e., $\varepsilon(t+1) = f[\hat{\theta}(t), \phi(t), \varepsilon^{\circ}(t+1)]$.

Equation (4.27) can be rewritten as:

$$\varepsilon(t+1) = y(t+1) - \hat{\theta}^T(t)\phi(t) - [(\hat{\theta}(t+1) - \hat{\theta}(t)]^T\phi(t) \tag{4.30}$$

The first two terms of the right hand side correspond to $\varepsilon^{\circ}(t+1)$, and from (4.29), one obtains:

$$\hat{\theta}(t+1) - \hat{\theta}(t) = F\phi(t)\varepsilon(t+1) \tag{4.31}$$

which enables to rewrite (4.30) as:

$$\varepsilon(t+1) = \varepsilon^{\circ}(t+1) - \phi^T(t)F\phi(t)\varepsilon(t+1) \tag{4.32}$$

from which the desired relation between $\varepsilon(t+1)$ and $\varepsilon^{\circ}(t+1)$ is obtained:

$$\varepsilon(t+1) = \frac{\varepsilon^{\circ}(t+1)}{1 + \phi^T(t)F\phi(t)} \tag{4.33}$$

and the algorithm of (4.29) becomes:

$$\hat{\theta}(t+1) = \hat{\theta}(t) + \frac{F\phi(t)\varepsilon^{\circ}(t+1)}{1 + \phi^T(t)F\phi(t)} \tag{4.34}$$

which is a *stable algorithm* irrespective of the value of the gain matrix F (positive definite). For a stability analysis see Sect. 4.4.

The division by $1+\phi^T(t)F\phi(t)$ introduces a normalization with respect to F and $\phi(t)$ which reduces the sensitivity of the algorithm with respect to F and $\phi(t)$.

4.3.2.2 Output Error Predictor

We will turn now towards the use of the improved gradient algorithm for the output error predictor described by (4.12) and (4.13). To apply the improved gradient approach to the output error predictor we will need first an expression of the *a posteriori* prediction error featuring explicitly the difference between the unknown and the estimated parameter vector.

$$\begin{aligned}\varepsilon(t+1) &= y(t+1) - \hat{y}(t+1)\\ &= -a_1 y(t) + b_1 u(t) - [-\hat{a}_1(t+1)\hat{y}(t) + \hat{b}_1(t+1)u(t)] \pm a_1\hat{y}(t)\\ &= -a_1\varepsilon(t) - [a_1 - \hat{a}_1(t+1)]\hat{y}(t) + [b_1 - \hat{b}_1(t+1)]u(t)\\ &= -a_1\varepsilon(t) + [\theta - \hat{\theta}(t+1)]^T\psi(t)\end{aligned} \tag{4.35}$$

Passing the term $-a_1\varepsilon(t)$ on the left side one gets:

$$(1 + a_1 q^{-1})\varepsilon(t+1) = [\theta - \hat{\theta}(t+1)]^T\psi(t) \tag{4.36}$$

Defining:

$$A(q^{-1}) = 1 + a_1 q^{-1} \tag{4.37}$$

one gets:

$$\varepsilon(t+1) = \frac{1}{A(q^{-1})}[\theta - \hat{\theta}(t+1)]^T\psi(t) \tag{4.38}$$

The gradient for the improved gradient algorithm is given by (4.26)

$$\frac{1}{2}\frac{\partial J(t+1)}{\partial\hat{\theta}(t+1)} = \frac{\partial\varepsilon(t+1)}{\partial\hat{\theta}(t+1)}\varepsilon(t+1) \tag{4.39}$$

and using (4.38) one gets

$$\frac{\partial\varepsilon(t+1)}{\partial\hat{\theta}(t+1)} = -\frac{1}{A(q^{-1})}\psi(t) = -\psi_f(t) \tag{4.40}$$

Then the parameter adaptation algorithm becomes:

$$\hat{\theta}(t+1) = \hat{\theta}(t) + F\psi_f(t)\varepsilon(t+1) \tag{4.41}$$

The PAA (4.41) cannot be implemented as it is since $A(q^{-1})$ is unknown. Several approximations are currently used; however, the conditions assuring the asymptotic stability of the resulting algorithm have to be established. The various approximations are detailed below.

(1) Output Error Algorithm (OE)

In this algorithm one simply approximates $(1/A(q^{-1}))\psi(t)$ by $\psi(t)$, i.e,

$$\psi_f(t) = \psi(t) \tag{4.42}$$

and one gets:

$$\hat{\theta}(t+1) = \hat{\theta}(t) + F\psi(t)\varepsilon(t+1) \tag{4.43}$$

(2) Output Error with Filtered Observations (FOLOE)

Define a filter $L(q^{-1})$ and suppose that it is close to $A(q^{-1})$. Neglecting the non-commutativity of the time-varying operators, one can rewrite the equation of the *a posteriori* prediction as:

$$\varepsilon(t+1) = \frac{1}{A(q^{-1})}[\theta - \hat{\theta}(t+1)]^T \psi(t) \tag{4.44}$$

$$= \frac{L(q^{-1})}{A(q^{-1})}[\theta - \hat{\theta}(t+1)]^T \psi_f(t) \tag{4.45}$$

where:

$$\psi_f(t) = \frac{1}{L(q^{-1})}\psi(t) \tag{4.46}$$

and the gradient of the criterion becomes

$$\frac{\partial \nu(t+1)}{\partial \hat{\theta}(t+1)} = -\frac{L(q^{-1})}{A(q^{-1})}\psi_f(t) \simeq -\psi_f(t) \tag{4.47}$$

which, taking into account the proximity of L and A, will be approximated by $-\psi_f(t)$.

(3) Output Error with Adaptive Filtered Observations (AFOLOE)

Since during the evolution of the adaptation algorithms the estimations of $\hat{A}(t, q^{-1})$ will approach $A(q^{-1})$, one replaces the fixed filter L by

$$L(t, q^{-1}) = \hat{A}(t, q^{-1}) \tag{4.48}$$

For all these algorithms, the *a posteriori* prediction error is computed as:

$$\varepsilon(t+1) = \frac{\varepsilon^{\circ}(t+1)}{1 + \psi_f^T(t) F \psi_f(t)} \tag{4.49}$$

(4) Output Error with Fixed Compensator (OEFC)

In this algorithm one defines an adaptation error as a filtered prediction error:

$$\nu(t+1) = D(q^{-1})\varepsilon(t+1) \tag{4.50}$$

where:

$$D(q^{-1}) = 1 + \sum_{i=1}^{n_D} d_i q^{-i} \tag{4.51}$$

is an asymptotically stable polynomial with $n_D \leq n_A$ (n_A—degree of the polynomial A) and we would like to minimize a criterion as in (4.25) but for $\nu(t+1)$, i.e.,

$$\min_{\hat{\theta}(t+1)} J(t+1) = [\nu(t+1)]^2 \tag{4.52}$$

In this case

$$\frac{\partial \nu(t+1)}{\partial \hat{\theta}(t+1)} = -\frac{D(q^{-1})}{A(q^{-1})}\psi(t) = -\psi_f(t) \tag{4.53}$$

Provided that $D(q^{-1})$ and $A(q^{-1})$ are close[2] one can use the following approximation:

$$\psi_f(t) \simeq \psi(t) \tag{4.54}$$

and the PAA takes the form

$$\hat{\theta}(t+1) = \hat{\theta}(t) + F\psi(t)\nu(t+1) \tag{4.55}$$

To make the algorithm (4.55) implementable, one has to give a relation between the *a posteriori* adaptation error given in (4.50) and the *a priori* adaptation error defined as:

$$\nu^\circ(t+1) = \varepsilon^\circ(t+1) + \sum_{i=1}^{n_D} d_i \varepsilon(t+1-i) \tag{4.56}$$

Note that the *a posteriori* prediction errors $\varepsilon(t), \varepsilon(t-1), \ldots$ are available at $t+1$. Subtracting (4.56) from (4.50), one gets:

$$\begin{aligned} \nu(t+1) - \nu^\circ(t+1) &= \varepsilon(t+1) - \varepsilon^\circ(t+1) \\ &= -\left[\hat{\theta}(t+1) - \hat{\theta}(t)\right]^T \psi(t) \end{aligned} \tag{4.57}$$

But, from (4.55), one obtains:

$$\hat{\theta}(t+1) - \hat{\theta}(t) = F\psi(t)\nu(t+1) \tag{4.58}$$

and (4.57) becomes:

$$\nu(t+1) + (F\psi(t)\nu(t+1))^T \psi(t) = \nu^\circ(t+1) \tag{4.59}$$

[2] As it will be shown in Sect. 4.4, this closeness is characterized in terms of the property that $D(z^{-1})/A(z^{-1})$ should be a strictly positive real transfer function.

from which one obtains:

$$\nu(t+1) = \frac{\nu^\circ(t+1)}{1 + \psi^T(t) F \psi(t)} \tag{4.60}$$

4.3.3 Recursive Least Squares Algorithm

When using the Improved Gradient Algorithm, $\varepsilon^2(t+1)$ is minimized at each step or, to be more precise, one moves in the quickest decreasing direction of the criterion, with a step depending on F. The minimization of $\varepsilon^2(t+1)$ at each step does not necessarily lead to the minimization of:

$$\sum_{i=1}^{t} \varepsilon^2(i+1)$$

on a time horizon, as is illustrated in Fig. 4.7. In fact, in the vicinity of the optimum, if the gain is not low enough, oscillations may occur around the minimum. On the other hand, in order to obtain a satisfactory convergence speed at the beginning when the optimum is far away, a high adaptation gain is preferable. In fact, the least squares algorithm offers such a variation profile for the adaptation gain. The same equations as in the gradient algorithm for the equation error configuration are considered for the plant, the prediction model, and the prediction errors, namely Eqs. (4.1)–(4.8).

The aim is to find a recursive algorithm of the form of (4.9) which minimizes the *least squares* criterion:

$$\min_{\hat{\theta}(t)} J(t) = \sum_{i=1}^{t} [y(i) - \hat{\theta}^T(t)\phi(i-1)]^2 \tag{4.61}$$

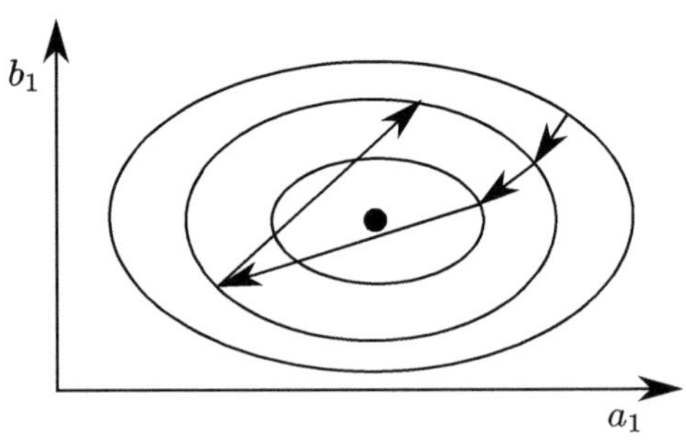

Fig. 4.7 Evolution of an adaptation algorithm of the gradient type

The term $\hat{\theta}(t)^T\phi(i-1)$ corresponds to:

$$\hat{\theta}^T(t)\phi(i-1) = -\hat{a}_1(t)y(i-1) + \hat{b}_1(t)u(i-1) = \hat{y}[i \mid \hat{\theta}(t)] \tag{4.62}$$

Therefore, this is the prediction of the output at instant i $(i \leq t)$ based on the parameter estimate at instant t obtained using t measurements.

First, a parameter θ must be estimated at instant t such that it minimizes the sum of the squares of the differences between the output of the plant and the output of the prediction model on a horizon of t measurements. The value of $\hat{\theta}(t)$, which minimizes the criterion (4.61), is obtained by seeking the value that cancels $\partial J(t)/\partial\hat{\theta}(t)$:

$$\frac{\partial J(t)}{\partial \hat{\theta}(t)} = -2\sum_{i=1}^{t}[y(i) - \hat{\theta}^T(t)\phi(i-1)]\phi(i-1) = 0 \tag{4.63}$$

From (4.63), taking into account that:

$$[\hat{\theta}^T(t)\phi(i-1)]\phi(i-1) = \phi(i-1)\phi^T(i-1)\hat{\theta}(t)$$

one obtains:

$$\left[\sum_{i=1}^{t}\phi(i-1)\phi^T(i-1)\right]\hat{\theta}(t) = \sum_{i=1}^{t} y(i)\phi(i-1)$$

and left multiplying by[3]:

$$\left[\sum_{i=1}^{t}\phi(i-1)\phi^T(i-1)\right]^{-1}$$

one obtains:

$$\hat{\theta}(t) = \left[\sum_{i=1}^{t}\phi(i-1)\phi^T(i-1)\right]^{-1}\sum_{i=1}^{t} y(i)\phi(i-1) = F(t)\sum_{i=1}^{t} y(i)\phi(i-1) \tag{4.64}$$

in which:

$$F(t)^{-1} = \sum_{i=1}^{t}\phi(i-1)\phi^T(i-1) \tag{4.65}$$

[3]It is assumed that the matrix $\sum_{i=1}^{t}\phi(i-1)\phi^T(i-1)$ is invertible. This corresponds to an *excitation* condition.

This estimation algorithm is not recursive. In order to obtain a recursive algorithm, the estimation of $\hat{\theta}(t+1)$ is considered:

$$\hat{\theta}(t+1) = F(t+1)\sum_{i=1}^{t+1} y(i)\phi(i-1) \tag{4.66}$$

$$F(t+1)^{-1} = \sum_{i=1}^{t+1} \phi(i-1)\phi^T(i-1) = F(t)^{-1} + \phi(t)\phi^T(t) \tag{4.67}$$

We can now express $\hat{\theta}(t+1)$ as a function of $\hat{\theta}(t)$:

$$\hat{\theta}(t+1) = \hat{\theta}(t) + \Delta\hat{\theta}(t+1) \tag{4.68}$$

From (4.66) one has:

$$\hat{\theta}(t+1) = F(t+1)\left[\sum_{i=1}^{t} y(i)\phi(i-1) + y(t+1)\phi(t)\right] \tag{4.69}$$

Taking into account (4.64), (4.69) can be rewritten as:

$$\hat{\theta}(t+1) = F(t+1)[F(t)^{-1}\hat{\theta}(t) + y(t+1)\phi(t)] \tag{4.70}$$

From (4.67), after post-multiplying both sides by $\hat{\theta}(t)$, one gets:

$$F(t)^{-1}\hat{\theta}(t) = F(t+1)^{-1}\hat{\theta}(t) - \phi(t)\phi^T(t)\hat{\theta}(t) \tag{4.71}$$

and (4.70) becomes:

$$\hat{\theta}(t+1) = F(t+1)\left(F(t+1)^{-1}\hat{\theta}(t) + \phi(t)[y(t+1) - \hat{\theta}^T(t)\phi(t)]\right) \tag{4.72}$$

Taking into account the expression of $\varepsilon^\circ(t+1)$ given by (4.21), the result is:

$$\hat{\theta}(t+1) = \hat{\theta}(t) + F(t+1)\phi(t)\varepsilon^\circ(t+1) \tag{4.73}$$

The adaptation algorithm of (4.73) has a recursive form similar to the basic gradient algorithm given in (4.23) except that the gain matrix $F(t+1)$ is now time-varying since it depends on the measurements (it automatically corrects the gradient direction and the step length). A recursive formula for $F(t+1)$ remains to be given from the recursive formula $F^{-1}(t+1)$ given in (4.67). This is obtained by using the *matrix inversion* lemma.

Matrix Inversion Lemma *Let F be a $(n \times n)$ dimensional nonsingular matrix, R a $(m \times m)$ dimensional nonsingular matrix and H a $(n \times m)$ dimensional matrix of maximum rank, then the following identity holds:*

$$(F^{-1} + HR^{-1}H^T)^{-1} = F - FH(R + H^T FH)^{-1}H^T F \tag{4.74}$$

Proof By direct multiplication one finds that:

$$[F - FH(R + H^T FH)^{-1}H^T F][F^{-1} + HR^{-1}H^T] = I$$

which ends the proof. □

For the case of (4.67), one chooses $H = \phi(t)$, $R = 1$ and one obtains from Eqs. (4.67) and (4.74):

$$F(t+1) = F(t) - \frac{F(t)\phi(t)\phi^T(t)F(t)}{1 + \phi^T(t)F(t)\phi(t)} \tag{4.75}$$

Putting together the different equations, a first formulation of the *Recursive Least Squares (RLS)* parameter adaptation algorithm (PAA) is given next:

$$\hat{\theta}(t+1) = \hat{\theta}(t) + F(t+1)\phi(t)\varepsilon^\circ(t+1) \tag{4.76}$$

$$F(t+1) = F(t) - \frac{F(t)\phi(t)\phi^T(t)F(t)}{1 + \phi^T(t)F(t)\phi(t)} \tag{4.77}$$

$$\varepsilon^\circ(t+1) = y(t+1) - \hat{\theta}^T(t)\phi(t) \tag{4.78}$$

In practice, the algorithm is started up at $t = 0$ by choosing:

$$F(0) = \frac{1}{\delta}I = (GI)I;\ 0 < \delta \ll 1 \tag{4.79}$$

An equivalent form of this algorithm is obtained by introducing the expression of $F(t+1)$ given by (4.77) in (4.76), where:

$$\hat{\theta}(t+1) - \hat{\theta}(t) = F(t+1)\phi(t)\varepsilon^\circ(t+1) = F(t)\phi(t)\frac{\varepsilon^\circ(t+1)}{1 + \phi^T(t)F(t)\phi(t)} \tag{4.80}$$

Nevertheless, from (4.7), (4.8) and (4.80), one obtains:

$$\begin{aligned}\varepsilon(t+1) &= y(t+1) - \hat{\theta}^T(t+1)\phi(t) = y(t+1) - \hat{\theta}(t)\phi(t) \\ &\quad - [\hat{\theta}(t+1) - \hat{\theta}(t)]^T\phi(t) = \varepsilon^\circ(t+1) \\ &\quad - \phi^T(t)F(t)\phi(t)\frac{\varepsilon^\circ(t+1)}{1 + \phi^T(t)F(t)\phi(t)} = \frac{\varepsilon^\circ(t+1)}{1 + \phi^T(t)F(t)\phi(t)}\end{aligned} \tag{4.81}$$

which expresses the relation between the *a posteriori* prediction error and the *a priori* prediction error. Using this relation in (4.80), an equivalent form of the parameter adaptation algorithm for the recursive least squares is obtained:

$$\hat{\theta}(t+1) = \hat{\theta}(t) + F(t)\phi(t)\varepsilon(t+1) \tag{4.82}$$

$$F(t+1)^{-1} = F(t)^{-1} + \phi(t)\phi^T(t) \tag{4.83}$$

$$F(t+1) = F(t) - \frac{F(t)\phi(t)\phi^T(t)F(t)}{1+\phi^T(t)F(t)\phi(t)} \tag{4.84}$$

$$\varepsilon(t+1) = \frac{y(t+1) - \hat{\theta}^T(t)\phi(t)}{1+\phi^T(t)F(t)\phi(t)} \tag{4.85}$$

The recursive least squares algorithm is an algorithm with a decreasing adaptation gain. This is clearly seen if the estimation of a single parameter is considered. In this case, $F(t)$ and $\phi(t)$ are scalars, and (4.84) becomes:

$$F(t+1) = \frac{F(t)}{1+\phi(t)^2F(t)} \leq F(t); \quad \phi(t),\ F(t) \in R^1$$

The same conclusion is obtained by observing that $F(t+1)^{-1}$ is the output of an integrator which has as input $\phi(t)\phi^T(t)$. Since $\phi(t)\phi^T(t) \geq 0$, one concludes that if $\phi(t)\phi^T(t) > 0$ in the average, then $F(t)^{-1}$ will tend towards infinity, i.e., $F(t)$ will tend towards zero.

The recursive least squares algorithm in fact gives less and less weight to the new prediction errors and thus to the new measurements. Consequently, this type of variation of the adaptation gain is not suitable for the estimation of time-varying parameters, and other variation profiles for the adaptation gain must therefore be considered.

The least squares algorithm presented up to now for $\hat{\theta}(t)$ and $\phi(t)$ of dimension 2 may be generalized for any dimensions resulting from the description of discrete-time systems of the form:

$$y(t) = \frac{q^{-d}B(q^{-1})}{A(q^{-1})}u(t) \tag{4.86}$$

where:

$$A(q^{-1}) = 1 + a_1q^{-1} + \cdots + a_{n_A}q^{-n_A} \tag{4.87}$$

$$B(q^{-1}) = b_1q^{-1} + \cdots + b_{n_B}q^{-n_B} \tag{4.88}$$

Equation (4.86) can be written in the form:

$$y(t+1) = -\sum_{i=1}^{n_A} a_i y(t+1-i) + \sum_{i=1}^{n_B} b_i u(t-d-i+1) = \theta^T\phi(t) \tag{4.89}$$

in which:

$$\theta^T = [a_1, \ldots, a_{n_A}, b_1, \ldots, b_{n_B}] \tag{4.90}$$

$$\phi^T(t) = [-y(t), \ldots, -y(t-n_A+1), u(t-d), \ldots, u(t-d-n_B+1)] \tag{4.91}$$

The *a priori* adjustable predictor is given in the general case by:

$$\hat{y}^\circ(t+1) = -\sum_{i=1}^{n_A} \hat{a}_i(t)y(t+1-i) + \sum_{i=1}^{n_B} \hat{b}_1(t)u(t-d-i+1) = \hat{\theta}^T(t)\phi(t) \tag{4.92}$$

in which:

$$\hat{\theta}^T(t) = [\hat{a}_1(t), \ldots, \hat{a}_{n_A}(t), \hat{b}_1(t), \ldots, \hat{b}_{n_B}(t)] \tag{4.93}$$

and for the estimation of $\hat{\theta}(t)$, the algorithm given in (4.82) through (4.85) is used, with the appropriate dimension for $\hat{\theta}(t)$, $\phi(t)$, and $F(t)$.

4.3.4 Choice of the Adaptation Gain

The recursive formula for the inverse of the adaptation gain $F(t+1)^{-1}$ given by (4.83) is generalized by introducing two weighting sequences $\lambda_1(t)$ and $\lambda_2(t)$, as indicated below:

$$\begin{gathered} F(t+1)^{-1} = \lambda_1(t)F(t)^{-1} + \lambda_2(t)\phi(t)\phi^T(t) \\ 0 < \lambda_1(t) \leq 1\ ;\ 0 \leq \lambda_2(t) < 2\ ;\ F(0) > 0 \end{gathered} \tag{4.94}$$

Note that $\lambda_1(t)$ and $\lambda_2(t)$ in (4.94) have the opposite effect. $\lambda_1(t) < 1$ tends to increase the adaptation gain (the inverse of the gain decreases); $\lambda_2(t) > 0$ tends to decrease the adaptation gain (the inverse of the gain increases). For each choice of sequences, $\lambda_1(t)$ and $\lambda_2(t)$ corresponds a *variation profile* of the adaptation gain and an interpretation in terms of the error criterion, which is minimized by the PAA. Equation (4.94) allows to interpret the inverse of the adaptation gain as the output of a filter $\lambda_2/(1-\lambda_1 q^{-1})$ having as input $\phi(t)\phi^T(t)$ and $F(0)^{-1}$ as an initial condition.

Using the *matrix inversion lemma* given by (4.74), one obtains from (4.94):

$$F(t+1) = \frac{1}{\lambda_1(t)}\left[F(t) - \frac{F(t)\phi(t)\phi^T(t)F(t)}{\frac{\lambda_1(t)}{\lambda_2(t)} + \phi^T(t)F(t)\phi(t)}\right] \tag{4.95}$$

Next, a certain number of choices for $\lambda_1(t)$ and $\lambda_2(t)$ and their interpretations will be given.

A1. Decreasing (Vanishing) Gain (Basic RLS)

In this case:

$$\lambda_1(t) = \lambda_1 = 1; \ \lambda_2(t) = 1 \tag{4.96}$$

and $F(t+1)^{-1}$ is given by (4.83), which leads to a decreasing adaptation gain. The minimized criterion is that of (4.61). This type of profile is suited to the estimation of the parameters of stationary systems or for the self-tuning operation of adaptive controllers or adaptive feedforward compensators.

A2. Constant Forgetting Factor

In this case:

$$\lambda_1(t) = \lambda_1; \ 0 < \lambda_1 < 1; \ \lambda_2(t) = \lambda_2 = 1 \tag{4.97}$$

The typical values for λ_1 are:

$$\lambda_1 = 0.95 \text{ to } 0.99$$

The criterion to be minimized will be:

$$J(t) = \sum_{i=1}^{t} \lambda_1^{(t-i)} [y(i) - \hat{\theta}^T(t)\phi(i-1)]^2 \tag{4.98}$$

The effect of $\lambda_1(t) < 1$ is to introduce increasingly weaker weighting on the old data $(i < t)$. This is why λ_1 is known as the *forgetting factor*. The maximum weight is given to the most recent error.

This type of profile is suited to the estimation of the parameters of slowly time-varying systems.

Remark The use of a constant forgetting factor without the monitoring of the maximum value of $F(t)$ causes problems if the $\{\phi(t)\phi^T(t)\}$ sequence becomes null in the average (steady state case) and the **adaptation gain will tend towards infinity**. In this case:

$$F(t+i)^{-1} = (\lambda_1)^i F(t)^{-1}$$

and

$$F(t+i) = (\lambda_1)^{-i} F(t).$$

For $\lambda_1 < 1$, $\lim_{i \to \infty} (\lambda_1)^{-i} = \infty$ and $F(t+i)$ will become asymptotically unbounded.

A3. Variable Forgetting Factor

In this case:

$$\lambda_2(t) = \lambda_2 = 1 \tag{4.99}$$

and the forgetting factor $\lambda_1(t)$ is given by:

$$\lambda_1(t) = \lambda_0\lambda_1(t-1) + 1 - \lambda_0 \; ; \; 0 < \lambda_0 < 1 \tag{4.100}$$

the typical values being:

$$\lambda_1(0) = 0.95 \text{ to } 0.99 \; ; \; \lambda_0 = 0.5 \text{ to } 0.99$$

($\lambda_1(t)$ can be interpreted as the output of a first order filter $(1-\lambda_0)\,/\left(1-\lambda_0 q^{-1}\right)$ with a unitary steady state gain and an initial condition $\lambda_1(0)$).

Relation (4.100) leads to a forgetting factor that asymptotically tends towards 1 (the adaptation gain tends towards a decreasing gain).

This type of profile is recommended for the model identification of stationary systems, since it avoids a too rapid decrease of the adaptation gain, thus generally resulting in an acceleration of the convergence (by maintaining a high gain at the beginning when the estimates are at a great distance from the optimum).

A4. Constant Trace

In this case, $\lambda_1(t)$ and $\lambda_2(t)$ are automatically chosen at each step in order to ensure a constant trace of the gain matrix (constant sum of the diagonal terms):

$$trF(t+1) = trF(t) = trF(0) = nGI \tag{4.101}$$

in which n is the number of parameters and GI the initial gain (typical values: $GI = 0.01$ to 4), the matrix $F(0)$ having the form:

$$F(0) = \begin{bmatrix} GI & & 0 \\ & \ddots & \\ 0 & & GI \end{bmatrix} \tag{4.102}$$

Using this technique, at each step there is a movement in the optimal direction of the RLS, but the gain is maintained approximately constant. The values of $\lambda_1(t)$ and $\lambda_2(t)$ are determined from the equation:

$$trF(t+1) = \frac{1}{\lambda_1(t)} tr\left[F(t) - \frac{F(t)\phi(t)\phi^T(t)F(t)}{\alpha(t) + \phi^T(t)F(t)\phi(t)}\right] \tag{4.103}$$

fixing the ratio $\alpha(t) = \lambda_1(t)/\lambda_2(t)$ ((4.103) is obtained from (4.95)).

This type of profile is suited to the model identification of systems with time-varying parameters and for adaptive control with non-vanishing adaptation.

A5. Decreasing Gain + Constant Trace

In this case, *A1* is switched to *A4* when:

$$trF(t) \le nG \; ; \; G = 0.01 \text{ to } 4 \tag{4.104}$$

in which G is chosen in advance. This profile is suited to the model identification of time variable systems and for adaptive control in the absence of initial information on the parameters.

A6. Variable Forgetting Factor + Constant Trace

In this case *A3* is switched to *A4* when:

$$trF(t) \leq nG \tag{4.105}$$

The use is the same as for *A5*.

A7. Constant Gain (Improved Gradient Algorithm)

In this case:

$$\lambda_1(t) = \lambda_1 = 1 \ ; \ \lambda_2(t) = \lambda_2 = 0 \tag{4.106}$$

and thus from (4.95), one obtains:

$$F(t+1) = F(t) = F(0) \tag{4.107}$$

The improved gradient adaptation algorithm given by (4.29) or (4.34) is then obtained.

This type of adaptation gain results in performances that are in general inferior to those provided by the *A1, A2, A3*, and *A4* profiles, but it is simpler to implement.

Choice of the Initial Gain $F(0)$

The initial gain $F(0)$ is usually chosen as a diagonal matrix of the form given by (4.79) and, respectively, (4.102).

In the absence of initial information upon the parameters to be estimated (typical value of initial estimates $=$ 0), a high initial gain (GI) is chosen.[4] A typical value is $GI = 1000$ (but higher values can be chosen). If an initial parameter estimation is available (resulting for example from a previous identification), a low initial gain is chosen. In general, in this case $GI \leq 0.1$. In adaptive regulation schemes in general, the initial trace of the adaptation gain ($n \times GI$, $n =$ number of parameters) is chosen larger but of the same order of magnitude as the desired constant trace.

4.3.4.1 Parameter Adaptation Algorithms with Scalar Adaptation Gain

This concerns an extension of PAA with constant adaptation gain of the form $F = \alpha I$ for $\alpha > 1$ (see the improved gradient algorithm Sect. 4.3.2) for the cases where $\alpha(t) = 1/\beta(t)$, i.e.,

$$F(t) = \alpha(t)I = \frac{1}{\beta(t)}I \tag{4.108}$$

[4] It can be shown that the size of the adaptation gain is related to the parameter error [1].

Some PAA's of this type are mentioned next:
(1) Improved Gradient

$$\beta(t) = const = 1/\alpha > 0 \Longrightarrow F(t) = F(0) = \alpha I \tag{4.109}$$

(2) Stochastic Approximation

$$\beta(t) = t \Longrightarrow F(t) = \frac{1}{t} I \tag{4.110}$$

This is the simplest PAA with time decreasing adaptation gain (very useful for the analysis of PAA in the presence of stochastic disturbances).
(3) Controlled Adaptation Gain

$$\begin{aligned} \beta(t+1) &= \lambda_1(t)\beta(t) + \lambda_2(t)\phi^T(t)\phi(t) \\ \beta(0) > 0 \ ; \ 0 &< \lambda_1(t) \leq 1 \ ; \ 0 \leq \lambda_2(t) < 2 \end{aligned} \tag{4.111}$$

The principal interest in using these algorithms is a simpler implementation than those using a matrix adaptation gain updating. Their disadvantage is that their performances are in general lower than those of the PAA using a matrix adaptation gain.

4.3.5 An Example

The influence of the choice of the adaptation gain profile will be illustrated in the applications considered in Chaps. 12, 13, 15 and 16. The subsequent example will try just to illustrate the influence of the adaptation gain profile upon the estimation of unknown but constant parameters in the presence of measurement noise (situation encountered in system identification). The simulated system is of the form:

$$G(q^{-1}) = \frac{q^{-1}(b_1 q^{-1} + b_2 q^{-2})}{1 + a_1 q^{-1} + a_2 q^{-2}} \tag{4.112}$$

The file *T2.mat* (available on the book website) contains 256 input/output data. The output of the system is disturbed by a measurement noise. The parameters will be estimated using an Equation Error Predictor combined with the Parameter Adaptation Algorithm given in Eqs. (4.76)–(4.78) but using various profiles of the adaptation gain generated with Eq. (4.94). It is known that using this predictor the estimation of the parameters will be biased in the presence of noise (there will be an error) [2],[5]

[5]In Chap. 5, it will be shown how the predictor should be modified in order to obtain unbiased estimated parameters.

however the objective here is just to illustrate the influence of the adaptation gain upon the evolution of the estimated parameters.

Figure 4.8 shows the evolution of the estimated parameters when using a *decreasing adaptation gain* (A.1) which corresponds to the classical RLS algorithm. One can see that despite the presence of measurement noise the parameters converge toward constant values. The evolution of the trace of the adaptation gain is illustrated in Fig. 4.9. One can also see in this figure the evolution of the trace of the adaptation gain when using a *variable forgetting factor* (A3). It keeps a slightly higher gain transiently which will influence slightly the speed of evolution of the estimated parameters.

Figure 4.10 shows the evolution of the estimated parameters when using a *decreasing adaptation gain + constant trace* (A.5). Since the adaptation gain never goes to zero the parameters will not converge toward a constant value in the presence of noise but the capability of tracking parameters variation is assured. The evolution of the trace of the adaptation gain in this case is also shown in Fig. 4.9.

Figure 4.11 shows the evolution of the estimated parameters when using a *diagonal constant adaptation gain matrix* (A.7) which corresponds to the improved gradient algorithm. One can see that the adaptation transient is longer than for the *decreasing adaptation gain + constant trace* and of course the estimated parameters will not converge toward a constant value.

4.4 Stability of Parameter Adaptation Algorithms

From the point of view of real-time identification and adaptive control, the parameter adaptation algorithms are supposed to operate on a very large number of measure-

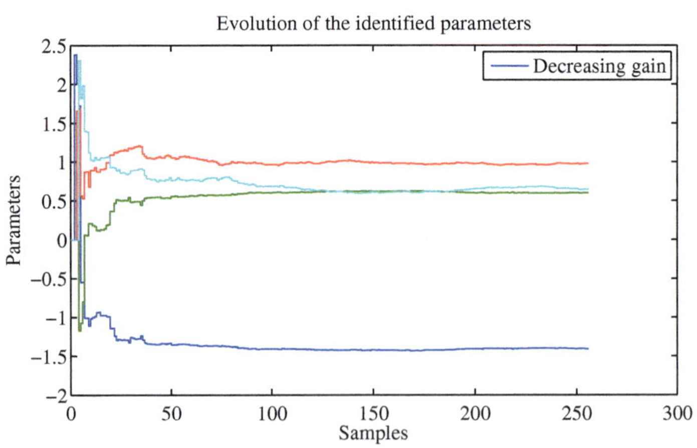

Fig. 4.8 Parameters evolution for decreasing adaptation gain algorithm

ments ($t \rightarrow \infty$). Therefore, it is necessary to examine the properties of parameter adaptation algorithms as $t \rightarrow \infty$. Specifically, one should study the conditions which guarantee:

$$\lim_{t\rightarrow\infty} \varepsilon(t+1) = 0. \tag{4.113}$$

This corresponds to the study of the stability of parameter adaptation algorithms. Conversely, other PAA can be derived starting from the stability condition.

Equivalent feedback representation of the full parameter estimation schemes is extremely helpful on one hand for deriving stability conditions and on the other hand for understanding the sense of the stability conditions.

4.4.1 Equivalent Feedback Representation of the Adaptive Predictors

To illustrate this approach we will consider the output error algorithm. The basic description of the adjustable output error predictor has been presented in Sect. 4.3, Eqs. (4.35)–(4.41).

The objective is now to write an equation for the *a posteriori* prediction error as a function of the parameter error. From (4.38) one obtains:

$$\varepsilon(t+1) = -\frac{1}{A(q^{-1})}\left(\hat{\theta}(t+1) - \theta\right)^T \psi^T(t) \tag{4.114}$$

$$= \frac{1}{A(q^{-1})}\left(-\psi^T(t)\tilde{\theta}(t+1)\right) \tag{4.115}$$

where:

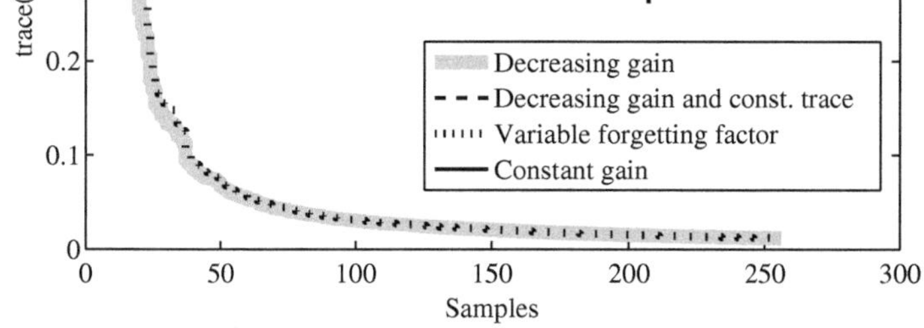

Fig. 4.9 Evolution of the trace of the adaptation gain matrix (vertical zoom within 0 and 0.5)

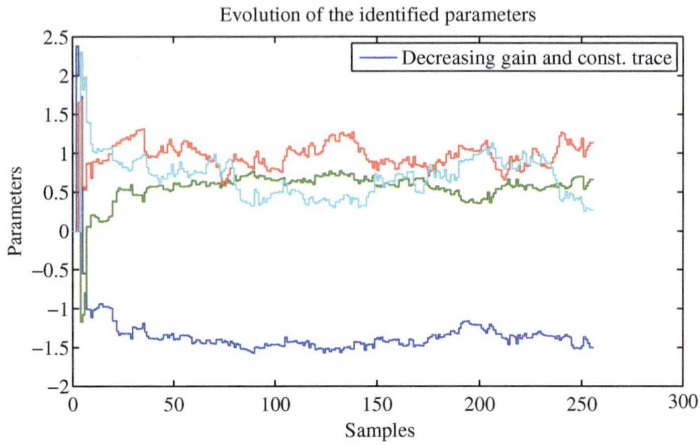

Fig. 4.10 Parameters evolution for decreasing adaptation gain combined with constant trace algorithm

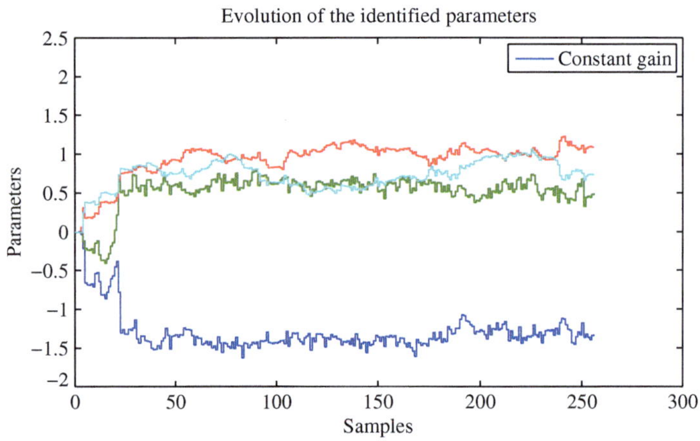

Fig. 4.11 Parameters evolution for constant adaptation gain algorithm

$$\tilde{\theta}(t+1) = \hat{\theta}(t+1) - \theta \tag{4.116}$$

This result remains valid even for higher order predictors where a_1 is replaced by $A^*(q-1) = a_1 + a_2 q^{-1} + \cdots + a_{n_A} q^{-n_A}$. In other words, the *a posteriori* prediction error is the output of a linear block characterized by a transfer function $1/A(z^{-1})$, whose input is $-\psi^T(t)\tilde{\theta}(t+1)$.

Once the equation for the *a posteriori* prediction error has been derived, the PAA synthesis problem can be formulated as a stability problem:

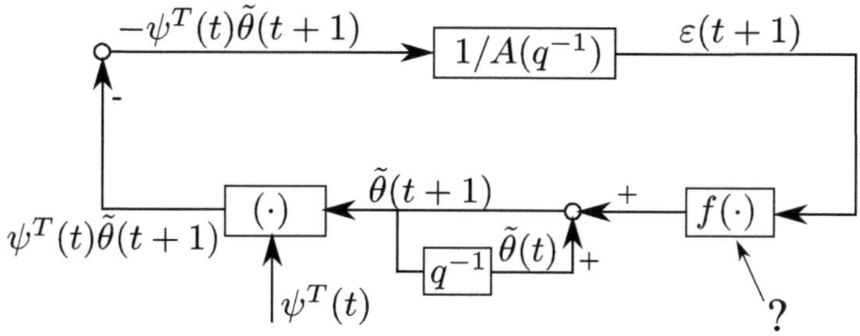

Fig. 4.12 Equivalent feedback system associated to the output error predictor

Find a PAA of the form:

$$\hat{\theta}(t+1) = \hat{\theta}(t) + f_\theta[\psi(t), \hat{\theta}(t), \varepsilon(t+1)] \tag{4.117}$$

$$\varepsilon(t+1) = f_\varepsilon[\psi(t), \hat{\theta}(t), \varepsilon^\circ(t+1)] \tag{4.118}$$

such that $\lim_{t\to\infty} \varepsilon(t+1) = 0$ *for all initial conditions* $\varepsilon(0), \hat{\theta}(0)$ *(or* $\tilde{\theta}(0)$*).*

Note that the structure of (4.117) assures the memory of the PAA (integral form), but other structures can be considered. The structure of (4.118) assures the causality of the algorithm.

From (4.117) subtracting θ from both sides, one gets:

$$\tilde{\theta}(t+1) = \tilde{\theta}(t) + f_\theta[\psi(t), \hat{\theta}(t), \varepsilon(t+1)] \tag{4.119}$$

and multiplying both sides by $\psi^T(t)$, yields:

$$\psi^T(t)\tilde{\theta}(t+1) = \psi^T(t)\tilde{\theta}(t) + \psi^T(t) f_\theta[\psi(t), \hat{\theta}(t), \varepsilon(t+1)] \tag{4.120}$$

Eqs. (4.115), (4.119), and (4.120) define an equivalent feedback system associated to the *output error* predictor as shown in Fig. 4.12.

The stability of the output predictor estimator is directly related to the stability of the equivalent nonlinear time-varying feedback system shown in Fig. 4.12. The major complication in the analysis of the stability of this configuration comes from the presence of a linear transfer function different from 1 (unity) on the feedforward path. The same analysis in the case of an equation error predictor shows that the feedforward block will have a transfer function equal to 1.

Therefore, both for taking into account various time-varying profiles for the adaptation gain and the possible presence of a transfer function in the feedforward path of the equivalent system, it is necessary to take a stability approach both for synthesis and analysis of the adaptation algorithms for various predictor configurations.

4.4.2 A General Structure and Stability of PAA

Two elements are crucial for the analysis of the adaptive schemes:

- the structure of the parameter adaptation algorithm; and
- the structure of the equation governing the generation of the adaptation error.

One can consider as a general structure for the PAA (integral type):

$$\hat{\theta}(t+1) = \hat{\theta}(t) + F(t)\Phi(t)\nu(t+1) \tag{4.121}$$

$$\nu(t+1) = \frac{\nu^\circ(t+1)}{1+\Phi^T(t)F(t)\Phi(t)} \tag{4.122}$$

$$F(t+1)^{-1} = \lambda_1(t)F(t)^{-1} + \lambda_2(t)\Phi(t)\Phi^T(t) \tag{4.123}$$

$$0 < \lambda_1(t) \leq 1; \quad 0 \leq \lambda_2(t) < 2$$

$$F(0) > 0; F^{-1}(t) > \alpha F^{-1}(0)\ ; \ \infty > \alpha > 0$$

where $\hat{\theta}(t)$ is the adjustable parameter vector, $F(t)$ is the adaptation gain, $\Phi(t)$ is the regressor (observation) vector, $\nu^\circ(t+1)$ is the *a priori* adaptation error and $\nu(t+1)$ is the *a posteriori* adaptation error (it is a function of the prediction error). The *a priori* adaptation error $\nu^\circ(t+1)$ depends only on the adjustable parameter vector $\hat{\theta}(i)$ up to and including $i = t$. $\nu^\circ(t+1)$ is in fact the prediction of $\nu(t+1)$ based on these $\hat{\theta}(i)$, i.e.:

$$\nu^\circ(t+1) = \nu(t+1|\hat{\theta}(t), \hat{\theta}(t-1), \ldots)$$

The adaptation gain matrix $F(t)$ is computed recursively using the *matrix inversion lemma* and (4.123) becomes:

$$F(t+1) = \frac{1}{\lambda_1(t)}\left[F(t) - \frac{F(t)\Phi(t)\Phi^T(t)F(t)}{\frac{\lambda_1(t)}{\lambda_2(t)} + \Phi^T(t)F(t)\Phi(t)}\right] \tag{4.124}$$

The real-time implementation of this algorithm requires to use a numerical robust updating formula guaranteeing the positive definiteness of the matrix $F(t)$. Such a solution is provided by the U-D factorization of the matrix $F(t)$. The details are given in Appendix B.

Associated with the PAA of Eqs. (4.121)–(4.123), one considers the class of adaptive systems for which the *a posteriori* adaptation error satisfies an equation of the form:

$$\nu(t+1) = H(q^{-1})\left[\theta - \hat{\theta}(t+1)\right]^T \Phi(t) \tag{4.125}$$

where:

$$H(q^{-1}) = \frac{H_1(q^{-1})}{H_2(q^{-1})} \tag{4.126}$$

with:

$$H_i(q^{-1}) = 1 + q^{-1}H_j^*(q^{-1}) = 1 + \sum_{i=1}^{n_j} h_i^j q^{-i} \; ; \; j = 1, 2 \tag{4.127}$$

and θ is a fixed value of the unknown parameter vector.

The relationship between *a priori* and *a posteriori* adaptation errors given in (4.122), can be alternatively expressed using (4.121) as:

$$\nu(t+1) = \left[\hat{\theta}(t) - \hat{\theta}(t+1)\right]^T \Phi(t) + \nu^\circ(t+1) \tag{4.128}$$

From (4.125) and (4.126) one gets:

$$\begin{aligned}\nu(t+1) = {} & \left[\theta - \hat{\theta}(t+1)\right]^T \Phi(t) - H_2^*(q^{-1})\nu(t) + \\ & + H_1^*(q^{-1})\left[\theta - \hat{\theta}(t)\right]^T \Phi(t-1)\end{aligned} \tag{4.129}$$

Adding and subtracting the term $\hat{\theta}^T(t)\Phi(t)$ in the right hand side of (4.129), one gets:

$$\begin{aligned}\nu(t+1) = {} & \left[\hat{\theta}(t) - \hat{\theta}(t+1)\right]^T \Phi(t) \\ & + \left(\left[\theta - \hat{\theta}(t)\right]^T \Phi(t) + H_1^*(q^{-1})\left[\theta - \hat{\theta}(t)\right]^T \Phi(t-1) - H_2^*(q^{-1})\nu(t)\right)\end{aligned} \tag{4.130}$$

Comparing (4.128) and (4.130), one observes that:

$$\begin{aligned}\nu^\circ(t+1) = {} & \left[\theta - \hat{\theta}(t)\right]^T \Phi(t) + H_1^*(q^{-1})\left[\theta - \hat{\theta}(t)\right]^T \Phi(t-1) \\ & - H_2^*(q^{-1})\nu(t)\end{aligned} \tag{4.131}$$

and it can be clearly seen that $\nu^\circ(t+1)$ depends upon $\hat{\theta}(i)$ for $i \leq t$.

The PAA of Eqs. (4.121)–(4.123), together with (4.125), define an equivalent feedback system with a linear time-invariant feedforward block and a time-varying and/or nonlinear feedback block (see Fig. 4.13a). For constant adaptation gain ($\lambda_2 = 0$), the feedback path in Fig. 4.13a is passive[6]; however, for time-varying adaptation gain ($\lambda_2 > 0$), one has to consider the equivalent feedback representation shown in Fig. 4.13b, where the new equivalent feedback path is passive.

[6] A passive system is characterized by the fact that the sum of the input/output products over any time horizon is larger than a finite negative constant.

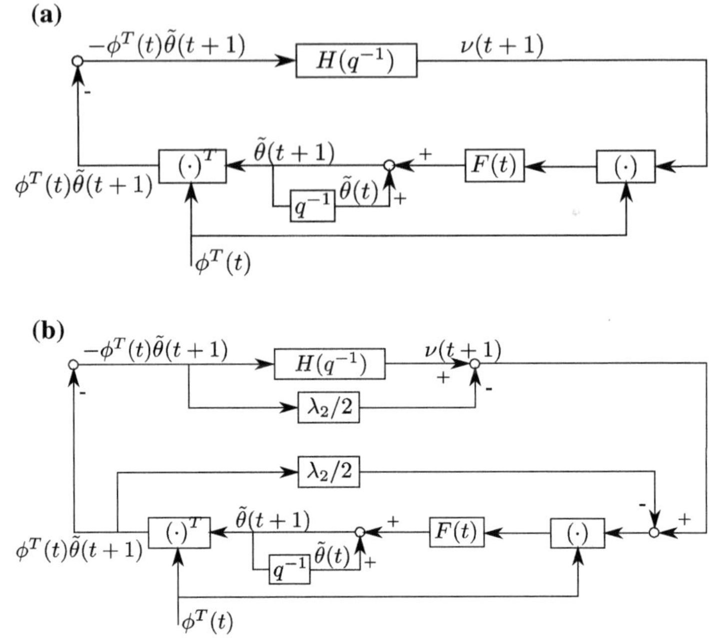

Fig. 4.13 Equivalent feedback system associated to the PAA with time-varying gain. **a** standard representation, **b** transformed equivalent feedback system

Exploiting the input-output properties of the equivalent feedback and feedforward block, one has the following general result (see [1] for the proof):

Theorem 4.1 *Consider the parameter adaptation algorithm given by Eqs. (4.121)–(4.123). Assume that the* a posteriori *adaptation error satisfies (4.125) where $\phi(t)$ is a bounded or unbounded vector sequence, $H(z^{-1})$ is a rational discrete transfer function (ratio of monic polynomials) and θ is a constant vector. Then, if:*

$$H'(z^{-1}) = H(z^{-1}) - \frac{\lambda_2}{2} \tag{4.132}$$

is strictly positive real *(SPR), where*

$$\max_t\,[\lambda_2(t)] \le \lambda_2 < 2, \tag{4.133}$$

one has for any bounded $\nu(0), \hat{\theta}(0)$:

$$1)\ \lim_{t_1\to\infty} \sum_{t=0}^{t_1} \nu^2(t+1) < C\left(\nu(0), \hat{\theta}(0)\right); 0 < C < \infty \tag{4.134}$$

$$2)\ \lim_{t\to\infty} \nu(t+1) = 0 \tag{4.135}$$

$$3)\ \lim_{t\to\infty}\left[\theta-\hat{\theta}(t+1)\right]^T\Phi(t)=0 \tag{4.136}$$

$$4)\ \lim_{t\to\infty}\left[\hat{\theta}(t+1)-\hat{\theta}(t)\right]^T F(t)^{-1}\left[\hat{\theta}(t+1)-\hat{\theta}(t)\right]=0 \tag{4.137}$$

4.4.2.1 Interpretation of the Results

1. A *strictly positive real* (SPR) transfer function is characterized by the following basic properties (there are others [1]):
   - It is asymptotically stable.
   - The real part of the transfer function is positive at all frequencies.

   The concept of SPR transfer functions is illustrated for continuous-time systems in the upper part of the Fig. 4.14 and for the discrete-time systems in the lower part of Fig. 4.14.
2. $\lim_{t\to\infty}\nu(t+1)=0$ can be interpreted as the output of the equivalent linear block which asymptotically tends to 0.
3. Since $H(z^{-1})$ is SPR, it is asymptotically stable as well as its inverse. Therefore, its input will also tend to zero which, taking into account the notations, leads to (4.136).
4. Equation (4.137) indicates that the asymptotic variations of the adjustable parameters tend to zero if $F(t)>0$.
5. For constant adaptation gain ($\lambda_1=1,\lambda_2=0$) condition (4.132) becomes:

$$H'(z^{-1})=H(z^{-1}) \tag{4.138}$$

   should be *strictly positive real* (SPR).
6. The explanation of the presence of the term $-\lambda_2/2$ in (4.132) is related to the loss of the passivity of the equivalent feedback path when using time-varying adaptation gain. This requires to consider the transformed equivalent feedback system shown in Fig. 4.13b.

Remark Note that $\lim_{t\to\infty}\nu(t+1)=0$ does not imply $\lim_{t\to\infty}\nu^{\circ}(t+1)=0$ since:

$$\nu(t+1)=\frac{\nu^{\circ}(t+1)}{1+\Phi^T(t)F(t)\Phi(t)}.$$

If $\Phi(t)$ is unbounded, then $\nu(t+1)$ can be zero with $\nu^{\circ}(t+1)\neq 0$. To conclude that $\lim_{t\to\infty}\nu^{\circ}(t+1)=0$, one should show that $\Phi(t)$ is bounded (assuming that $F(t)$ is bounded).

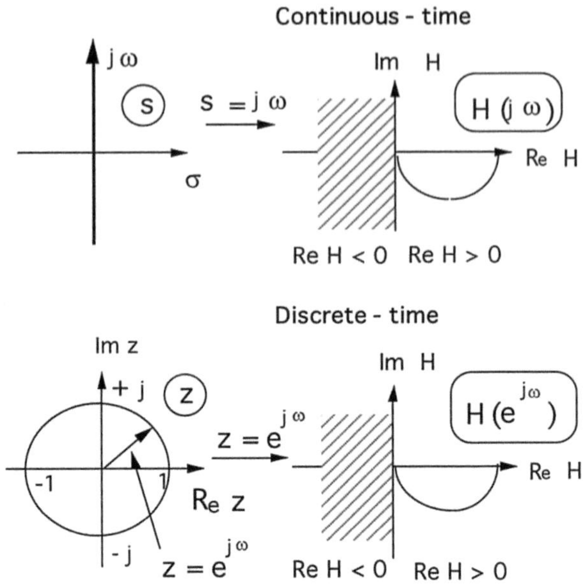

Fig. 4.14 Strictly positive real transfer functions

4.4.3 Output Error Algorithms—Stability Analysis

Applying this theorem to the output error configurations (with constant adaptation gain) presented in Sect. 4.3.2.2, will lead to the following stability conditions.

Output Error

In this case $\nu(t+1) = \varepsilon(t+1)$, $\Phi(t) = \psi(t)$ and the discrete-time transfer function:

$$H(z^{-1}) = \frac{1}{A(z^{-1})} \tag{4.139}$$

should be SPR. This condition may be restrictive in certain situations. In order to overcome this difficulty, one may consider filtering the *a posteriori* prediction error before its use in the PAA or to filter the observation vector (see Sect. 4.3.2).

Output Error with Fixed Compensator

Using (4.115), $\nu(t+1)$ can be expressed as:

$$\nu(t+1) = \frac{D(q^{-1})}{A(q^{-1})}[\theta - \hat{\theta}(t+1)]^T \psi(t) \tag{4.140}$$

and $\Phi(t) = \psi(t)$. In this case:

$$H(z^{-1}) = \frac{D(z^{-1})}{A(z^{-1})} \tag{4.141}$$

should be SPR.

Output Error with Filtered Observations

In this case $\nu(t+1) = \varepsilon(t+1)$ and $\Phi(t) = \psi_f(t)$. The equation of the *a posteriori* prediction error (neglecting the non-commutativity of time-varying operators) is given in (4.38)

$$\varepsilon(t+1) = \frac{1}{A(q^{-1})}[\theta - \hat{\theta}(t+1)]^T \psi(t) = \frac{L(q^{-1})}{A(q^{-1})}[\theta - \hat{\theta}(t+1)]^T \psi_f(t) \tag{4.142}$$

and applying the Theorem 4.1 for stability of PAA one concludes that

$$H(z^{-1}) = \frac{L(z^{-1})}{A(z^{-1}} \tag{4.143}$$

should be SPR. An exact algorithm for the output error with filtered observations is given in [1, Sect. 5.5.3].

4.4.3.1 A Swapping Result

The above developments on output error with filtered observations have used the following relationship:

$$\nu(t+1) = H(q^{-1})\left[\theta - \hat{\theta}(t+1)\right]^T \phi(t) = \left[\theta - \hat{\theta}(t+1)\right]^T \phi_f(t) + O \tag{4.144}$$

where

$$\phi_f(t) = H(q^{-1})\phi(t) \tag{4.145}$$

and the error term O is assumed to be negligible. Exact algorithms can be developed [1] but they are not necessary in practice. Therefore systematically relationship (4.144) will be used neglecting the error term (swapping error) in order to apply Theorem 4.1 for the synthesis of stable adaptation algorithms.

4.5 Parametric Convergence

4.5.1 The Problem

As will be shown, the convergence toward zero of the adaptation or prediction error does not imply in every case that the estimated parameters will converge toward the true parameters. The objective will be to determine under what conditions the convergence of the adaptation (prediction) error will imply the convergence toward the true parameters. We will make the hypothesis that such a value of parameter vector exists.

In order to illustrate the influence of the excitation signal for the parametric convergence, let us consider the discrete-time system model described by:

$$y(t+1) = -a_1 y(t) + b_1 u(t) \tag{4.146}$$

and consider an estimated model described by:

$$\hat{y}(t+1) = -\hat{a}_1 y(t) + \hat{b}_1 u(t) \tag{4.147}$$

in which $\hat{y}(t+1)$ is the output predicted by the estimation model with the constant parameters $\hat{a}_1, \hat{b}_1$.

Now assume that $u(t) = constant$ and that the parameters $a_1, b_1, \hat{a}_1, \hat{b}_1$ verify the following relation:

$$\frac{b_1}{1+a_1} = \frac{\hat{b}_1}{1+\hat{a}_1} \tag{4.148}$$

i.e., the steady state gains of the system and of the estimated model are equal even if $\hat{b}_1 \neq b_1$ and $\hat{a}_1 \neq a_1$. Under the effect of the constant input $u(t) = u$, the plant output will be given by:

$$y(t+1) = y(t) = \frac{b_1}{1+a_1} u \tag{4.149}$$

and the output of the estimated prediction model will be given by:

$$\hat{y}(t+1) = \hat{y}(t) = \frac{\hat{b}_1}{1+\hat{a}_1} u \tag{4.150}$$

Nevertheless, taking into account (4.148), it results that:

$$\varepsilon(t+1) = y(t+1) - \hat{y}(t+1) = 0 \tag{4.151}$$
$$\text{for } u(t) = \text{constant} \ ; \ \hat{a}_1 \neq a_1 \ ; \ \hat{b}_1 \neq b_1$$

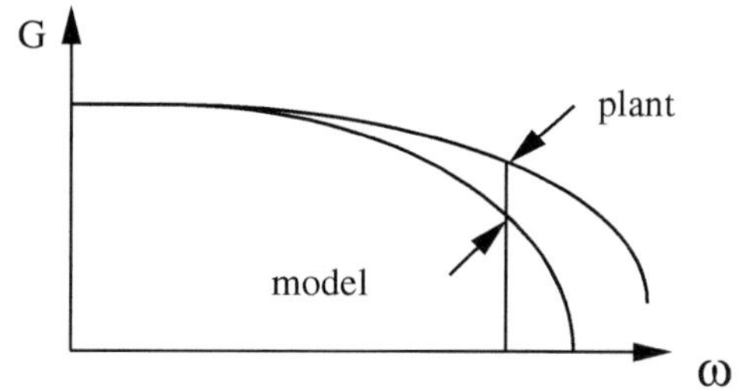

Fig. 4.15 Gain frequency characteristics of two systems with the same steady state gain

It can thus be concluded from this example that the application of a constant input does not allow to distinguish these two models, since they both have the same steady state gain.

If the frequency characteristics of both systems are represented, they will superpose each other at zero frequency and the difference between them will appear for frequencies other than zero since the poles of the two systems are different. Such a situation is shown in Fig. 4.15.

Figure 4.15 indicates that in order to highlight the difference between the two models (i.e., between the parameters) a signal $u(t) = \sin \omega t$ $(\omega \neq 0)$ must be applied instead of signal $u(t) = constant$.

Let us analyze the phenomenon in more detail. From (4.146) and (4.147), one obtains:

$$\varepsilon(t+1) = y(t+1) - \hat{y}(t+1) = (a_1 - \hat{a}_1)y(t) + (b_1 - \hat{b}_1)u(t) = 0 \quad (4.152)$$

From (4.146), $y(t)$ can be expressed as a function of $u(t)$ using the system transfer operator:

$$y(t) = \frac{b_1 q^{-1}}{1 + a_1 q^{-1}} u(t) \quad (4.153)$$

Introducing the expression $y(t)$ given by (4.153) in (4.152) and after multiplying by $(1 + a_1 q^{-1})$, one obtains:

$$\begin{aligned} \varepsilon(t+1) &= \left[(a_1 - \hat{a}_1)b_1 q^{-1} + (b_1 - \hat{b}_1)(1 + a_1 q^{-1})\right] u(t) \\ &= \left[(b_1 - \hat{b}_1) + q^{-1}(b_1 \hat{a}_1 - a_1 \hat{b}_1)\right] u(t) = 0 \end{aligned} \quad (4.154)$$

We are concerned with finding the characteristics of $u(t)$ so that (4.154) implies zero parametric errors. Denoting:

$$b_1 - \hat{b}_1 = \alpha_0 \; ; \; b_1 \hat{a}_1 - a_1 \hat{b}_1 = \alpha_1 \quad (4.155)$$

Equation (4.154) is thus written as:

$$(\alpha_0 + \alpha_1 q^{-1})u(t) = 0 \tag{4.156}$$

which is a difference equation having a solution of the discretized exponential type.

Let us take:

$$u(t) = z^t = e^{sT_s t} \tag{4.157}$$

where T_s is the sampling period. Equation (4.156) is then written:

$$(\alpha_0 + z^{-1}\alpha_1)z^t = (z\alpha_0 + \alpha_1)z^{t-1} = 0 \tag{4.158}$$

and it will be verified for z, which is the solution of the characteristic equation:

$$z\alpha_0 + \alpha_1 = 0 \tag{4.159}$$

One obtains:

$$z = -\frac{\alpha_1}{\alpha_0} = e^{\sigma T_s} \tag{4.160}$$

$$\sigma = \text{real} \; ; \; \left(\frac{\alpha_1}{\alpha_0} < 0\right)$$

and the nonperiodic solution:

$$u(t) = e^{\sigma T_s t} \tag{4.161}$$

leads to the verification of (4.156) and (4.154) respectively without $\hat{b}_1 = b_1$ and $\hat{a}_1 = a_1$. Indeed the signal $u(t) = constant$ previously considered, corresponds to $\sigma = 0$, i.e., $-\alpha_1 = \alpha_0$; however,

$$-\alpha_1 = \alpha_0 \implies b_1 - \hat{b}_1 = a_1\hat{b}_1 - b_1\hat{a}_1 \implies \frac{b_1}{1+a_1} = \frac{\hat{b}_1}{1+\hat{a}_1}. \tag{4.162}$$

In conclusion, if $u(t) = constant$, only the steady state gain of the system is correctly estimated. In order to correctly estimate the system model parameters, $u(t)$ must thus be found such that $\varepsilon(t) = 0$ implies $\hat{b}_1 = b_1$ and $\hat{a}_1 = a_1$. This will be obtained if $u(t)$ is not a possible solution of (4.156).

Let

$$u(t) = e^{j\omega T_s t} \text{ or } e^{-j\omega T_s t} \tag{4.163}$$

For $u(t) = e^{j\omega T_s t}$, (4.156) becomes:

$$\left(e^{j\omega T_s}\alpha_0 + \alpha_1\right) e^{j\omega T_s(t-1)} = 0 \tag{4.164}$$

Since α_0 and α_1 are real, $e^{j\omega T_s t}$ cannot be a root of the characteristic equation (4.164) and therefore $\varepsilon(t+1) = 0$ will be obtained only if:

$$\alpha_0 = \alpha_1 = 0 \implies \hat{b}_1 = b_1;\ \hat{a}_1 = a_1 \tag{4.165}$$

It was this type of input that was previously proposed ($\sin \omega t = (e^{j\omega t} - e^{-j\omega t})/2j$) when the frequency characteristics of the two models were examined. A non zero frequency sinusoid is thus required in order to identify two parameters. The signal $u(t)$ which in this case is a sinusoid, is a *persistently exciting signal* of order 2 (allowing to estimate 2 parameters).

This approach for determining the input $u(t)$ allowing satisfactory model parameter identification may also be applied to systems of the general form

$$y(t) = -\sum_{i=1}^{n_A} a_i y(t-i) + \sum_{i=1}^{n_B} b_i u(t-d-i) \tag{4.166}$$

for which the total number of parameters to be identified is:

$$\text{number of parameters} = n_A + n_B.$$

In this case $u(t)$ can be chosen as a sum of p-sinusoids of different frequencies:

$$u(t) = -\sum_{i=1}^{p} \sin(\omega_i T_e t) \tag{4.167}$$

and the value p, allowing good parameter identification, is given by

$$\begin{cases} p \geq \frac{n_A+n_B}{2} & \text{for } n_A + n_B \text{ even} \\ p \geq \frac{n_A+n_B+1}{2} & \text{for } n_A + n_B \text{ odd} \end{cases} \tag{4.168}$$

In other words, in order to identify a correct model it is necessary to apply a frequency rich input. The standard solution in practice is provided by the use of "pseudo-random binary sequences". Pseudo-random binary sequences are sequences of rectangular pulses, modulated in width, which approximate a discrete-time white noise and thus have a spectral content rich in frequencies (see Sect. 5.2).

The above result will be used also for analysing the parameter convergence in adaptive feedback control schemes used for attenuation of narrow-band disturbances (see Chap. 12).

4.6 The LMS Family of Parameter Adaptation Algorithms

The least mean squares (LMS) family of PAA has its origin in the paper of Widrow and Hoff [3]. This algorithm corresponds to the "basic gradient algorithm" discussed in Sect. 4.3.1 particularized for the case of an FIR structure ($a_i \equiv 0$) and constant diagonal adaptation gain matrix. Starting from this initial algorithm an impressive number of developments and applications have been done by the signal processing community concerned with adaptive filtering and by the community concerned with active noise (vibration) control. Efforts to analyze the behaviour of the resulting algorithms have been also done. These developments have been made ignoring till the mid nineties (with some exceptions as for example [4]) the parameter adaptation algorithms developed by the control community in the context of system identification and adaptive control.

To make a bridge between the family of LMS algorithms and those developed by the control community (presented in this book) one has to consider several aspects which characterize a PAA:

- Structure of the estimated model.
- Type of adaptation error (*a priori* or *a posteriori*).
- Type of adaptation gain.
- Generation of the regressor vector.
- Properties of the algorithm (stability in deterministic context, convergence in a stochastic context).

Most applications in adaptive filtering and adaptive feedforward noise/vibration compensation are related to the "output error" structure considered in Sect. 4.3.2.2. As long as an FIR structure is considered, there is no difference between an "output error" configuration and an "equation error" configuration. The extension of the LMS approach when an IIR structure is considered lead to an "output error" configuration. This extension attributed to Feintuch (1976) [5] is called ULMS or RLMS [6] (the standard LMS for FIR configurations being called XLMS); however, in the field of adaptive parameter estimation an algorithm for an IIR output error configuration designed form stability considerations (but which can be interpreted also from the improved gradient point of view) has been already proposed in 1971 [7]. It uses the concept of *a posteriori* adaptation error and constant adaptation gain while in [5] the *a priori* adaptation error is used. For small adaptation gain the Feintuch ULMS algorithm can be viewed as an approximation of the algorithm given in [7]. The algorithm given in [7] has been extended for the case of time-varying adaptation gain in [8] and compared with other algorithms in [9]. The asymptotic unbiasedness in the mean has been shown in [10, 11]. Conditions for convergence with probability 1 have been established in [12].

The use of the *a posteriori* prediction (adaptation) error is crucial for the stability of the adaptation algorithms. Once this option is considered, the stability analysis becomes much easier and conditions upon the the *strict real positivity* of some transfer functions in order to guarantee stability come in view.

Concerning a further comparison between LMS type algorithms and those presented in this chapter one has to consider two cases:

- scalar constant adaptation gain; and
- matrix time-varying adaptation gain.

In other terms, in addition to the structure (which gives the name to an algorithm) and the type of adaptation error used one has to specify what type of adaptation gain is used. Both LMS type algorithms and those given in this chapter can be operated with this two types of adaptation gain. In developing the output error algorithms in this chapter, it was pointed from the beginning that "output error" can be used for adaptive feedforward compensation for the case when the transfer function of the secondary path (see Chap. 1) is equal to one (or very close), i.e., $G = 1$ (or very close). Nevertheless, in practice $G \neq 1$ and this complicates the analysis of the algorithms (in particular stability aspects). This problem is discussed in detail in Chap. 15. One of the popular solutions adopted in the LMS family of algorithms is to filter the regressor vector by G. This solution generated the FXLMS algorithm [13, 14] when using an FIR structure and the FULMS algorithm when an IIR structure is considered [15]. These corresponds in fact to particular cases of the family of algorithms which will be proposed in Chap. 15 for adaptive feedforward compensation. The FXLMS and FULMS have a serious drawback in terms of stability and convergence in the presence of internal positive feedback [16, 17]. Experimental comparison of FULMS with algorithms proposed in the book will be presented in Chap. 15. Comparison with other related algorithms can also be found in Sect. 15.5.

4.7 Concluding Remarks

In this chapter we have presented discrete-time *parameter adaptation algorithms* (PAA) and we have examined their properties.

We wish to emphasize the following basic ideas:

1. The PAA has in general the following recursive form (integral adaptation):

$$\hat{\theta}(t+1) = \hat{\theta}(t) + F(t)\phi(t)\nu(t+1)$$

   where $\hat{\theta}(t)$ is the adjustable parameter vector. At each step the correcting term is formed by the product of the adaptation error $\nu(t+1)$, the regressor (observation) vector $\phi(t)$ and the adaptation gain matrix $F(t)$. The adaptation error $\nu(t+1)$ is computed from the measurements up to and including $t+1$ and the estimated parameters up to t.
2. Several approaches can be used for the derivation of PAA among which we have considered:

- recursive minimization of a criterion in term of the adaptation error;
- transformation of an off-line parameter estimation into a recursive parameter estimation; and
- stability considerations.

Nevertheless, since the resulting system is nonlinear, a stability analysis is mandatory.

3. An equivalent feedback system (EFR) can be associated with the PAA in the cases where the adaptation error equation features the parameter error explicitly. The use of the EFR simplifies drastically the stability analysis (or synthesis) via the use of the properties of passive systems connected in feedback.
4. For general classes of adaptation error equations and PAA, stability conditions for the resulting adaptive systems have been given.
5. A variety of choices for the time profile of the adaptation gain are possible. The choice depends upon the specific application.

4.8 Notes and References

The chapter focused on PAA with an integral structure which is the most used in practice. Other structure exists [1]. In particular the integral+proportional PAA is interesting for active vibration control. Details are given in Appendix E.

The books [1, 18–22] give an extensive coverage of the discrete time PAA from the stability point of view in a deterministic environment and from the convergence point of view in a stochastic environment. The book [6] gives a presentation of the PAA starting from the LMS algorithm. Development and analysis of LMS can be found in [13, 14] (Filtered-X LMS), [15, 23] (Filtered-U LMS), [24] (full gradient algorithm) [16, 25].

References

1. Landau ID, Lozano R, M'Saad M, Karimi A (2011) Adaptive control, 2nd edn. Springer, London
2. Landau I, Zito G (2005) Digital control systems - design, identification and implementation. Springer, London
3. Widrow B, Hoff M (1960) Adaptive swithching circuits. Oric IRE WESCON Conv Rec 4(Session 16):96–104
4. Johnson C Jr (1979) A convergence proof for a hyperstable adaptive recursive filter (corresp.). IEEE Trans Inf Theory 25(6):745–749. doi:10.1109/TIT.1979.1056097
5. Feintuch P (1976) An adaptive recursive LMS filter. Proc IEEE 64(11):1622–1624. doi:10.1109/PROC.1976.10384
6. Elliott S (2001) Signal processing for active control. Academic Press, San Diego
7. Landau I (1972) Synthesis of hyperstable discrete model reference adaptive systems. In: Proceedings of 5th Asilomar conference on circuits and systems (8–10 November 1971), pp 591–595

8. Landau I (1973) Algorithme pour l'identification à l'aide d'un modèle ajustable parallèle. Comptes Rendus de l'Académie des Sciences 277(Séria A):197–200
9. Landau I (1974) A survey of model reference adaptive techniques - theory and applications. Automatica 10:353–379
10. Landau I (1974) An asymptotic unbiased recursive identifier for linear systems. In: Proceedings of IEEE conference on decision and control, pp 288–294
11. Landau I (1976) Unbiased recursive identification using model reference adaptive techniques. IEEE Trans Autom Control 21(2):194–202. doi:10.1109/TAC.1976.1101195
12. Ljung L (1977) Analysis of recursive stochastic algorithms. IEEE Trans Autom Control 22(4):551–575. doi:10.1109/TAC.1977.1101561
13. Burgess J (1981) Active adaptive sound control in a duct: a computer simulation. J Acoust Soc Am 70:715–726
14. Widrow B, Shur D, Shaffer S (1981) On adaptive inverse control. In: Proceedings of the 15th Asilomar conference on circuits, systems and computers. Pacific Grove, CA, USA
15. Eriksson L (1991) Development of the filtered-U LMS algorithm for active noise control. J Acoust Soc Am 89(1):257–261
16. Wang A, Ren W (2003) Convergence analysis of the filtered-U algorithm for active noise control. Signal Process 83:1239–1254
17. Landau I, Alma M, Airimitoaie T (2011) Adaptive feedforward compensation algorithms for active vibration control with mechanical coupling. Automatica 47(10):2185–2196. doi:10.1016/j.automatica.2011.08.015
18. Goodwin G, Sin K (1984) Adaptive filtering prediction and control. Prentice Hall, New Jersey
19. Ljung L, Söderström T (1983) Theory and practice of recursive identification. The MIT Press, Cambridge
20. Landau I (1979) Adaptive control: the model reference approach. Marcel Dekker, New York
21. Astrom KJ, Hagander P, Sternby J (1984) Zeros of sampled systems. Automatica 20(1):31–38. doi:10.1109/CDC.1980.271968
22. Anderson B, Bitmead R, Johnson C, Kokotovic P, Kosut R, Mareels I, Praly L, Riedle B (1986) Stability of adaptive systems. The MIT Press, Cambridge
23. Eriksson L, Allie M, Greiner R (1987) The selection and application of an IIR adaptive filter for use in active sound attenuation. IEEE Trans Acoust Speech Signal Process 35(4):433–437. doi:10.1109/TASSP.1987.1165165
24. Crawford D, Stewart R (1997) Adaptive IIR filtered-V algorithms for active noise control. J Acoust Soc Am 101(4)
25. Fraanje R, Verhaegen M, Doelman N (1999) Convergence analysis of the iltered-U LMS algorithm for active noise control in case perfect cancellation is not possible. Signal Process 73:255–266

Chapter 5
Identification of the Active Vibration Control Systems—The Bases

5.1 Introduction

To design an active control one needs the dynamical model of the compensator systems (from the control to be applied to the measurement of the residual acceleration or force).[1] Model identification from experimental data is a well established methodology [1, 2]. Identification of dynamic systems is an experimental approach for determining a dynamic model of a system. It includes four steps:

(1) Input–output data acquisition under an experimental protocol.
(2) Estimation of the model complexity (structure).
(3) Estimation of the model parameters.
(4) Validation of the identified model (structure of the model and values of the parameters).

A complete identification operation must comprise the four stages indicated above. System identification should be viewed as an iterative process as illustrated in Fig. 5.1 which has as objective to obtain a model which passes the model validation test and then can be used safely for control design.

The typical input excitation sequence is a PRBS (pseudorandom binary sequence). The type of model which will be identified is a discrete-time parametric model allowing later to directly design a control algorithm straightforwardly implementable on a computer. Model validation is the final key point. The estimation of the model parameters is done in a noisy environment. It is important to emphasize that it does not exist one single algorithm that can provide in all the cases a good model (i.e., which passes the model validation tests). Therefore, the appropriate algorithm which allows to obtain a model which passes the validation tests has to be used.

[1]Linear feedback regulator design will require also the model of the disturbance. Linear feedforward compensator design will require in addition a model of the primary path. Design of adaptive regulators or of feedforward compensators require only the model of the secondary path.

I.D. Landau et al., *Adaptive and Robust Active Vibration Control*,
Advances in Industrial Control, DOI 10.1007/978-3-319-41450-8_5

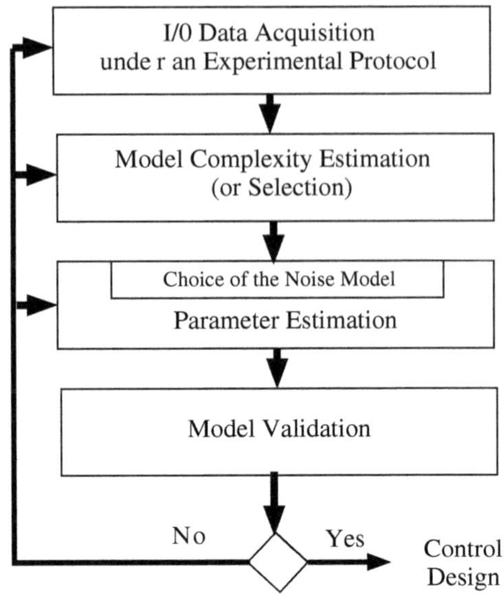

Fig. 5.1 The identification methodology

In what follows, we would like to summarize some of the basic facts in system identification. For a detailed coverage of the subject, see [1, 3].

Figure 5.2 shows the principle of parameter estimation of a discrete-time model. An adjustable model of the discretized plant is built. Its parameters are driven by a parameter adaptation algorithm such that the prediction error (the difference between the true output and the predicted output by the model) is minimized in the sense of a certain criterion.

Several assumptions are implicitly made when one uses this approach

- the order of the discrete-time model representing the system is known;
- in the absence of noise the adaptation algorithm will drive the prediction error towards zero;

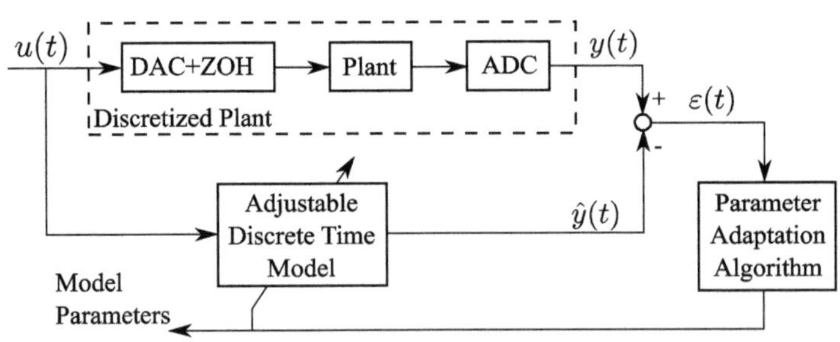

Fig. 5.2 Principle of model parameters estimation

- in the presence of noise, the estimated parameters will be asymptotically unbiased[2] (i.e., the noise does not influence asymptotically the precision of the parameter estimation); and
- the input to the system (the testing signal) is such that null prediction error implies null parameter errors (persistent excitation property).

The various steps indicated in Fig. 5.1 tries to assure that the parameter estimation algorithm will provide the good parameter estimates.

5.2 Input–Output Data Acquisition and Preprocessing

5.2.1 Input–Output Data Acquisition Under an Experimental Protocol

The experimental protocol should assure persistent excitation for the number of parameters to be estimated. It can be shown (see Chap. 4, Sect. 4.5 and [1]) that for identifying $2n$ parameters, the excitation signal should contain at least $n + 1$ sinusoids of distinct frequencies. To go beyond this constraints one usually uses Pseudorandom Binary Sequences (PRBS) since they contain a large number of sinusoids with energy equally distributed over the frequency domain. In addition the magnitude of the signal is constant allowing and easy selection with respect to the magnitude constraints on the plant input.

5.2.2 Pseudorandom Binary Sequences (PRBS)

Pseudorandom binary sequences are sequences of rectangular pulses, modulated in width, that approximate a discrete-time white noise and thus have a spectral content *rich* in frequencies. They owe their name *pseudo random* to the fact that they are characterized by a *sequence length* within which the variations in pulse width vary randomly, but that over a large time horizon, they are periodic, the period being defined by the length of the sequence. In the practice of system identification, one generally uses just one complete sequence and we should examine the properties of such a sequence.

The PRBS are generated by means of shift registers with feedback (implemented in hardware or software).[3] The maximum length of a sequence is $L = 2^N - 1$, where N is the number of cells of the shift register.

[2]The parameter estimation error induced by the measurement noise is called "bias".

[3]Routines for generating PRBS can be downloaded from the websites: http://www.landau-adaptivecontrol.org and http://www.gipsa-lab.grenoble-inp.fr/~ioandore.landau/identificationandcontrol/.

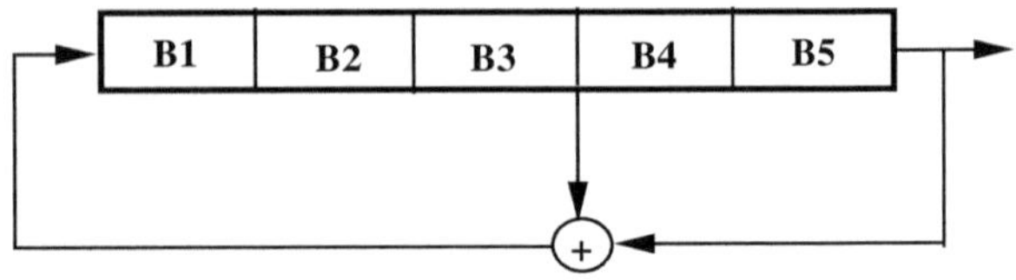

Fig. 5.3 Generation of a PRBS of length $2^5 - 1 = 31$ sampling periods

Table 5.1 Generation of maximum length PRBS

Number of cells N	Sequence length $L = 2^N - 1$	Bits added B_i and B_j
5	31	3 and 5
6	63	5 and 6
7	127	4 and 7
8	255	2, 3, 4 and 8
9	511	5 and 9
10	1023	7 and 10

Figure 5.3 presents the generation of a PRBS of length $31 = 2^5 - 1$ obtained by means of a $N = 5$-cells shift register. Note that at least one of the N cells of the shift register should have an initial logic value different from zero (one generally takes all the initial values of the N cells equal to the logic value 1).

Table 5.1 gives the structure enabling maximum length PRBS to be generated for different numbers of cells. Note also a very important characteristic element of the PRBS: *the maximum duration of a PRBS impulse is equal to N (number of cells).* This property is to be considered when choosing a PRBS for system identification.[4]

In order to cover the entire frequency spectrum generated by a particular PRBS, the length of a test must be at least equal to the length of the sequence. In a large number of cases, the duration of the test L is chosen equal to the length of the sequence. Through the use of a frequency divider for the clock frequency of the PRBS, it is possible to shape the energy distribution in the frequency domain. This is why, in a large number of practical situations, a submultiple of the sampling frequency is chosen as the clock frequency for the PRBS. Note that dividing the clock frequency of the PRBS will reduce the frequency range corresponding to a constant spectral density in the high frequencies while augmenting the spectral density in the low frequencies. In general, this will not affect the quality of identification, either because in many cases when this solution is considered, the plant to be identified has a low band pass or because the effect or the reduction of the signal/noise ratio at high frequencies can be compensated by the use of appropriate identification techniques; however, it is recommended to choose $p \leq 4$ where p is the frequency divider.

[4]Functions *prbs.m* and *prbs.c* available on the websites: http://www.landau-adaptivecontrol.org and http://www.gipsa-lab.grenoble-inp.fr/ ioandore.landau/identificationandcontrol/ allow to generate PRBS of various lengths and magnitudes.

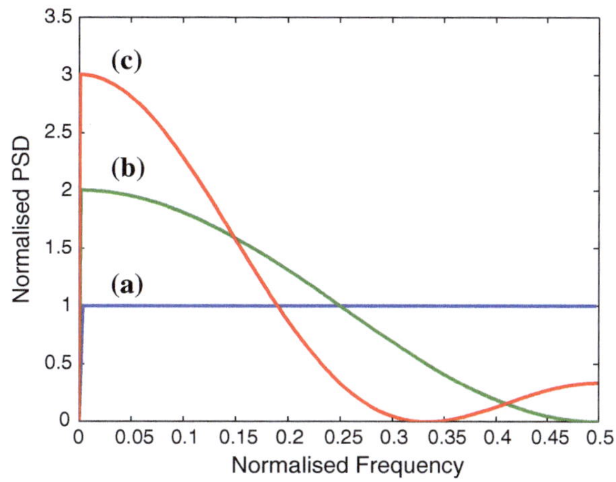

Fig. 5.4 Spectral density of a PRBS sequence, **a** $N = 8, p = 1$, **b** $N = 8, p = 2$, **c** $N = 8, p = 3$

Figure 5.4 shows the spectral density of PRBS sequences generated with $N = 8$ for $p = 1, 2, 3$. As one can see, the energy of the spectrum is reduced in the high frequencies and augmented in the lower frequencies. Furthermore, for $p = 3$ a hole occurs at $f_s/3$.

Until now, we have been concerned only with the choice of the length and clock frequency of the PRBS; however, the magnitude of the PRBS must also be considered. Although the magnitude of the PRBS may be very low, it should lead to output variations larger than the residual noise level. If the signal/noise ratio is too low, the length of the test must be augmented in order to obtain a satisfactory parameter estimation.

Note that in a large number of applications, the significant increase in the PRBS level may be undesirable in view of the nonlinear character of the plants to be identified (we are concerned with the identification of a linear model around an operating point).

5.2.3 Data Preprocessing

The first aspect is that one works with centred data (variations of the real data) so that the first operation is the centering of the input/output data by subtracting their mean value.

When identifying the compensator system in active vibration control systems, one has to take into account the double differentiator behaviour. This means that a part of the model is known and we should identify only the unknown part. To do this, the input applied to the real system is filtered by a double discrete-time differentiator filter

$$(1 - q^{-1})^2 = 1 - 2q^{-1} + q^{-2} \tag{5.1}$$

This new input/output sequence is then centred and used together with the measured output data for identifying the unknown part of the model. After the unknown part of the model will be identified, the double differentiator should be included in the final model (the two transfer operators will be multiplied).

5.3 Model Order Estimation from Data

It is extremely important to be able to estimate the order of the system from input/output data since it is hard from physical reasoning to get a reliable estimation of the order of the system. To introduce the problem of order estimation from data, we will start with an example. Assume that the plant model can be described by:

$$y(t) = -a_1 y(t-1) + b_1 u(t-1) \tag{5.2}$$

and that the data are noise free. The order of this model is $n = n_A = n_B = 1$.

Question: Is there any way to test from data if the order assumption is correct? To do so, construct the following matrix:

$$\begin{bmatrix} y(t) & \vdots\, y(t-1) & u(t-1) \\ y(t-1) & \vdots\, y(t-2) & u(t-2) \\ y(t-2) & \vdots\, y(t-3) & u(t-3) \end{bmatrix} = \left[Y(t) \,\vdots\, R(1) \right] \tag{5.3}$$

Clearly, if the order of the model given in Eq. (5.2) is correct, the vector $Y(t)$ will be a linear combination of the columns of $R(1)$ ($Y(t) = R(1)\theta$ with $\theta^T = [-a_1, b_1]$) and the rank of the matrix will be 2 (instead of 3). If the plant model is of order 2 or higher, the matrix in (5.3) will be full rank. Of course, this procedure can be extended for testing the order of a model by testing the rank of the matrix $[Y(t), R(\hat{n})]$ where:

$$R(\hat{n}) = [Y(t-1), U(t-1), Y(t-2), U(t-2) \ldots Y(t-\hat{n}), U(t-\hat{n})], \tag{5.4}$$

$$Y^T(t) = [y(t), y(t-1) \ldots], \qquad U^T(t) = [u(t), u(t-1) \ldots]. \tag{5.5}$$

Unfortunately, as a consequence of the presence of noise, this procedure cannot directly be applied in practice.

A more practical approach results from the observation that the rank test problem can be approached by the searching of $\hat{\theta}$ which minimizes the following criterion for an estimated value of the order $\hat{n}$.

$$V_{LS}(\hat{n}, N) = \min_{\hat{\theta}} \frac{1}{N} \| Y(t) - R(\hat{n})\hat{\theta} \|^2 \tag{5.6}$$

where N is the number of samples. But this criterion is nothing else than an equivalent formulation of the least squares [4]. If the conditions for unbiased estimation using least squares are fulfilled, (5.6) is an efficient way for assessing the order of the model since $V_{LS}(\hat{n}) - V_{LS}(\hat{n}+1) \rightarrow 0$ when $\hat{n} \geq n$.

In the meantime, the objective of the identification is to estimate lower order models (parsimony principle) and therefore, it is reasonable to add in the criterion (5.6) a term which penalizes the complexity of the model. Therefore, the *penalized criterion* for order estimation will take the form:

$$J_{LS}(\hat{n}, N) = V_{LS}(\hat{n}, N) + S(\hat{n}, N) \tag{5.7}$$

where typically

$$S(\hat{n}, N) = 2\hat{n}X(N) \tag{5.8}$$

and V_{LS} represents the *non penalized criterion*. $X(N)$ in (5.8) is a function that decreases with N. For example, in the so called $BIC_{LS}(\hat{n}, N)$ criterion, $X(N) = \frac{\log N}{N}$ (other choices are possible—see [3–5]) and the order $\hat{n}$ is selected as the one which minimizes J_{LS} given by (5.7). Unfortunately, the results are unsatisfactory in practice because in the majority of situations, the conditions for unbiased parameter estimation using least squares are not fulfilled.

In [5, 6], it is proposed to replace the matrix $R(\hat{n})$ by an instrumental variable matrix $Z(\hat{n})$ whose elements will not be correlated with the measurement noise. Such an instrumental matrix $Z(\hat{n})$ can be obtained by replacing in the matrix $R(\hat{n})$, the columns $Y(t-1),\ Y(t-2),\ Y(t-3)$ by delayed version of $U(t-L-i)$, i.e., where $L > n$:

$$Z(\hat{n}) = [U(t-L-1), U(t-1), U(t-L-2), U(t-2)\ldots] \tag{5.9}$$

and therefore, the following criterion is used for the order estimation:

$$J_{IV}(\hat{n}, N) = \min_{\hat{\theta}} \frac{1}{N} \|Y(t) - Z(\hat{n})\hat{\theta}\|^2 + \frac{2\hat{n}\log N}{N} \tag{5.10}$$

and

$$\hat{n} = \min_{\hat{n}} J_{IV}(\hat{n}). \tag{5.11}$$

A typical curve of the evolution of the criterion (5.10) as a function of $\hat{n}$ is shown in Fig. 5.5.

It is shown in [5] that using this criterion a consistent estimate of the order $\hat{n}$ is obtained under mild noise conditions (i.e., $\lim_{N\rightarrow\infty} Pr(\hat{n} = n) = 1$ where Pr denotes the probability). Comparisons with other order estimation criteria are also provided in this reference.

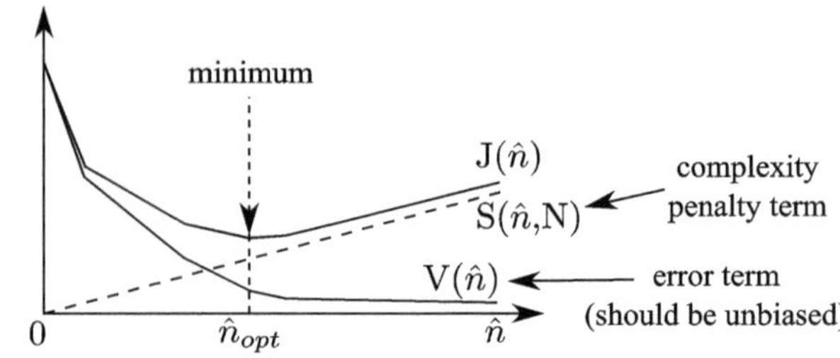

Fig. 5.5 Evaluation of the criterion for order estimation

Once an estimated order $\hat{n}$ is selected, one can apply a similar procedure to estimate $\hat{n}_A$, $\hat{n} - \hat{d}$, $\hat{n}_B + \hat{d}$, from which $\hat{n}_A$, $\hat{n}_B$ and $\hat{d}$ are obtained.[5]

5.4 Parameter Estimation Algorithms

The algorithms which will be used for parameter estimation will depend on the assumptions made on the noise disturbing the measurements, assumptions which have to be confirmed by the model validation.

It is important to emphasize that no one single *plant* + *noise* structure can describe all the situations encountered in practice. Furthermore, there is no a unique parameter estimation algorithm that may be used with all possible plant + noise structures such that the estimated parameters are always unbiased. The most typical structures for *plant* + *noise* are shown in Fig. 5.6.

The various "plant + noise" models shown in Fig. 5.6 can be described by:

$$y(t) = \frac{q^{-d}B(q^{-1})}{A(q^{-1})}u(t) + \eta(t) \tag{5.12}$$

For structure $S1$ one has:

$$\eta(t) = \frac{1}{A(q^{-1})}e(t) \tag{5.13}$$

where $e(t)$ is a discrete-time Gaussian white noise (zero mean and standard deviation σ).

For structure $S2$ one has:

$$\eta(t) = w(t) \tag{5.14}$$

a centred noise of finite power and uncorrelated with the input $u(t)$.

[5]Routines corresponding to this method in MATLAB (*estorderiv.m*) and Scilab (*estorderiv.sci*) can be downloaded from the websites: http://www.landau-adaptivecontrol.org and http://www.gipsa-lab.grenoble-inp.fr/ ioandore.landau/identificationandcontrol/.

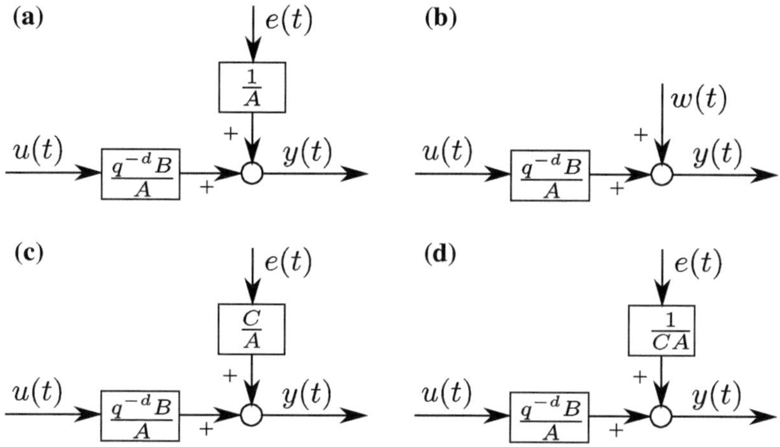

Fig. 5.6 Structure of the "plant + noise" models. **a S1**: $A(q^{-1})y(t) = q^{-d}B(q^{-1})u(t) + e(t)$. **b S2**: $A(q^{-1})y(t) = q^{-d}B(q^{-1})u(t) + A(q^{-1})w(t)$. **c S3**: $A(q^{-1})y(t) = q^{-d}B(q^{-1})u(t) + C(q^{-1})e(t)$. **d S4**: $A(q^{-1})y(t) = q^{-d}B(q^{-1})u(t) + (1/C(q^{-1}))e(t)$

For structure $S3$ one has

$$\eta(t) = \frac{C(q^{-1})}{A(q^{-1})}e(t) \tag{5.15}$$

and for structure $S4$ one has:

$$\eta(t) = \frac{1}{C(q^{-1})A(q^{-1})}e(t) \tag{5.16}$$

Based on the experience of the authors in identifying active vibration control systems, one can say that in most of the situations they are represented correctly by ARMAX models (structure $S3$). Therefore, most likely, algorithms for estimating parameters for ARMAX models will provide good results (should be confirmed by model validation). The simplest and in general most efficient algorithms for identifying active vibration control systems are the "recursive extended least squares" and the "output error with extend predictor model".[6] Details on these two type of algorithms will be given next. Nevertheless, there is no guarantee that ARMAX representation is the good one for all possible configuration which can be encountered in practice. Therefore one has to be prepared to use also other parameter estimation algorithms if the validation of the identified models using the above mentioned algorithms fails.[7]

---

[6]Routines for these algorithms can be downloaded from the websites: http://www.landau-adaptivecontrol.org and http://www.gipsa-lab.grenoble-inp.fr/~ioandore.landau/identificationandcontrol/.

[7]The interactive stand alone software iReg (http://tudor-bogdan.airimitoaie.name/ireg.html) provides parameter estimations algorithms for all the mentioned "plant + noise" structures as well as

All the recursive parameter estimation algorithms use the same parameter adaptation algorithm:

$$\hat{\theta}(t+1) = \hat{\theta}(t) + F(t)\Phi(t)\nu(t+1) \tag{5.17}$$

$$F(t+1)^{-1} = \lambda_1(t)F(t)^{-1} + \lambda_2(t)\Phi(t)\Phi^T(t) \tag{5.18}$$

$$0 < \lambda_1(t) \leq 1; \quad 0 \leq \lambda_2(t) < 2; \; F(0) > 0$$

$$F^{-1}(t) > \alpha F^{-1}(0); \quad 0 < \alpha < \infty$$

$$\nu(t+1) = \frac{\nu^{\circ}(t+1)}{1 + \Phi^T(t)F(t)\Phi(t)} \tag{5.19}$$

What changes from an identification algorithm to another is:

- the structure of the adjustable predictor;
- how the adaptation error is generated;
- how the regressor vector is generated;
- how the adaptation error is generated; and
- the size of the adjustable parameter vector (the number of parameters).

The various options for the selection of the time profile of the adaptation gain $F(t)$ in (5.19) have been discussed in Sect. 4.3.4. For system identification of a linear time invariant models, a decreasing adaptation gain type algorithm will be used or an algorithm with variable forgetting factor. We will present next the "recursive extended least squares" and the "output error with extended predictor".

5.4.1 Recursive Extended Least Squares (RELS)

This method has been developed in order to identify without bias *plant* + *noise* models of the form (ARMAX model):

$$A(q^{-1})y(t) = q^{-d}B(q^{-1})u(t) + C(q^{-1})e(t) \tag{5.20}$$

The idea is to simultaneously identify the plant model and the noise model, in order to obtain a prediction (adaptation) error which is asymptotically white.

The model generating the data can be expressed as:

$$\begin{aligned} y(t+1) &= -A^*(q^{-1})y(t) + B^*(q^{-1})u(t-d) + C^*(q^{-1})e(t) + e(t+1) \\ &= \theta^T\phi_0(t) + e(t+1) \end{aligned} \tag{5.21}$$

(Footnote 7 continued)
an automated identification procedure covering all the stages of system identification. It has been extensively used for identification of active vibration control systems.

where:

$$\theta^T = [a_1 \ldots a_{n_A}, b_1 \ldots b_{n_B}, c_1 \ldots c_{nc}] \tag{5.22}$$

$$\phi_0^T(t) = [-y(t) \cdots - y(t-n_A+1), u(t-d) \ldots u(t-d-n_B+1), e(t) \ldots e(t-n_c+1)] \tag{5.23}$$

Assume that the parameters are known and construct a predictor that will give a white prediction error:

$$\hat{y}(t+1) = -A^*(q^{-1})y(t) + B^*(q^{-1})u(t-d) + C^*(q^{-1})e(t) \tag{5.24}$$

Furthermore, this predictor will minimize $E\{[y(t+1) - \hat{y}(t+1)]^2\}$ [1].

The prediction error, in the case of known parameters, is given by:

$$\varepsilon(t+1) = y(t+1) - \hat{y}(t+1) = e(t+1) \tag{5.25}$$

This allows rewriting Eq. (5.24) in the form:

$$\hat{y}(t+1) = -A^*(q^{-1})y(t) + B^*(q^{-1})u(t-d) + C^*(q^{-1})\varepsilon(t) \tag{5.26}$$

Subtracting now (5.26) from (5.21), one gets:

$$\varepsilon(t+1) = -C^*(q^{-1})\,[\varepsilon(t) - e(t)] + e(t) \tag{5.27}$$

i.e.,

$$C(q^{-1})\,[\varepsilon(t+1) - e(t+1)] = 0 \tag{5.28}$$

Since $C(q^{-1})$ is an asymptotically stable polynomial, it results that $\varepsilon(t+1)$ will become white asymptotically.

The adaptive version of this predictor is as follows. The *a priori* adjustable predictor will take the form:

$$\hat{y}^{\circ}(t+1) = -\hat{A}^*(q^{-1},t)y(t) + \hat{B}^*(q^{-1},t)u(t) + \hat{C}^*(q^{-1},t)\varepsilon(t) = \hat{\theta}^T(t)\phi(t) \tag{5.29}$$

in which:

$$\hat{\theta}^T = [\hat{a}_1(t) \ldots \hat{a}_{n_A}(t), \hat{b}_1(t) \ldots \hat{b}_{n_A}(t), \hat{c}_1(t) \ldots \hat{c}_{n_A}(t)] \tag{5.30}$$

$$\phi^T(t) = [-y(t) \cdots - y(t-n_A+1), u(t-d) \ldots u(t-d-n_B+1), \varepsilon(t) \ldots \varepsilon(t-n_c+1)] \tag{5.31}$$

The *a posteriori* adjustable predictor will be given by:

$$\hat{y}(t+1) = \hat{\theta}^T(t+1)\phi(t) \tag{5.32}$$

The *a posteriori* prediction error $\varepsilon(t)$ which enters in the observation vector of the predictor is given by

$$\varepsilon(t) = y(t) - \hat{y}(t) \tag{5.33}$$

(where $\hat{y}(t)$ is now the *a posteriori* output of the adjustable predictor) and the *a priori* prediction error is given by:

$$\varepsilon^{\circ}(t+1) = y(t+1) - \hat{y}^{\circ}(t+1) \tag{5.34}$$

The *a posteriori* prediction equation is obtained subtracting (5.32) from (5.21) and observing that (5.21) can be alternatively expressed as:

$$y(t+1) = \theta^T \phi(t) - C^*(q^{-1})\varepsilon(t) + C(q^{-1})e(t) \tag{5.35}$$

(by adding and subtracting the term $\pm C^*(q^{-1})\varepsilon(t)$). One obtains:

$$\varepsilon(t+1) = -C^*(q^{-1})\varepsilon(t) + \left[\theta - \hat{\theta}(t+1)\right]^T \phi(t) + C(q^{-1})e(t) \tag{5.36}$$

from which it results that:

$$\varepsilon(t+1) = \frac{1}{C(q^{-1})}\left[\theta - \hat{\theta}(t+1)\right]^T \phi(t) + e(t) \tag{5.37}$$

In the deterministic case $C(q^{-1}) = 1$, $e(t) \equiv 0$ and it can be seen that (5.37) has the appropriate format corresponding to Theorem 4.1 given in Chap. 4. One immediately concludes, using the PAA given in (5.17) through (5.19), with $\Phi(t) = \phi(t)$, $\nu(t) = \varepsilon(t)$, and $\nu^{\circ}(t) = \varepsilon^{\circ}(t)$ that, in the deterministic case, global asymptotic stability is assured without any positive real condition. In stochastic environment, either using ODE or martingales, it can be shown [7] that the convergence is assured provided that (sufficient condition):

$$H'(z^{-1}) = \frac{1}{C(z^{-1})} - \frac{\lambda_2}{2} \tag{5.38}$$

is a strictly positive real transfer function for $2 > \lambda_2 \geq \max_t \lambda_2(t)$.

5.4.2 Output Error with Extended Prediction Model (XOLOE)

This algorithm can be used for identification of *plant + noise* models of the ARMAX form. It has been developed initially with the aim to remove the *positive real condition* required by the *output error* algorithm. It turns out that the XOLOE can be interpreted as a variant of the ELS. To see this, consider the *a priori* output of the adjustable predictor for ELS (5.29), which can be rewritten as follows by adding and subtracting the term $\pm\hat{A}^*(q^{-1},t)\hat{y}(t)$:

$$\hat{y}^\circ(t+1) = -\hat{A}^*(q^{-1},t)\hat{y}(t) + \hat{B}^*(q^{-1},t)u(t-d) + \left[\hat{C}^*(q^{-1},t)\varepsilon(t) - \hat{A}^*(q^{-1},t)[y(t)-\hat{y}(t)]\right] \quad (5.39)$$

Defining:

$$\hat{H}^*(q^{-1},t) = \hat{C}^*(q^{-1},t) - \hat{A}^*(q^{-1},t) = \hat{h}_1(t) + q^{-1}\hat{h}_2(t) + \cdots$$

with:

$$\hat{h}_i(t) = \hat{c}_i(t) - \hat{a}_i(t);\ \ i = 1,2\ldots \max(n_A, n_C)$$

Equation (5.39) can be rewritten as:

$$\hat{y}^\circ(t+1) = -\hat{A}^*(q^{-1},t)\hat{y}(t) + \hat{B}^*(q^{-1},t)u(t-d) + \hat{H}^*(q^{-1},t)\varepsilon(t) \quad (5.40)$$

$$= \hat{\theta}^T(t)\phi(t) \quad (5.41)$$

where now:

$$\hat{\theta}^T(t) = \left[\hat{a}_1(t)\ldots\hat{a}_{n_A}, \hat{b}_1(t)\ldots\hat{b}_{n_B}(t), \hat{h}_1(t)\ldots\hat{h}_{n_H}(t)\right]$$

$$\phi^T(t) = [-\hat{y}(t),\ldots\hat{y}(t-n_A+1), u(t-d)\ldots u(t-d-n_B+1), \varepsilon(t)\ldots\varepsilon(t-n_C+1)]$$

Equation (5.40) corresponds to the adjustable predictor for the output error with extended prediction model. One immediately concludes, using the PAA given in (5.17)–(5.19), with $\Phi(t)=\phi(t)$, $\nu(t)=\varepsilon(t)$, and $\nu^\circ(t)=\varepsilon^\circ(t)$ (defined in Eqs. (5.33) and (5.34), respectively) that, in the deterministic case, global asymptotic stability is assured without any positive real condition. In the stochastic context, one has the (sufficient) convergence condition: $H'(z^{-1}) = \frac{1}{C(z^{-1})} - \frac{\lambda_2}{2}$ should be SPR ($2 > \lambda_2 \geq \max_t \lambda_2(t)$) similar to that for ELS.

Despite their similar asymptotic properties, there is a difference in the first n_A components of the vector $\phi(t)$. While the RELS algorithm uses the measurements $y(t), y(t-1),\ldots$ directly affected by the noise, the XOLOE algorithm uses the predicted *a posteriori* outputs $\hat{y}(t), \hat{y}(t-1)$ which depend upon the noise only indirectly through the PAA. This explains why a better estimation is obtained with XOLOE than with RELS over short or medium time horizons (it removes the bias more quickly).

5.5 Validation of the Identified Models

The identification methods considered above (*recursive extended least squares* and *output error with extended predictor*) belongs to the class of methods based on the whitening of the residual errors, i.e., the identified ARMAX predictor is an optimal predictor if the residual error is a white noise. If the residual prediction error is a white noise sequence, in addition to obtaining unbiased parameter estimates, this also means that the identified model gives the best prediction for the plant output in the sense that it minimizes the variance of the prediction error. On the other hand, since the residual error is white and a white noise is not correlated with any other variable, then all the correlations between the input and the output of the plant are represented by the identified model and what remains unmodelled does not depend on the input.

The principle of the validation method is as follows:

- If the plant + noise structure chosen is correct, i.e., representative of reality.
- If an appropriate parameter estimation method for the structure chosen has been used.
- If the orders of the polynomials $A(q^{-1})$, $B(q^{-1})$, $C(q^{-1})$ and the value of d (delay) have been correctly chosen (the plant model is in the model set).

Then the prediction error $\varepsilon(t)$ asymptotically tends toward a white noise, which implies:

$$\lim_{t\to\infty} E\{\varepsilon(t)\varepsilon(t-i)\} = 0; \qquad i = 1, 2, 3\ldots;\ -1, -2, -3\ldots$$

The validation method implements this principle.[8] It is made up of several steps:

(1) Creation of an I/O file for the identified model (using the same input sequence as for the system).
(2) Creation of a residual prediction error file for the identified model.
(3) *Whiteness* (uncorrelatedness) test on the residual prediction errors sequence.

5.5.1 Whiteness Test

Let $\{\varepsilon(t)\}$ be the centred sequence of the residual prediction errors (centred: measured value–mean value). One computes:

[8]Routines corresponding to this validation method in MATLAB and Scilab can be downloaded from the websites: http://www.landau-adaptivecontrol.org and http://www.gipsa-lab.grenoble-inp.fr/ ioandore.landau/identificationandcontrol/.

$$R(0) = \frac{1}{N}\sum_{t=1}^{N}\varepsilon^2(t),\ RN(0) = \frac{R(0)}{R(0)} = 1, \tag{5.42}$$

$$R(i) = \frac{1}{N}\sum_{t=1}^{N}\varepsilon(t)\varepsilon(t-i),\ RN(i) = \frac{R(i)}{R(0)},\quad i = 1, 2, 3, \ldots n_A, \ldots \tag{5.43}$$

with $i_{\max} \geq n_A$ (degree of polynomial $A(q^{-1})$), which are estimations of the (normalized) autocorrelations.

If the residual prediction error sequence is perfectly white (theoretical situation), and the number of samples is very large ($N \rightarrow \infty$), then $RN(0) = 1$, $RN(i) = 0$, $i \geq 1$.[9]

In real situations, however, this is never the case (i.e., $RN(i) \neq 0;\ i \geq 1$), since on the one hand, $\varepsilon(t)$ contains residual structural errors (order errors, nonlinear effects, non-Gaussian noises), and on the other hand, the number of samples may be relatively small in some cases. Also, it should be kept in mind that one always seeks to identify *good* simple models (with few parameters).

One considers as a practical validation criterion (extensively tested on applications):

$$RN(0) = 1;\ \mid RN(i) \mid \leq \frac{2.17}{\sqrt{N}};\ i \geq 1 \tag{5.44}$$

where N is the number of samples.

This test has been defined taking into account the fact that for a white noise sequence $RN(i), i \neq 0$ has an asymptotically Gaussian (normal) distribution with zero mean and standard deviation:

$$\sigma = \frac{1}{\sqrt{N}}$$

The confidence interval considered in (5.44) corresponds to a 3 % level of significance of the hypothesis test for Gaussian distribution.

If $RN(i)$ obeys the Gaussian distribution $(0, 1/\sqrt{N})$, there is only a probability of 1.5 % that $RN(i)$ is larger than $2.17/\sqrt{N}$, or that $RN(i)$ is smaller than $-2.17/\sqrt{N}$. Therefore, if a computed value $RN(i)$ falls outside the range of the confidence interval, the hypothesis $\varepsilon(t)$ and $\varepsilon(t-i)$ are independent should be rejected, i.e., $\{\varepsilon(t)\}$ is not a white noise sequence.

The following remarks are important:

- If several identified models have the same complexity (number of parameters), one chooses the model given by the methods that lead to the smallest $|RN(i)|$;
- A *too good* validation criterion indicates that model simplifications may be possible.

[9] Conversely, for Gaussian data, uncorrelation implies independence. In this case, $RN(i) = 0, i \geq 1$ implies independence between $\varepsilon(t), \varepsilon(t-1)\ldots$, i.e., the sequence of residuals $\{\varepsilon(t)\}$ is a Gaussian white noise.

- To a certain extent, taking into account the relative weight of various non-Gaussian and modelling errors (which increases with the number of samples), the validation criterion may be slightly tightened for small N and slightly relaxed for large N. Therefore, for simplicity's sake, one can consider as a basic practical numerical value for the validation criterion value:

$$| RN(i) | \leq 0.15; \;\; i \geq 1.$$

Note also that a complete model validation implies, after the validation using the input/output sequence used for identification, a validation using a plant input/output sequence other than the one used for identification.

5.6 Concluding Remarks

Basic elements for the identification of discrete-time models for dynamical systems have been laid down in this chapter. The following facts have to be emphasized:

1. System identification includes four basic steps:
   - input/output data acquisition under an experimental protocol;
   - estimation or selection of the model complexity;
   - estimation of the model parameters; and
   - validation of the identified model (structure of the model and values of parameters).

   This procedure has to be repeated (with appropriate changes at each step) if the validation of the model fails.
2. Recursive or off-line parameter estimation algorithms can be used for identification of the plant model parameters.
3. The various recursive parameter estimation algorithms use the same structure for the PAA. They differ from each other in the following ways:
   - structure of the adjustable predictor;
   - nature of the components of the observation vector; and
   - the way in which the adaptation error is generated.
4. The stochastic noises, which contaminate the measured output, may cause errors in the parameter estimates (bias). For a specific type of noise, appropriate recursive identification algorithms providing asymptotically unbiased estimates are available.
5. A unique *plant + noise* model structure that describes all the situations encountered in practice does not exist, nor is there a unique identification method providing satisfactory parameter estimates (unbiased estimates) in all situations.

5.7 Notes and References

A more detailed discussion of the subject following the same pathway can be found in [1]. The associated website: http://www.gipsa-lab.grenoble-inp.fr/~ioandore.landau/identificationandcontrol/ provides MATLAB and scilab functions for system identification as well as simulated and real input/output data for training.

For a general coverage of system identification see [3, 4].

References

1. Landau I, Zito G (2005) Digital control systems - design, identification and implementation. Springer, London
2. Ljung L, Söderström T (1983) Theory and practice of recursive identification. The MIT Press, Cambridge
3. Ljung L (1999) System identification - theory for the user, 2nd edn. Prentice Hall, Englewood Cliffs
4. Soderstrom, T., Stoica, P.: System Identification. Prentice Hall (1989)
5. Duong HN, Landau ID (1996) An IV based criterion for model order selection. Automatica 32(6):909–914
6. Duong HN, Landau I (1994) On statistical properties of a test for model structure selection using the extended instrumental variable approach. IEEE Trans Autom Control 39(1):211–215. doi:10.1109/9.273371
7. Landau ID, Lozano R, M'Saad M, Karimi A (2011) Adaptive control, 2nd edn. Springer, London

Chapter 6
Identification of the Test Benches in Open-Loop Operation

6.1 Identification of the Active Hydraulic Suspension in Open-Loop Operation

The active suspension has been presented in Sect. 2.1. It will be used to enhance damping properties of the passive damper in the frequency region 25–50 Hz. In the same frequency region, active rejection of vibrations will be also considered. Outside this region, the passive part offers good vibration isolation properties. For active damping the specifications in the frequency domain will go up to 150 Hz. Above this frequency, the system should operate almost in open-loop. The sampling frequency is 800 Hz.

The block diagram for the primary and secondary path is shown in Fig. 6.1, where $u(t)$ will be the excitation of the secondary path and u_p the excitation of the primary path.

The linear time-invariant (LTI) discrete-time model of the secondary path used for controller design has the form

$$G(z^{-1}) = \frac{z^{-d}B(z^{-1})}{A(z^{-1})} \tag{6.1}$$

where

$$A(z^{-1}) = 1 + a_1 z^{-1} + \cdots + a_{n_A} z^{-n_A}, \tag{6.2}$$

$$B(z^{-1}) = b_1 z^{-1} + \cdots + b_{n_B} z^{-n_B}, \tag{6.3}$$

and d is the secondary path pure time delay in number of sampling periods.[1]

[1]The complex variable z^{-1} will be used to characterize the system's behaviour in the frequency domain and the delay operator q^{-1} will be used for the time domain analysis.

I.D. Landau et al., *Adaptive and Robust Active Vibration Control*,
Advances in Industrial Control, DOI 10.1007/978-3-319-41450-8_6

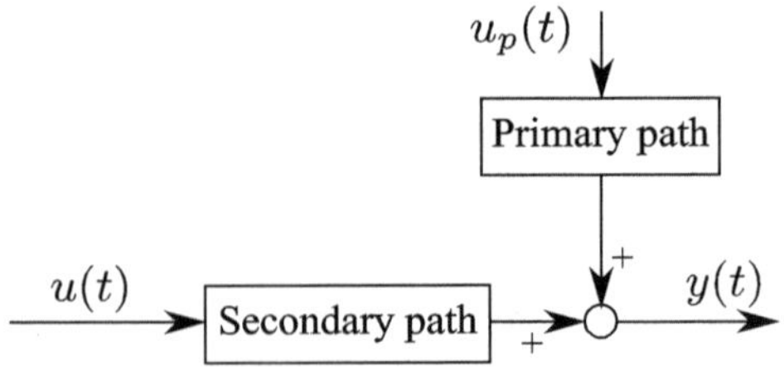

Fig. 6.1 Block diagram of the primary and secondary path

The linear time-invariant (LTI) discrete-time model of the primary path has the form

$$D(z^{-1}) = \frac{q^{-d_D} B_D(z^{-1})}{A_D(z^{-1})} \tag{6.4}$$

The identification of the secondary and primary paths will be presented next. The model of the secondary path will be used throughout the book for controller design while the model of the primary path will be used for simulation only.

6.1.1 *Identification of the Secondary Path*

6.1.1.1 Data Acquisition

Since the main frequency range for control is between 25 and 50 Hz, a Pseudo Random Binary Sequence (PRBS) with a clock frequency divider of 4 has been used in order to enhance the energy of the input excitation in low frequencies. The risk of missing some dynamics around 200 Hz is not important since the system will operate almost in open-loop at frequencies over 150 Hz. The PRBS used has the following characteristics

- magnitude $= 0.2\,\text{V}$;
- number of cells: $N = 9$ (sequence length: $L = 2^N - 1 = 511$);
- frequency divider: $p = 4$; and
- number of acquired samples: 2048.

Since the secondary path has a double differentiator behaviour (input: position, output: force) as indicated in Sect. 5.2.3, this will be considered as a "known" part of the system and the objective will be to identify the "unknown" part only. To do this, the input sequence will be filtered by a double discrete-time differentiator (i.e., $(1 - q^{-1})^2$) as shown in Fig. 6.2, i.e., $B(q^{-1}) = (1 - q^{-1})^2 \cdot B'(q^{-1})$.

Once B' will be identified, the discrete-time double differentiator will be included in the final complete model.

The input/output data file *data_identActiveSusp_SecPath.mat* is available on the book website (the input is already filtered through a discrete-time double differentiator).

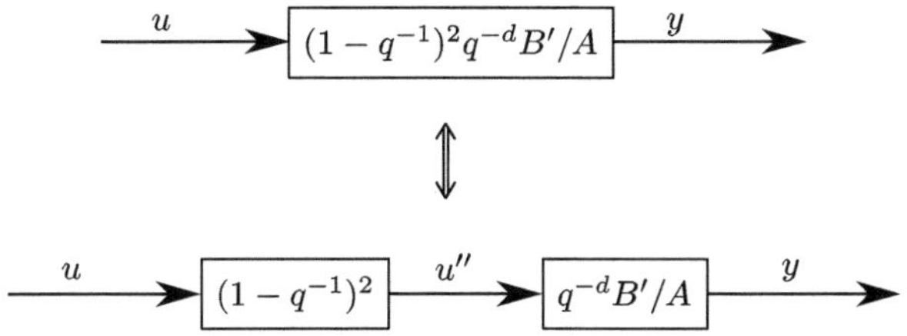

Fig. 6.2 Including the double differentiator for identification of the secondary path

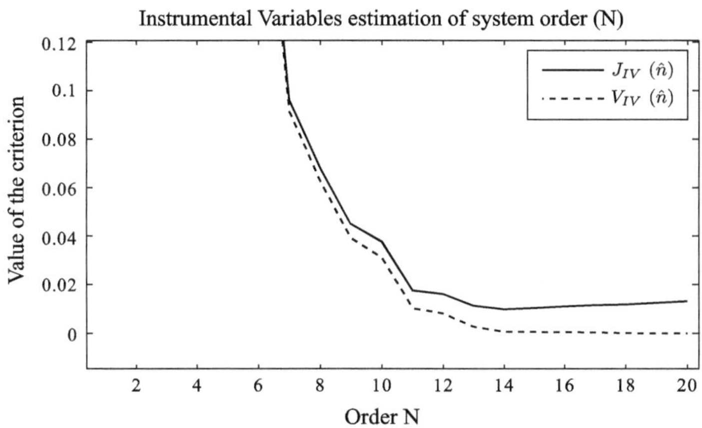

Fig. 6.3 Estimation of the system order *n* (active suspension)

6.1.1.2 Order Estimation

The algorithm described in Sect. 5.3, for order estimation and based on the use of instrumental variables has been used.[2] The non-penalized criterion V_{IV} (dashed line) and the complexity penalized criterion J_{IV} (solid line) for the estimation of n (order of the system) are shown in Fig. 6.3. As it can be seen the minimum of J_{IV} is not very sharp but however clear. $n = 14$ has been selected.

Proceeding further to the estimation of the orders of polynomials A, B' and delay d, the values obtained are $n_A = 13$, $n_{B'} = 11$, $d = 3$. From Fig. 6.4 (zoom) it can be seen that the criterion for the selection of n_A gives extremely close results for $n_A = 13$ and $n_A = 14$. It was found that $n_A = 14$ gives better results in terms of statistical model validation. For parameter estimation, since the complexity of the designed controller will depend on $n_B + d$, it was decided to take $n_{B'} = 14$, $d = 0$ (the model with $n_{B'} = 11$ and $d = 3$ gives very close results).

[2]See function *estorderiv.m* on the book website.

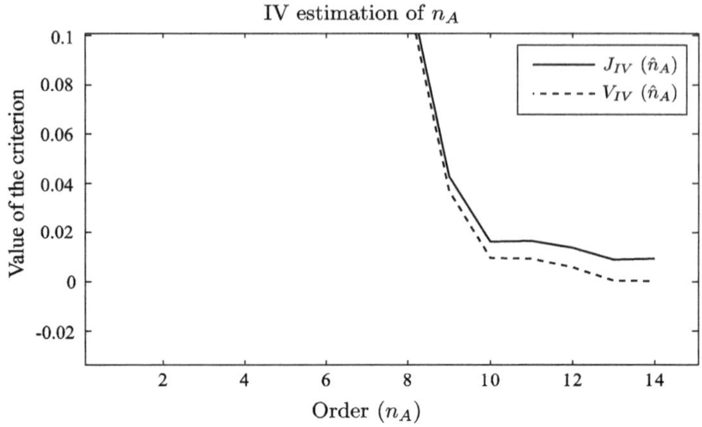

Fig. 6.4 Estimation of the order of polynomial A (n_A) (active suspension)

6.1.1.3 Parameter Estimation

Models with order $n_A = 13$, $n_A = 14$ and $d = 0$ have been identified. As indicated in Sect. 5.4 extensive validation tests on this system as well as on other AVC (see subsequent section) indicate that the ARMAX representation of the "plant + noise" provides the best results. Various methods dedicated to parameter estimation of ARMAX model have been used (Recursive extended least squares, Output error with extended prediction model, Recursive maximum likelihood [1]). These algorithms have been used with a *variable forgetting factor*

$$\lambda_1(t) = \lambda_0 \lambda_1(t-1) + 1 - \lambda_0; \quad 0 < \lambda_0 < 1, \tag{6.5}$$

with $\lambda_1(0) = 0.97$ and $\lambda_0 = 0.97$. The various obtained models have been validated and compared in terms of quality of validation.

6.1.1.4 Model Validation

Models with $n_A = 14$, $n_{B'} = 14$, $d = 0$ and with $n_A = 13$, $n_{B'} = 14$, $d = 0$ obtained with various parameter estimations have been compared using the whiteness test. The best results have been obtained using estimated models with $nA = 14$, $n_{B'} = 14$, $d = 0$ and estimating the parameters using Recursive Extended Least Squares (RELS) or Output Error with Extended Prediction Model (XOLOE). Both have been used with a variable forgetting factor. Figure 6.5 shows the validation results for the RELS model and Fig. 6.6 shows the validation results for the XOLOE model ($WRN(i)$ corresponds to the normalized autocorrelations defined in Chap. 5, Eq. (5.43)).

Table 6.1 summarizes the validation results.

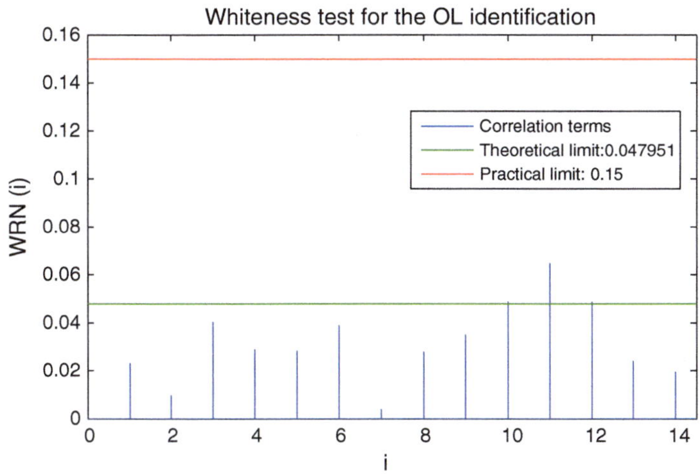

Fig. 6.5 Whiteness test for the RELS identified model (active suspension)

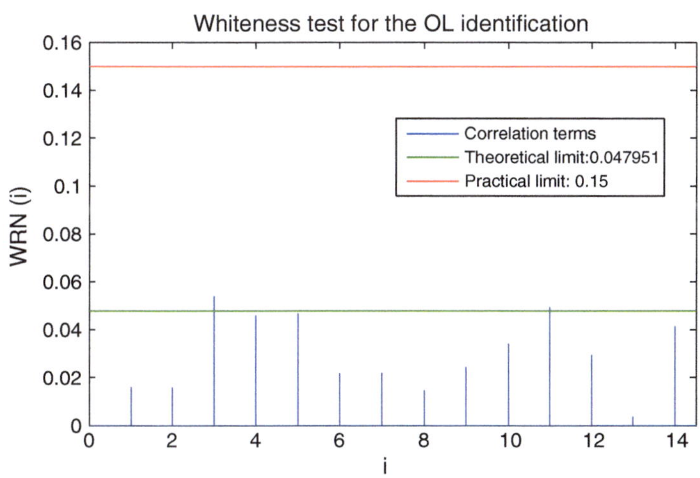

Fig. 6.6 Whiteness test for the XOLOE identified model (active suspension)

The frequency characteristics of the two models are indistinguishable except at very low frequencies. Finally the RELS model has been chosen since the RN(i) for lower values of i are smaller than those of the XOLOE model. The parameters of the model are given in Table 6.2 and they are stored in the file *SecPath_activeSusp.mat* available on the book website. Table 6.3 gives the frequency and the damping of the poles of the secondary path identified model (RELS).

Table 6.1 Summary of the whiteness test validations (active suspension)

Algorithm	Error energy	Maximum RN(i)	RN(i) over limit
RELS	0.0092	0.0647 (i = 11)	3
XOLOE	0.0090	0.0540 (i = 3)	2

Table 6.2 Parameters of the identified secondary path model (RELS) (active suspension)

Parameter	Value	Parameter	Value	Parameter	Value	Parameter	Value
a_0	1.0000	a_9	0.5008	b_0	0.0000	b_9	−2.3676
a_1	−0.0586	a_{10}	0.2481	b_1	0.0251	b_{10}	2.3658
a_2	0.4092	a_{11}	−0.4152	b_2	0.0647	b_{11}	2.5058
a_3	−0.9164	a_{12}	−0.0154	b_3	−0.1246	b_{12}	2.8960
a_4	−0.5737	a_{13}	−0.3473	b_4	−0.4606	b_{13}	−0.5826
a_5	−0.5834	a_{14}	−0.0795	b_5	2.7988	b_{14}	0.1619
a_6	−0.3110			b_6	1.2316	b_{15}	−2.5355
a_7	0.6052			b_7	−3.3935	b_{16}	0.4735
a_8	0.6965			b_8	−3.0591		

Table 6.3 Frequency and damping of the poles of the secondary path identified model (RELS) (active suspension)

Poles	Damping	Frequency (Hz)
$0.955993 + 0.000000i$	1.000000	5.730231
$0.950132 - 0.242697i$	0.077941	31.939243
$0.950132 + 0.242697i$	0.077941	31.939243
$0.265498 - 0.920456i$	0.033259	164.335890
$0.265498 + 0.920456i$	0.033259	164.335890
$0.162674 - 0.753066i$	0.188593	176.071865
$0.162674 + 0.753066i$	0.188593	176.071865
$-0.301786 - 0.925822i$	0.014095	240.144314
$-0.301786 + 0.925822i$	0.014095	240.144314
$-0.547208 - 0.798935i$	0.014803	276.492803
$-0.547208 + 0.798935i$	0.014803	276.492803
$-0.869136 - 0.255155i$	0.034615	363.860597
$-0.869136 + 0.255155i$	0.034615	363.860597
$-0.217701 + 0.000000i$	0.436606	444.616040

Figure 6.7 gives the poles zero map of the RELS model of the secondary path. The model of the secondary path is characterized by the presence of several very low damped complex poles and unstable zeros. The frequency characteristics of the secondary path is shown in Fig. 6.8.

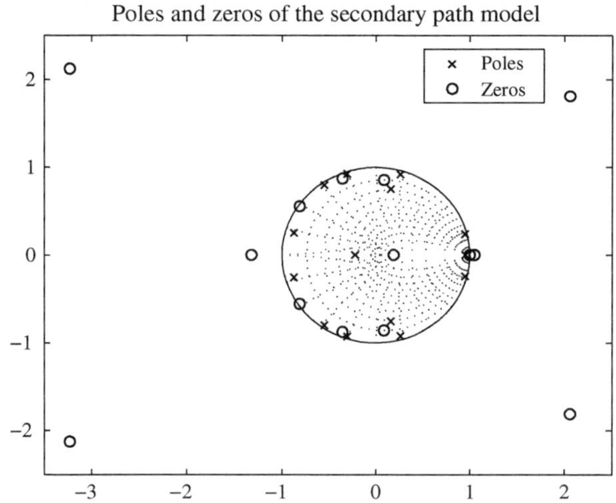

Fig. 6.7 Poles-zeros map for the secondary path identified model (RELS) (active suspension)

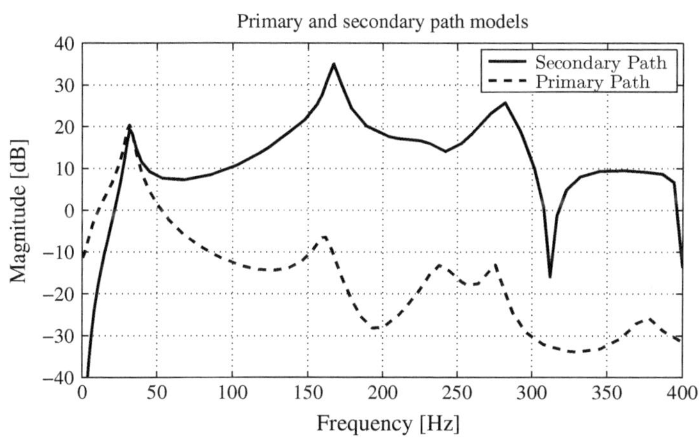

Fig. 6.8 Frequency characteristics of the identified model of the primary and secondary paths (active suspension)

6.1.2 Identification of the Primary Path

The same identification methodology used for the secondary path has been used for identifying the model of the primary path. The identification was done between the excitation of the shaker u_p and the residual acceleration in the absence of the compensation action. The estimated orders are: $n_{A_p} = 12$, $n_{B_p} = 9$, $d_p = 3$. The excitation was a PRBS sequence generated with a shift register with N = 9 cells and without frequency divider ($p = 1$). Like for the secondary path, the existence of the double differentiator has been taken in account. The best model in terms

Table 6.4 Parameters of the identified model of the primary path (active suspension)

Parameter	Value	Parameter	Value	Parameter	Value	Parameter	Value
a_0	1.0000	a_7	0.7709	b_0	0.0000	b_7	0.1325
a_1	−0.3862	a_8	0.2417	b_1	−0.1016	b_8	0.0552
a_2	−0.2391	a_9	−0.0932	b_2	−0.2085		
a_3	−0.6875	a_{10}	−0.1747	b_3	−0.1375		
a_4	−0.3052	a_{11}	−0.4845	b_4	−0.0393		
a_5	0.4003	a_{12}	0.2735	b_5	0.0985		
a_6	−0.1430			b_6	0.1536		

of validation has been obtained with RELS algorithm using adaptation gain with variable forgetting factor ($\lambda_1(0) = \lambda_0 = 0.97$). The frequency characteristics of the identified model for the primary path are shown in Fig. 6.8. The model of the primary path shows a very strong resonance at 31.59 Hz which needs to be damped. There also other very low damped complex zeros over 160 Hz. The parameters of the model are given in Table 6.4 and they are available in the file *PrimPath_activeSusp.mat* to be downloaded from the book website.

6.2 Identification of the AVC System Using Feedback Compensation Through an Inertial Actuator

The AVC system using an inertial actuator has been described in Sect. 2.2. The block diagram of the primary and secondary paths is the same as for the active suspension and is given in Fig. 6.1, where $u(t)$ will be the excitation of the secondary path and u_p the excitation of the primary path.

The identification of the primary and secondary paths in open-loop operation will be presented. The open-loop identification procedure is done in the absence of the controller and of the disturbance. The primary path is identified only for simulation purposes. The sampling frequency is 800 Hz.

6.2.1 Identification of the Secondary Path

6.2.1.1 Data Acquisition

As persistent excitation signal a PRBS generated with a shift register having $N = 10$ cells and a frequency divider $p = 2$ is used. The magnitude used is: 0.085 V. The input/output data file *data_identSAAI_SecPath.mat* is available on the book website.

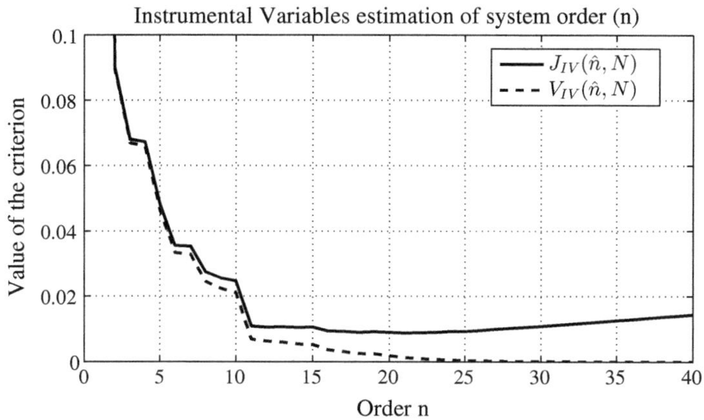

Fig. 6.9 Evaluation of the criterion for order estimation of the secondary path (global view) (inertial actuator AVC)

Since both paths present a double differentiator behaviour, this "known" dynamics do not need to be estimated in the open-loop identification procedure and the objective is to identify the "unknown" part only. This procedure has been already used in the previous Sect. 6.1 and is illustrated in Fig. 6.2. The model of the system without the double differentiator will be identified and the double differentiator will be included in the final complete model.

6.2.1.2 Order Estimation

The estimation of the order of the model for the secondary path is done using the procedure described in Sect. 5.3. Assuming that the measurements are affected by non-white noise, one uses for complexity estimation the *estororderiv.m*[3] algorithm which implements the estimation procedure using delayed instrumental variables.

Figure 6.9 shows the complexity estimation criterion for the order of the secondary path model $n = \max\{n_A, n_B + d\}$. The dashed line represents the V_{IV} criterion as a function of the order $\hat{n}$ and the solid line represents the complexity penalized criterion J_{IV}. A zoom of the penalized criterion is shown in Fig. 6.10. It can be observed that the penalized criterion is almost flat between 20 and 23, which suggests that any one of these values will give a good result.

After validation of the estimated models of various orders obtained for various $\hat{n}$ between 20 and 23 and comparing also with the power spectral density (PSD) of the output, it has been concluded that the best compromise between complexity and quality of the model for the secondary path is given by order $\hat{n} = 22$.

[3] Available on the book website.

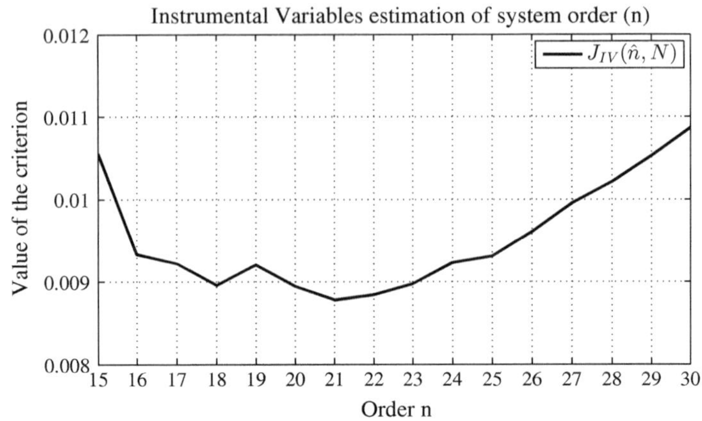

Fig. 6.10 Evaluation of the criterion for order estimation of the secondary path (zoom) (inertial actuator AVC)

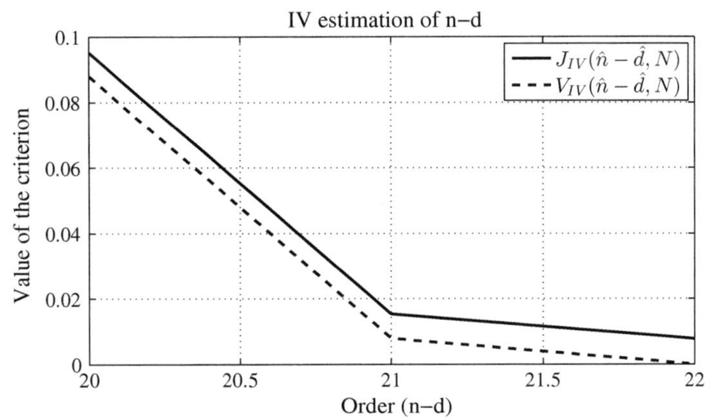

Fig. 6.11 Criterion for estimation of the order of the $n - d$ (inertial actuator AVC)

Once the order of the system is estimated, the estimation of $n - d$, $n_B + d$ and n_A follows using the same type of criterion as for the estimation of the order of the system. The estimated delay of the secondary path is obtained as $\hat{d} = 0$ from Fig. 6.11, since the minimum value of the criterion is obtained for $n - d = 22$ (with $n = 22$). From the estimation of $n_B + d$ in Fig. 6.12, and taking in account that $d = 0$ it results that the estimate of the order of the numerator is $\hat{n}_{B'} = 19$ (without the double differentiator). Finally the plant denominator order is estimated from Fig. 6.13 as $\hat{n}_A = 18$.

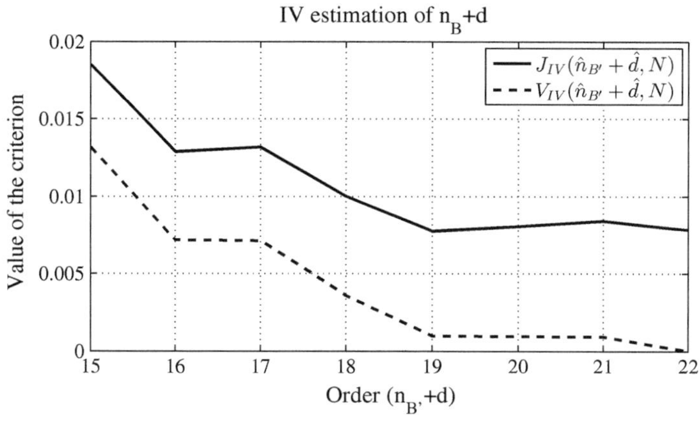

Fig. 6.12 Criterion for estimation of the order of the order $n_B + d$ (inertial actuator AVC)

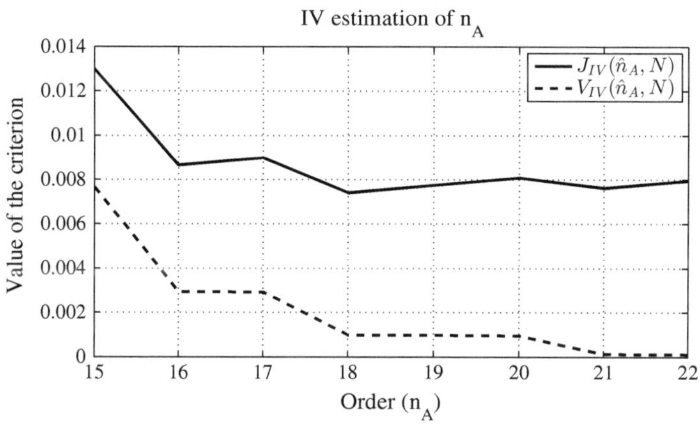

Fig. 6.13 Criterion for estimation of the order of the order of n_A (inertial actuator AVC)

6.2.1.3 Parameter Estimation

As for the active suspension, it was found that parameter estimation methods dedicated to ARMAX "model + noise" structure give the best validation results. These methods include Recursive Extended Least Squares (RELS), Output Error with Extended Estimation Model (XOLOE) and Recursive Maximum Likelihood (RML) [1]. These algorithms have been used with a decreasing adaptation gain.

6.2.1.4 Model Validation

The normalized autocorrelations for the methods RELS, XOLOE and RML are shown, respectively, in Figs. 6.14, 6.15 and 6.16 for the estimated models of the secondary path ($WRN(i)$) corresponds to the normalized autocorrelations defined in Chap. 5, Eq. (5.43). Table 6.5 gives the maximum normalized autocorrelation term and the error variance for each of these methods. RELS algorithm gives the best results for the identification of the secondary path.

The parameters of the secondary path are given in Table 6.6 and they can be found in the file *SecPath_SAAI.mat* available on the book website. The frequency

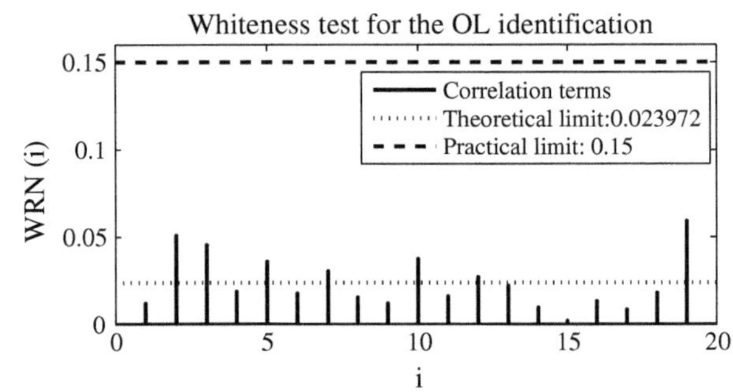

Fig. 6.14 Whiteness test for the RELS identified model of the secondary path (inertial actuator AVC)

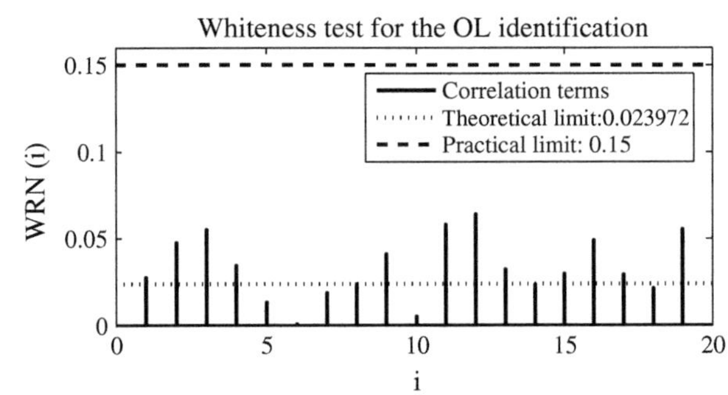

Fig. 6.15 Whiteness test for the XOLOE identified model of the secondary path (inertial actuator AVC)

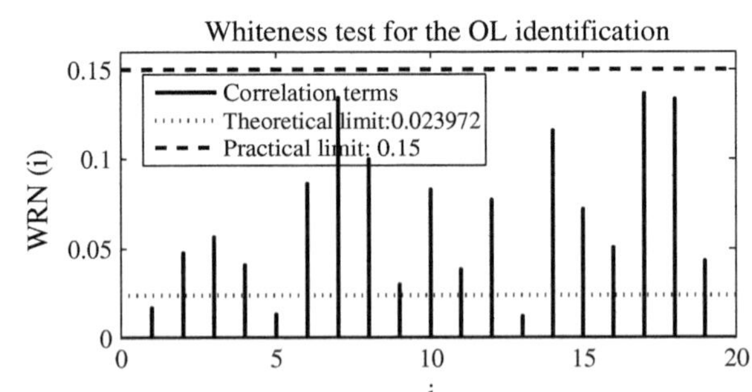

Fig. 6.16 Whiteness test for the RML identified model of the secondary path (inertial actuator AVC)

Table 6.5 Summary of the validation results for RELS, XOLOE and RML (inertial actuator AVC)

Algorithm	$\max_{WRN(i)}$	Error variance
RELS	0.059	9.7e-06
XOLOE	0.0642	1.0024e-05
RML	0.1367	7.9383e-06

Table 6.6 Identified parameters of the secondary path model (inertial actuator AVC)

Parameter	Value	Parameter	Value	Parameter	Value	Parameter	Value
a_0	1	a_{11}	−0.6107	b_0	0	b_{11}	−0.0179
a_1	−1.7074	a_{12}	0.5858	b_1	−0.0127	b_{12}	0.0164
a_2	1.2791	a_{13}	−0.2963	b_2	−0.0876	b_{13}	−0.0425
a_3	−0.8861	a_{14}	0.5336	b_3	0.0812	b_{14}	0.0031
a_4	1.2235	a_{15}	−0.9736	b_4	0.0157	b_{15}	0.0089
a_5	−1.1388	a_{16}	0.7849	b_5	0.0103	b_{16}	0.0166
a_6	0.6129	a_{17}	−0.3860	b_6	0.0380	b_{17}	0.0717
a_7	−0.7381	a_{18}	0.1902	b_7	−0.0580	b_{18}	−0.0508
a_8	1.0485			b_8	−0.0064	b_{19}	−0.0012
a_9	−0.3791			b_9	0.0195	b_{20}	−0.0093
a_{10}	0.2289			b_{10}	0.0188	b_{21}	−0.0139

Table 6.7 Poles of the open-loop secondary path identified model (inertial actuator AVC)

Poles	Damping	Frequency (Hz)
0.8982 ± 0.2008	0.3530	29.9221
0.9280 ± 0.3645	0.0079	47.6491
0.6642 ± 0.7203	0.0247	105.1909
0.5260 ± 0.8050	0.0395	126.4064
0.0623 ± 0.5832	0.3423	198.4441
-0.1229 ± 0.9689	0.0139	216.0790
-0.4533 ± 0.8394	0.0228	263.1134
-0.7524 ± 0.6297	0.0078	311.2826
-0.8965 ± 0.2856	0.0215	360.8163

and damping of the poles and zeros for the secondary path are given in Tables 6.7 and 6.8. The poles-zeros map of the secondary path is given in Fig. 6.17. The frequency characteristic of the secondary path is shown in Fig. 6.18. There are several low damped complex poles and zeros in the open-loop identified model. There are also very close resonances and antiresonances. The operational zone is defined between 50 and 95 Hz.

Table 6.8 Zeros of the open-loop secondary path identified model (inertial actuator AVC)

Zeros	Damping	Frequency (Hz)
1	0	0
1	0	0
0.9292 ± 0.3559	0.0135	46.5756
0.7097 ± 0.7037	0.0008	99.4567
0.2359 ± 0.9201	0.0389	168.1703
0.1054 ± 0.6063	0.3279	188.5135
-0.1100 ± 1.0087	-0.0087	213.8383
-0.4362 ± 0.8273	0.0325	261.9247
-0.8085 ± 0.5713	0.0040	321.6776
-0.9753 ± 0.1243	0.0056	383.8647
-0.4908	0.2209	410.1361
-7.7152	-0.5452	477.1543

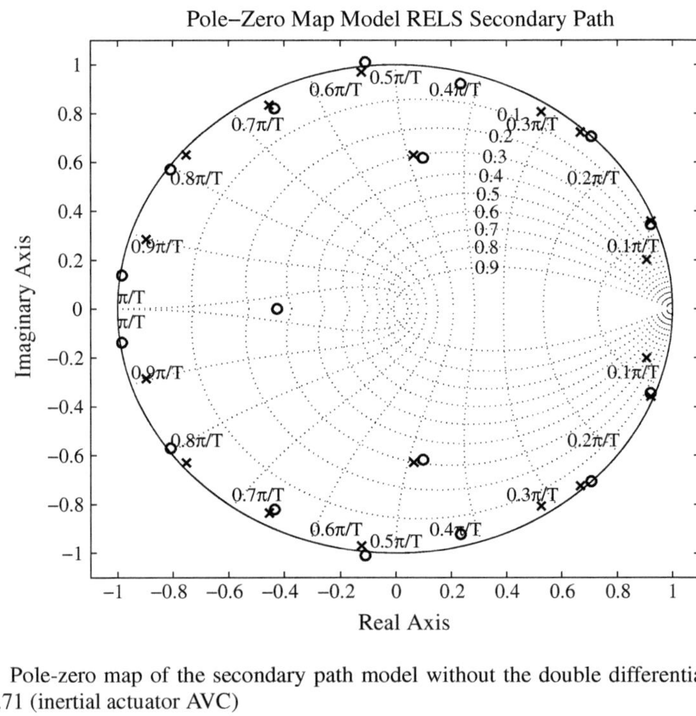

Fig. 6.17 Pole-zero map of the secondary path model without the double differentiator and the zero at -7.71 (inertial actuator AVC)

6.2.2 *Identification of the Primary Path*

Similar analysis has been done also for the identified model of the primary path. The orders of the primary path are: $\hat{n} = 14$, $\hat{n}_{A_D} = 13$, and $\hat{n}_{B'_D} = 14$ with a plant delay

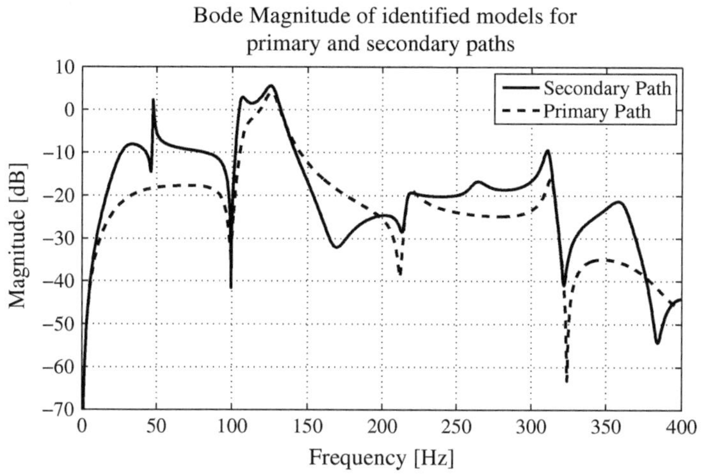

Fig. 6.18 Frequency characteristics of the primary and secondary paths (inertial actuator AVC)

of $d_D = 0$. The parameters of the identified primary path model are given in file *PrimPath_SAAI.mat*. The frequency characteristic of the primary path is shown in Fig. 6.18.

6.3 Identification of the Active Distributed Flexible Mechanical Structure Using Feedforward–Feedback Compensation

The AVC system using feedforward–feedback compensation has been described in Sect. 2.3. In this section, open-loop identification results for the secondary, positive feedback coupling (reverse) and primary paths will be presented. Note also that, although for adaptive control it is sufficient to estimate the secondary and reverse paths, for simulation and model-based controller design it is also necessary to identify the primary path.

The primary path is characterized by the asymptotically stable transfer function

$$D(z^{-1}) = \frac{B_D(z^{-1})}{A_D(z^{-1})} \tag{6.6}$$

where

$$B_D(z^{-1}) = b_1^D z^{-1} + \cdots + b_{n_{B_D}}^D z^{-n_{B_D}} \tag{6.7}$$

$$A_D(z^{-1}) = 1 + a_1^D z^{-1} + \cdots + a_{n_{A_D}}^D z^{-n_{A_D}} \tag{6.8}$$

The unmeasurable value of the output of the primary path (when the compensation is active) is denoted $x(t)$. The secondary path is characterized by the asymptotically stable transfer function:

$$G(z^{-1}) = \frac{B_G(z^{-1})}{A_G(z^{-1})} \tag{6.9}$$

where

$$B_G(z^{-1}) = b_1^G z^{-1} + \cdots + b_{n_{B_G}}^G z^{-n_{B_G}} = z^{-1} B_G^*(z^{-1}) \tag{6.10}$$

$$A_G(z^{-1}) = 1 + a_1^G z^{-1} + \cdots + a_{n_{A_G}}^G z^{-n_{A_G}} \tag{6.11}$$

The positive feedback coupling (the reverse path) is characterized by the asymptotically stable transfer function

$$M(z^{-1}) = \frac{B_M(z^{-1})}{A_M(z^{-1})} \tag{6.12}$$

where

$$B_M(z^{-1}) = b_1^M z^{-1} + \cdots + b_{n_{B_M}}^M z^{-n_{B_M}} = q^{-1} B_M^*(q^{-1}) \tag{6.13}$$

$$A_M(z^{-1}) = 1 + a_1^M z^{-1} + \cdots + a_{n_{A_M}}^M z^{-n_{A_M}} \tag{6.14}$$

The integer delay, if any, is included in the polynomials B_X.

The methodology used for parametric system identification is similar to that presented in the previous sections. The sampling frequency is 800 Hz.

The identification of the primary, secondary and reverse paths has been done in the absence of the compensator (see Fig. 2.11). For the secondary and reverse paths a PRBS excitation signal with a shift register having $N = 12$ cells (without frequency divider, $p = 1$) has been applied[4] at the input of the inertial actuator II, where the control signal $\hat{u}(t)$ is applied (see Figs. 2.9 and 2.10). For the primary path, a different PRBS signal with $N = 10$ and frequency divider $p = 4$ has been applied at the input of the inertial actuator I.

For the secondary path, $G(q^{-1})$, the output is the residual acceleration measurement, $e^\circ(t)$ in Fig. 2.11b. The input/output data necessary for the identification is given in file *data_identif_G.mat* and is available on the book website. Given that the input is the position of the inertial actuator II and that the output is an acceleration, it can be concluded that a double differentiator exists. As described also in Sects. 6.1 and 6.2, the *a priori* known properties of the system can be considered by filtering the input $u(t)$ through a filter that represents the *a priori* known dynamics. Then the resulting signal $u''(t)$ will be used as input for the identification procedure. At the end, the double differentiator will be included in the model.

[4]In previous publications [2, 3], models identified with $N = 10$ and $p = 4$ have been used. The differences in the identified models frequency characteristics are negligible.

The estimated orders of the model for the secondary path (without the double differentiator) are $n_{B_G} = 12$, $n_{A_G} = 14$. The best results, in terms of validation, have been obtained with the *Recursive Extended Least Square* method using decreasing adaptation gain. The result of the whiteness test validation is shown in Fig. 6.19 ($WRN(i)$ corresponds to the normalized autocorrelations defined in Chap. 5, Eq. (5.43)). The parameters of the estimated model are given in Table 6.9 and are also given in the file *SecPathModel.mat* available from the book website. The frequency characteristics of the secondary path are shown in Fig. 6.20, solid line. It features several very low damped vibration modes and antiresonances, as can be noticed from Tables 6.10 and 6.11, respectively. As a consequence of the double differentiator behaviour, a double zero at $z = 1$ is also present.

For the reverse path, $M(q^{-1})$, the output is the signal delivered by the primary transducer (accelerometer) $\hat{y}_1(t)$. The input/output data necessary for identification is given in file *data_identif_M.mat* and is available on the book website. Similarly to the secondary path, the input to the reverse path is a position while the output is an acceleration. It is clear that a double differentiator is present. The model's complexity has been estimated to be $n_{B_M} = 11$, $n_{A_M} = 13$ (without the double differentiator). The

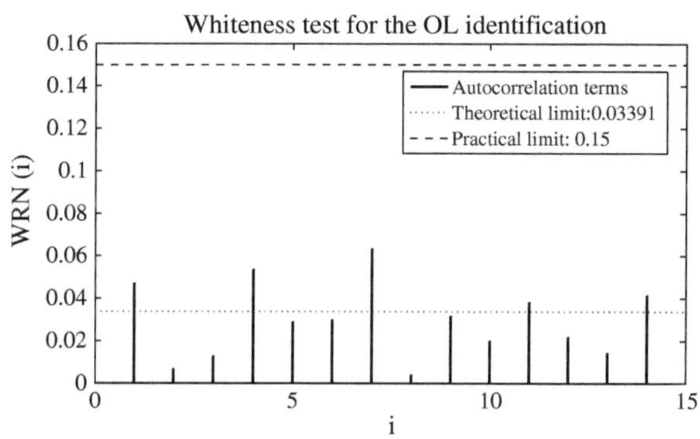

Fig. 6.19 Whiteness test for the RELS identified secondary path model (AVC with feedforward)

Table 6.9 Parameters for the identified model of the secondary path (AVC with feedforward)

Parameter	Value	Parameter	Value	Parameter	Value	Parameter	Value
a_0	1	a_8	0.0212	b_0	0	b_8	−0.7000
a_1	−2.6416	a_9	0.0761	b_1	−0.1923	b_9	0.7212
a_2	3.4603	a_{10}	1.0527	b_2	0.2225	b_{10}	0.0451
a_3	−2.4405	a_{11}	−1.3628	b_3	0.4228	b_{11}	−0.4273
a_4	1.5221	a_{12}	0.7597	b_4	−0.9161	b_{12}	−0.0306
a_5	−1.8122	a_{13}	−0.1076	b_5	0.4604	b_{13}	0.4383
a_6	2.3666	a_{14}	0.0462	b_6	0.2332	b_{14}	−0.2270
a_7	−1.3779			b_7	−0.0502		

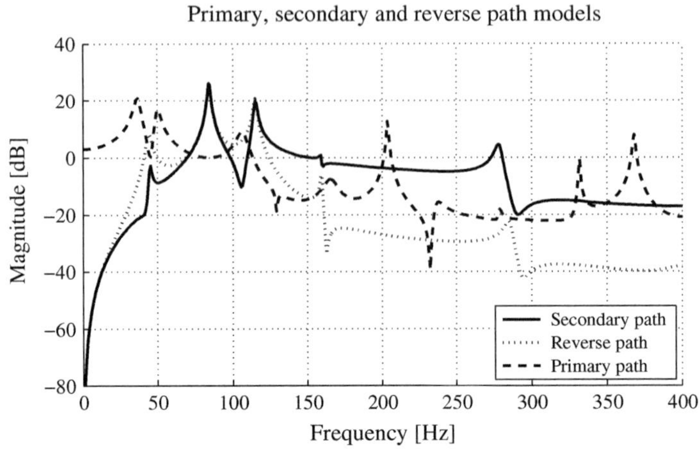

Fig. 6.20 Frequency characteristics of the primary, secondary and reverse identified paths (AVC with feedforward)

Table 6.10 Frequency and damping of the poles of the secondary path identified model (AVC with feedforward)

Poles	Damping	Frequency (Hz)
0.9323 ± 0.3443	0.0176	45.0468
0.7850 ± 0.6099	0.0090	84.1065
0.6131 ± 0.7794	0.0093	115.1355
0.3128 ± 0.9443	0.0042	159.2716
0.0097 ± 0.2646	0.6547	258.4384
-0.5680 ± 0.8006	0.0085	278.5733
-0.7640 ± 0.3690	0.0609	343.3584

Table 6.11 Frequency and damping of the zeroes of the secondary path identified model (AVC with feedforward)

Zeroes	Damping	Frequency (Hz)
1	0	0
1	0	0
0.9272 ± 0.3245	0.0528	42.9288
0.6624 ± 0.7295	0.0176	106.1476
0.3105 ± 0.9452	0.0040	159.5853
-0.6275 ± 0.7404	0.0131	289.5403
-0.7728 ± 0.3688	0.0574	343.8781
-1.8425	-0.1910	407.4983

best results, in terms of validation, have been obtained with the *Recursive Extended Least Square* method with decreasing adaptation gain (see Fig. 6.21). The parameters of the estimated model numerator and denominator are given in the file *ReversePathModel.mat* available from the book website. The frequency characteristic of the reverse path is shown in Fig. 6.20 (dotted line). There are several very low damped vibration modes and antiresonances as can be seen in Tables 6.12 and 6.13. There are also two zeros on the unit circle corresponding to the double differentiator behaviour. The gain of the reverse path is of the same order of magnitude as the gain of the secondary path up to 150 Hz, indicating a strong positive feedback in this frequency zone.

The primary path is identified between $w(t)$ and $e^{\circ}(t)$ in the absence of the compensator (see Fig. 2.11). The signal $w(t)$ is the result of the excitation $s(t)$ (PRBS designed with $N = 10$ bits shift register and frequency divider $p = 4$) passed through the transfer function $W(z^{-1})$.

The estimated orders of the primary path model are $n_{B_D} = 20$, $n_{A_D} = 20$. The best results in terms of validation have been obtained using FOLOE algorithm with

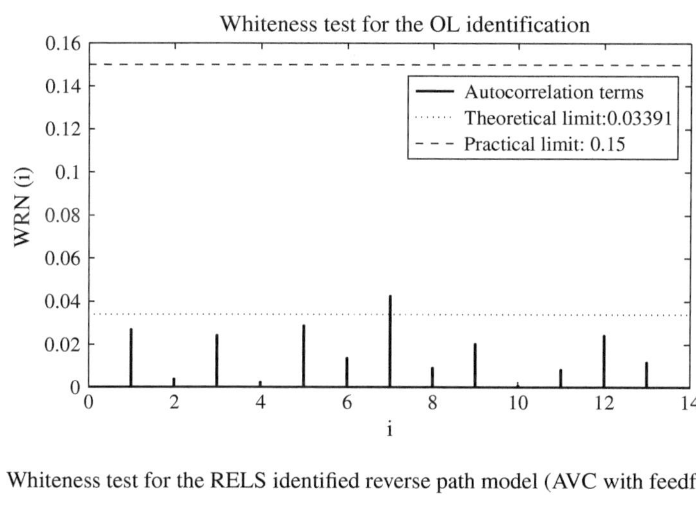

Fig. 6.21 Whiteness test for the RELS identified reverse path model (AVC with feedforward)

Table 6.12 Poles of the reverse path identified model (AVC with feedforward)

Poles	Damping	Frequency (Hz)
0.9287 ± 0.3361	0.0357	44.2370
0.7863 ± 0.6087	0.0086	83.8780
0.6139 ± 0.7784	0.0096	114.9852
0.3112 ± 0.9453	0.0039	159.5034
-0.6093 ± 0.7671	0.0092	285.4759
-0.3781 ± 0.3018	0.2822	327.5534
-0.8967	0.0347	400.2411

Table 6.13 Zeroes of the reverse path identified model (AVC with feedforward)

Zeroes	Damping	Frequency (Hz)
1	0	0
1	0	0
0.3853	1.0000	121.4376
1.0198 ± 1.5544	−0.5307	148.7535
0.2883 ± 0.9522	0.0040	162.5682
-0.6527 ± 0.7248	0.0108	293.3561
−0.8467	0.0529	400.5609
−0.6375	0.1418	404.0855
−3.6729	−0.3826	432.9424

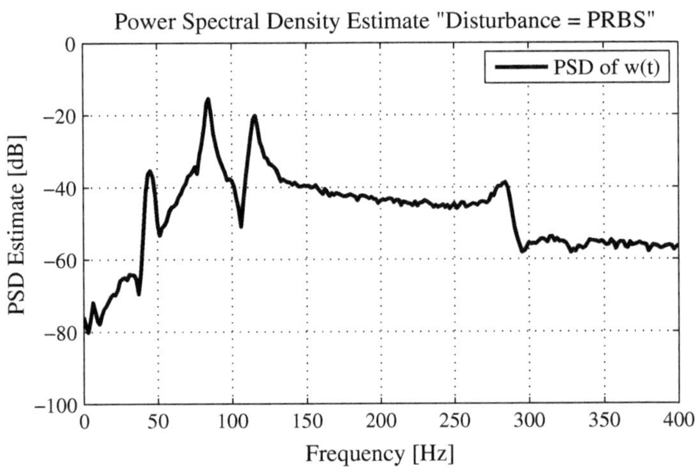

Fig. 6.22 Power spectral density of the disturbance $w(t)$ (AVC with feedforward)

variable forgetting factor with $\lambda_1(0) = \lambda_0 = 0.95$. The fixed filter used in FOLOE ($L = \hat{A}$) has been obtained by running first the AFOLOE algorithm with the same adaptation gain (see Sect. 4.3.2.2 in this book and also [1, 4] for more details on the FOLOE and AFOLOE algorithms). The parameters of the identified primary path model are given in file *PrimPathModel.mat* available from the book website. The data file *data_identif_D.mat* used to obtain these parameters is also available on the book website. The frequency characteristic is presented in Fig. 6.20 (dashed line).

The primary path model is used for simulations, detailed performance evaluation and for the design of linear feedforward compensators (see Chap. 14). Note that the primary path features a strong resonance at 106 Hz, exactly where the secondary path has a pair of low damped complex zeros (almost no gain). Therefore, one cannot expect a good attenuation around this frequency.

For identification purposes, it is also of interest to characterize the spectrum of the disturbance $w(t)$. Looking at the power spectral density of the signal $w(t)$ in Fig. 6.22, it can be observed that it has enough energy in the frequency band from 40 to 275 Hz. This corresponds to the frequency band where also the secondary path has enough gain. As such, the identified model of the primary path will be relevant and the compensation signal can effectively influence the residual acceleration.

6.4 Concluding Remarks

- The methodology discussed in Chap. 5 has been used successfully for identifying the dynamical models of the test benches in open-loop operation.
- The criterion for order estimation has for all three test benches a relatively flat minimum. This requires a comparative test of models of different orders around the value corresponding to the minimum of the criterion.
- Based on the identification of several test benches one can say that dynamical models for AVC can be relevantly represented in most of the cases by ARMAX models.
- Among the various algorithms available for the ARMAX model structure, it was found that RELS and XOLOE algorithms provide the best results for the specific problems considered.

6.5 Notes and References

The book website provides input/output data and the models for all three test benches.

The models of the test benches have been used in [2, 5–7], as well as in other papers. For the identification of another test bench see [8].

References

1. Landau I, Zito G (2005) Digital control systems - design, identification and implementation. Springer, London
2. Landau I, Alma M, Airimitoaie T (2011) Adaptive feedforward compensation algorithms for active vibration control with mechanical coupling. Automatica 47(10):2185–2196. doi:10.1016/j.automatica.2011.08.015
3. Landau ID, Airimitoaie TB, Alma M (2013) IIR Youla-Kučera parameterized adaptive feedforward compensators for active vibration control with mechanical coupling. IEEE Trans Control Syst Technol 21(3):765–779
4. Landau ID, Lozano R, M'Saad M, Karimi A (2011) Adaptive control, 2nd edn. Springer, London
5. Landau I, Constantinescu A, Rey D (2005) Adaptive narrow band disturbance rejection applied to an active suspension - an internal model principle approach. Automatica 41(4):563–574

6. Landau I, Karimi A, Constantinescu A (2001) Direct controller order reduction by identification in closed loop. Automatica 37:1689–1702
7. Landau ID, Silva AC, Airimitoaie TB, Buche G, Noé M (2013) Benchmark on adaptive regulation - rejection of unknown/time-varying multiple narrow band disturbances. Eur J Control 19(4):237–252. doi:10.1016/j.ejcon.2013.05.007
8. Landau I, Alma M, Martinez J, Buche G (2011) Adaptive suppression of multiple time-varying unknown vibrations using an inertial actuator. IEEE Trans Control Syst Technol 19(6):1327–1338. doi:10.1109/TCST.2010.2091641

Chapter 7
Digital Control Strategies for Active Vibration Control—The Bases

7.1 The Digital Controller

The basic equation for the polynomial digital controller to be used in active vibration control (subsequently called RS controller) is (see Fig. 7.1)

$$S(q^{-1})u(t) = -R(q^{-1})y(t) \tag{7.1}$$

where $u(t)$ is the plant input, $y(t)$ is the measured plant output, and

$$S(q^{-1}) = s_0 + s_1 q^{-1} + \cdots + s_{n_S} q^{-n_S} = s_0 + q^{-1} S^*(q^{-1}), \tag{7.2}$$

$$R(q^{-1}) = r_0 + r_1 q^{-1} + \cdots + r_{n_R} q^{-n_R}, \tag{7.3}$$

are, respectively, the denominator and numerator of the controller

$$K(q^{-1}) = \frac{R(q^{-1})}{S(q^{-1})}. \tag{7.4}$$

Equation (7.1) can also be written as

$$u(t) = \frac{1}{s_0}\left[-S^*(q^{-1})u(t-1) - R(q^{-1})y(t)\right] \tag{7.5}$$

Note that for a number of control algorithms (like pole placement) $s_0 = 1$ in (7.2).

Consider

$$G(q^{-1}) = \frac{q^{-d} B(q^{-1})}{A(q^{-1})} \tag{7.6}$$

as the pulse transfer operator of the cascade DAC + ZOH + continuous-time system + ADC, then the transfer function of the open-loop system is written as

I.D. Landau et al., *Adaptive and Robust Active Vibration Control*,
Advances in Industrial Control, DOI 10.1007/978-3-319-41450-8_7

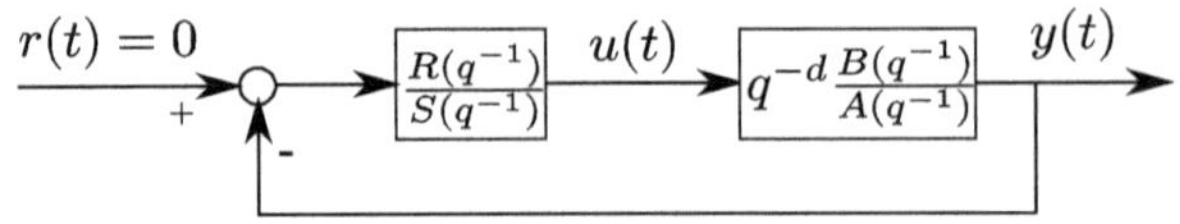

Fig. 7.1 Discrete feedback RS controller

$$H_{OL}(z^{-1}) = K(z^{-1})G(z^{-1}) = \frac{B(z^{-1})R(z^{-1})}{A(z^{-1})S(z^{-1})} \tag{7.7}$$

and the closed-loop transfer function between the reference signal $r(t)$ and the output $y(t)$, using controller (7.4), has the expression

$$S_{yr}(z^{-1}) = \frac{KG}{1+KG} = \frac{B(z^{-1})R(z^{-1})}{A(z^{-1})S(z^{-1}) + B(z^{-1})R(z^{-1})} = \frac{B(z^{-1})R(z^{-1})}{P(z^{-1})}, \tag{7.8}$$

where

$$P(z^{-1}) = A(z^{-1})S(z^{-1}) + z^{-d-1}B^{\star}(z^{-1})R(z^{-1}) \tag{7.9}$$

$$= A(z^{-1})S(z^{-1}) + z^{-d}B(z^{-1})R(z^{-1}) \tag{7.10}$$

is the denominator of the closed-loop transfer function that defines the closed-loop system poles. S_{yr} is known also as the *complementary sensitivity function*.

In the presence of disturbances (see Fig. 7.2), there are other important transfer functions to consider, relating the disturbance to the output and the input of the plant.

The transfer function between the disturbance $p(t)$ and the output $y(t)$ (*output sensitivity function*) is given by

$$S_{yp}(z^{-1}) = \frac{1}{1+KG} = \frac{A(z^{-1})S(z^{-1})}{P(z^{-1})}. \tag{7.11}$$

The transfer function between the disturbance $p(t)$ and the input of the plant $u(t)$ (*input sensitivity function*) is given by

$$S_{up}(z^{-1}) = -\frac{K}{1+KG} = -\frac{A(z^{-1})R(z^{-1})}{P(z^{-1})}. \tag{7.12}$$

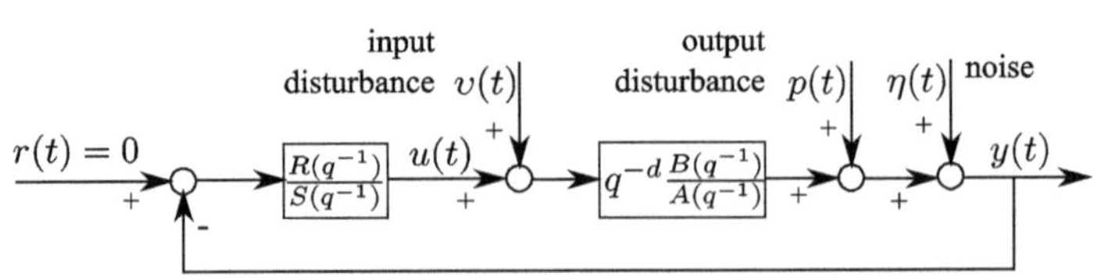

Fig. 7.2 Discrete feedback RS controller with input/output disturbances and measurement noise

Another important transfer function describes the influence on the output of a disturbance $v(t)$ on the plant input. This sensitivity function (*input disturbance-output sensitivity function*) is given by

$$S_{yv}(z^{-1}) = \frac{G}{1+KG} = \frac{B(z^{-1})S(z^{-1})}{P(z^{-1})}. \tag{7.13}$$

The feedback system presented in Fig. 7.2 is asymptotically stable if and only if all the four sensitivity functions S_{yr}, S_{yp}, S_{up} and S_{yv} are asymptotically stable.

As it will be shown soon, the perfect rejection of disturbances with known characteristics or conversely opening of the loop for certain disturbances will require the introduction of some fixed pre-specified polynomials in S and R. The general structure of R and S will be of the form:

$$S(z^{-1}) = S'(z^{-1})H_S(z^{-1}) \tag{7.14}$$

$$R(z^{-1}) = R'(z^{-1})H_R(z^{-1}) \tag{7.15}$$

where $H_S(z^{-1})$ and $H_R(z^{-1})$ are monic fixed polynomials, which are introduced in the controller for achieving certain performances with respect to disturbances. Using this parametrization, the closed-loop poles will be given by

$$P(z^{-1}) = A(z^{-1})H_S(z^{-1})S'(z^{-1}) + z^{-d-1}B^{\star}(z^{-1})H_R(z^{-1})R'(z^{-1}) \tag{7.16}$$

Note that $H_S(z^{-1})$ and $H_R(z^{-1})$ can be interpreted as an "augmentation" of the plant model (for computation purposes).

The design of the RS controller can be done in the frequency domain using transfer functions (operators).

7.2 Pole Placement

The pole placement strategy is applicable to plant models of the form of Eq. (7.6). We will make the following hypothesis upon the plant model of Eq. (7.6):

(H1) No restrictions upon the orders of the polynomials $A(z^{-1})$, $B(z^{-1})$ and the value of the delay d.
(H2) The orders n_A, n_B, the delay d and the coefficients of $A(z^{-1})$ and $B(z^{-1})$ are known.
(H3) The zeros of $B(z^{-1})$ can be inside or outside the unit circle.
(H4) $A(z^{-1})$ and $B(z^{-1})$ (or AH_S and BH_R) do not have any common factors.
(H5) The zeros of $A(z^{-1})$ can be inside or outside the unit circle.

The control law is of the form (7.1) and the polynomials $R(z^{-1})$ and $S(z^{-1})$ have the structure of Eqs. (7.14) and (7.15).

The closed-loop behaviour is defined by

- the desired closed-loop poles;
- the choice of the fixed parts $H_R(z^{-1})$ and $H_S(z^{-1})$.

The desired closed-loop poles are chosen under the form as follows:

$$P(z^{-1}) = P_D(z^{-1}) \cdot P_F(z^{-1}) \tag{7.17}$$

where $P_D(z^{-1})$ defines the *dominant poles* and $P_F(z^{-1})$ defines the *auxiliary poles*.

Often $P_D(z^{-1})$ is chosen to include all the stable poles of the plant in open-loop with the option of eventually modifying the damping of the complex poles.

The role of $P_F(z^{-1})$ is on one hand to introduce a filtering effect at certain frequencies and on the other hand to improve the robustness of the controller.

With the notations:

$$\begin{aligned} n_A &= \deg A \; ; \quad n_B = \deg B \\ n_{H_S} &= \deg H_S \; ; \quad n_{H_R} = \deg H_R \end{aligned}$$

and under the hypotheses H1 to H5, (7.16) has a unique solution for S' and R', of minimal degree for

$$n_P = \deg P(z^{-1}) \leq n_A + n_{H_S} + n_B + n_{H_R} + d - 1 \tag{7.18}$$

$$n_{S'} = \deg S'(z^{-1}) = n_B + n_{H_R} + d - 1 \tag{7.19}$$

$$n_{R'} = \deg R'(z^{-1}) = n_A + n_{H_S} - 1 \tag{7.20}$$

with

$$S'(z^{-1}) = 1 + s'_1 z^{-1} + \cdots + s'_{n_S} z^{-n_S} \tag{7.21}$$

$$R'(z^{-1}) = r'_0 + r'_1 z^{-1} + \cdots + r'_{n_R} z^{-n_R} \tag{7.22}$$

For a proof see [1, 2]. Various methods for solving this equation are available.[1]

7.2.1 Choice of H_R and H_S—Examples

Opening the Loop

In a number of applications, the measured signal may contain specific frequencies which should not be attenuated by the regulator. In such cases the system should be in open-loop at these frequencies.

[1]See functions *bezoutd.m* (MATLAB®) or *bezoutd.sci* (Scilab) on the book website.

From (7.12) in the absence of the reference, the input to the plant is given by

$$u(t) = S_{up}(q^{-1})p(t) = \frac{A(q^{-1})H_R(q^{-1})R'(q^{-1})}{P(q^{-1})}p(t) \tag{7.23}$$

and therefore in order to make the input sensitivity function zero at a given frequency f, one should introduce a pair of undamped zeros in $H_R(q^{-1})$, i.e.,:

$$H_R(q^{-1}) = (1 + \beta q^{-1} + q^{-2}) \tag{7.24}$$

where

$$\beta = -2\cos(\omega T_S) = -2\cos(2\pi\frac{f}{f_S})$$

In many cases it is desired that the controller does not react to signals of frequencies close to $0.5 f_S$ (where the gain of the system is in general very low). In such cases, one uses

$$H_R(q^{-1}) = (1 + \beta q^{-1}) \tag{7.25}$$

where

$$0 < \beta \leq 1$$

Note that $(1 + \beta q^{-1})^2$ corresponds to a second order with a damped resonance frequency equal to $\omega_S/2$ as follows:

$$\omega_0\sqrt{1-\zeta^2} = \frac{\omega_S}{2}$$

and the corresponding damping ζ is related to β by

$$\beta = e^{-\frac{\zeta}{\sqrt{1-\zeta^2}}\pi}$$

For $\beta = 1$, the system will operate in open-loop at $f_S/2$.

In active vibration control systems, the gain of the secondary path at 0 Hz is zero (double differentiator behaviour). It is therefore not reasonable to send a control signal at this frequency. The system should operate in open-loop at this frequency. To achieve this, one uses

$$H_R(q^{-1}) = (1 - q^{-1}) \tag{7.26}$$

Perfect Rejection of an Harmonic Disturbance

The disturbance $p(t)$ can be represented as the result of a Dirac function $\delta(t)$ passed through a filter $D(q^{-1})$ (called the model of the disturbance)

$$D(q^{-1})p(t) = \delta(t) \tag{7.27}$$

In the case of an harmonic disturbance, the model is

$$(1 + \alpha q^{-1} + q^{-2})p(t) = \delta(t) \tag{7.28}$$

with

$$\alpha = -2\cos(\omega T_S) = -2\cos(2\pi \frac{f}{f_S}) \tag{7.29}$$

From (7.11) in the absence of a reference, one has

$$y(t) = \frac{A(q^{-1})H_S(q^{-1})S'(q^{-1})}{P(q^{-1})} p(t) \tag{7.30}$$

The problem can be viewed as choosing $H_S(q^{-1})$ such that the gain of the transfer function between $p(t)$ and $y(t)$ be zero at this frequency.

To achieve this one should choose

$$H_S(q^{-1}) = (1 + \alpha q^{-1} + q^{-2}) \tag{7.31}$$

In this case the expression of $y(t)$ taking into account (7.28), (7.30) and (7.31) becomes

$$y(t) = \frac{A(q^{-1})S'(q^{-1})}{P(q^{-1})} \delta(t) \tag{7.32}$$

and it results that asymptotically $y(t)$ tends to zero since $P(q^{-1})$ is asymptotically stable. This result is nothing else than the *internal model principle* which will be stated next.

7.2.2 *Internal Model Principle (IMP)*

Suppose that $p(t)$ is a deterministic disturbance, so it can be written as

$$p(t) = \frac{N_p(q^{-1})}{D_p(q^{-1})} \cdot \delta(t), \tag{7.33}$$

where $\delta(t)$ is a Dirac impulse and $N_p(z^{-1})$, $D_p(z^{-1})$ are coprime polynomials in z^{-1}, of degrees n_{N_p} and n_{D_p}, respectively (see also Fig. 7.1). In the case of stationary disturbances, the roots of $D_p(z^{-1})$ are on the unit circle. The energy of the disturbance is essentially represented by D_p. The contribution of the terms of N_p is weak asymptotically compared to the effect of D_p, so one can neglect the effect of N_p for a steady-state analysis of the effect of the disturbance upon the system.

Internal Model Principle: The effect of the disturbance given in (7.33) upon the output:

$$y(t) = \frac{A(q^{-1})S(q^{-1})}{P(q^{-1})} \cdot \frac{N_p(q^{-1})}{D_p(q^{-1})} \cdot \delta(t), \tag{7.34}$$

where $D_p(z^{-1})$ is a polynomial with roots on the unit circle and $P(z^{-1})$ is an asymptotically stable polynomial, converges asymptotically towards zero if and only if the polynomial $S(z^{-1})$ in the RS controller has the following form:

$$S(z^{-1}) = D_p(z^{-1})S'(z^{-1}). \tag{7.35}$$

In other terms, the pre-specified part of $S(z^{-1})$ should be chosen as $H_S(z^{-1}) = D_p(z^{-1})$ and the controller is computed using (7.16), where P, D_p, A, B, H_R and d are given.[2]

The IMC principle says that *in order to completely reject a disturbance asymptotically (i.e., in steady state), the controller should include the model of the disturbance.*

7.2.3 Youla–Kučera Parametrization

Using the Youla–Kučera parametrization (Q-parametrization) of all stable controllers ([3, 4]), the controller polynomials $R(z^{-1})$ and $S(z^{-1})$ get the following form:

$$R(z^{-1}) = R_0(z^{-1}) + A(z^{-1})Q(z^{-1}) \tag{7.36}$$
$$S(z^{-1}) = S_0(z^{-1}) - z^{-d}B(z^{-1})Q(z^{-1}) \tag{7.37}$$

where (R_0, S_0) is the so-called central controller and Q is the YK or Q filter which can be a FIR or an IIR filter. Figure 7.3 gives a representation of the Youla–Kučera parametrization of the R–S controllers. The *central* controller (R_0, S_0) can be computed by pole placement (but any other design technique can be used). Given the plant model (A, B, d) and the desired closed-loop poles specified by the roots of $P(z^{-1})$ one has to solve

$$P(z^{-1}) = A(z^{-1})S_0(z^{-1}) + z^{-d}B(z^{-1})R_0(z^{-1})\,. \tag{7.38}$$

If $Q(z^{-1})$ is considered to be a polynomial of the form (FIR filter):

$$Q(z^{-1}) = q_0 + q_1 z^{-1} + \cdots + q_{n_Q} z^{-n_Q}\,. \tag{7.39}$$

Equations (7.36) and (7.37) characterize the set of all stabilizable controllers assigning the closed-loop poles as defined by $P(z^{-1})$. It can be easily verified by simple computation, that the poles of the closed-loop remain unchanged; however, the par-

[2]Of course it is assumed that D_p and BH_R do not have common factors.

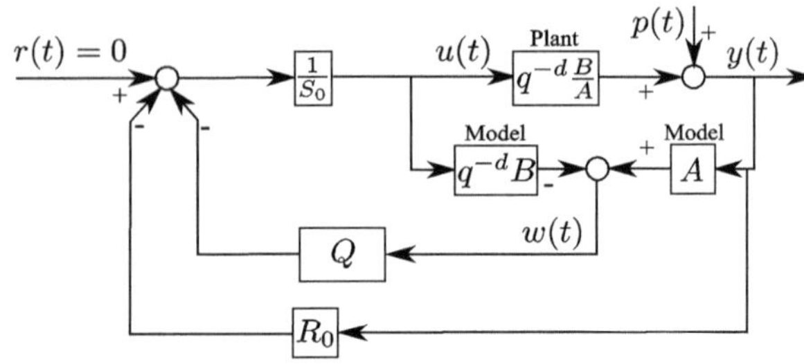

Fig. 7.3 The Youla–Kučera parametrized RS digital controller

ticular interest of the YK parametrization is the fact that the internal model of the disturbance can be incorporated in the controller by an appropriate choice of the filter Q. This filter should be such that the resulting polynomial S has the form $S = S'D_p$, i.e.,:

$$S'(z^{-1})D_p(z^{-1}) = S_0(z^{-1}) - z^{-d}B(z^{-1})Q(z^{-1}), \tag{7.40}$$

To compute $Q(z^{-1})$ in order that the polynomial $S(z^{-1})$ given by (7.37) incorporates the internal model of the disturbance (7.33) one has to solve the following diophantine equation:

$$S'(z^{-1})D_p(z^{-1}) + z^{-d}B(z^{-1})Q(z^{-1}) = S_0(z^{-1}), \tag{7.41}$$

where $D_p(z^{-1})$, d, $B(z^{-1})$ and $S_0(z^{-1})$ are known and $S'(z^{-1})$ and $Q(z^{-1})$ are unknown. Equation (7.41) has a unique solution for $S'(z^{-1})$ et $Q(z^{-1})$ with: $n_{S_0} \leq n_{D_p} + n_B + d - 1$, $n_{S'} = n_B + d - 1$, $n_Q = n_{D_p} - 1$. One sees that the order n_Q of the polynomial Q depends upon the structure of the disturbance model.

Consider now the case of a Q filter as ratio of rational polynomials (IIR filter) with an asymptotically stable denominator as follows:

$$Q(z^{-1}) = \frac{B_Q(z^{-1})}{A_Q(z^{-1})} \tag{7.42}$$

The YK controller will have the structure:

$$R(z^{-1}) = A_Q(z^{-1})R_0(z^{-1}) + A(z^{-1})B_Q(z^{-1}) \tag{7.43}$$

$$S(z^{-1}) = A_Q(z^{-1})S_0(z^{-1}) - z^{-d}B(z^{-1})B_Q(z^{-1}) \tag{7.44}$$

but in this case the poles of the closed-loop will be given by

$$P(z^{-1})_{QIIR} = P(z^{-1})A_Q(z^{-1}) \tag{7.45}$$

In the case of IIR Q filters, the poles of the denominator of Q will appear as additional poles of the closed-loop. This parametrization will be discussed in detail in Sects. 7.4 and 12.2 together with the preservation of the pre-specified fixed part of the controller H_R and H_S.

7.2.4 Robustness Margins

The Nyquist plot of the open-loop transfer function allows one to assess the influence of the modelling errors and to derive appropriate specifications for the controller design in order to assure the *robust stability* of the closed-loop system for certain classes of plant model uncertainties.

The open-loop transfer function corresponding to the use of an RS controller is:

$$H_{OL}(z^{-1}) = \frac{z^{-d}B(z^{-1})R(z^{-1})}{A(z^{-1})S(z^{-1})} \tag{7.46}$$

By making $z = e^{j\omega}$, where ω is the normalized frequency ($\omega = \omega T_s = 2\pi f/f_s$, f_s sampling frequency, T_s sampling period), the Nyquist plot of the open-loop transfer function $H_{OL}(e^{-j\omega})$ can be drawn. In general, one considers for the normalized frequency ω the domain between 0 and π (i.e., between 0 and $0.5f_s$). Note that the Nyquist plot between π and 2π is symmetric with respect to the real axis of the Nyquist plot between 0 and π. An example of a Nyquist plot is given in Fig. 7.4.

The vector connecting a point of the Nyquist plot with the origin corresponds to $H_{OL}(e^{-j\omega})$ for a certain normalized frequency. The point $[-1, j0]$ on the diagram of Fig. 7.4 corresponds to the *critical point*. From Fig. 7.4, it results that the vector connecting the *critical point* with the Nyquist plot of $H_{OL}(e^{-j\omega})$ has the following expression:

$$1 + H_{OL}(z^{-1}) = \frac{A(z^{-1})S(z^{-1}) + z^{-d}B(z^{-1})R(z^{-1})}{A(z^{-1})S(z^{-1})} = S_{yp}^{-1}(z^{-1}) \tag{7.47}$$

This vector corresponds to the inverse of the output sensitivity function $S_{yp}(z^{-1})$ given by Eq. (7.11) and the zeros of S_{yp}^{-1} are the poles of the closed-loop system. In order that the closed-loop system be asymptotically stable, it is necessary that all the zeros of S_{yp}^{-1} lie inside the unit circle.

The necessary and sufficient conditions in the frequency domain for the asymptotic stability of the closed-loop system are given by the Nyquist criterion. For the case of

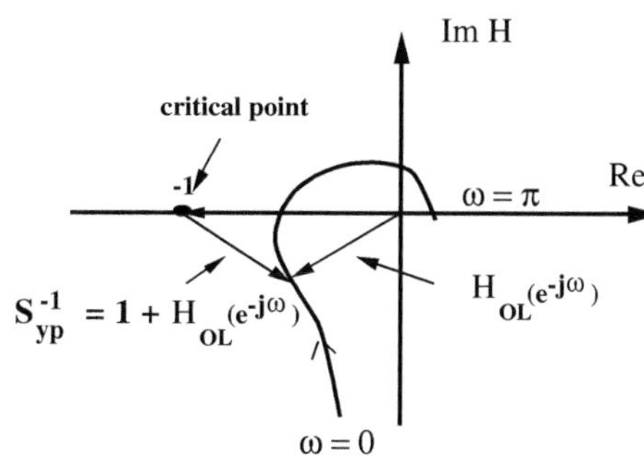

Fig. 7.4 The Nyquist plot of a discrete-time transfer function and the critical point

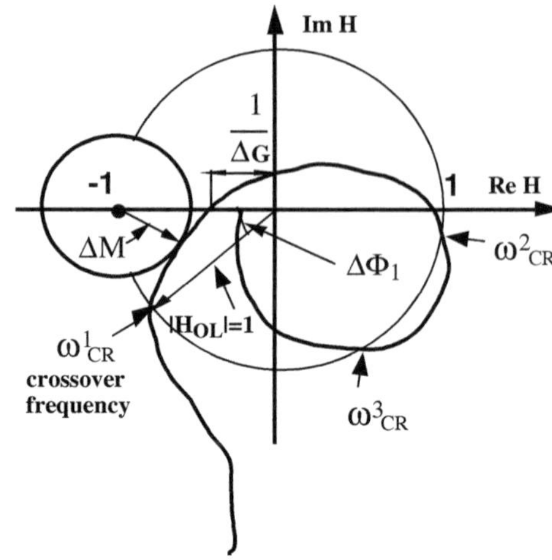

Fig. 7.5 Modulus, gain and phase margins

open-loop stable systems (in our case this corresponds to $A(z^{-1}) = 0$ and $S(z^{-1}) = 0 \implies |z| < 1$), the Nyquist criterion is expressed as:

Stability Criterion (Open-Loop Stable Systems)

The Nyquist plot of $H_{OL}(z^{-1})$ traversed in the sense of growing frequencies (from $\omega = 0$ to $\omega = \pi$ leaves the critical point $[-1, j0]$ on the left.

Using *pole placement*, the Nyquist criterion will be satisfied for the *nominal* plant model because $R(z^{-1})$ and $S(z^{-1})$ are computed using Eq. 7.10 for an asymptotically stable polynomial $P(z^{-1})$ defining the desired closed-loop poles ($P(z^{-1}) = 0 \implies |z| < 1$). Of course, we are assuming at this stage that the resulting $S(z^{-1})$ is also stable.[3]

The minimal distance between the Nyquist plot of $H_{OL}(z^{-1})$ and the critical point will define a *stability margin*. This minimal distance according to Eq. (7.47) will depend upon the maximum of the modulus of the output sensitivity function.

This stability margin which we will call subsequently the *modulus margin* could be linked to the uncertainties upon the plant model.

The following indicators serve for characterizing the distance between the Nyquist plot of $H_{OL}(z^{-1})$ and the critical point $[-1, j0]$ (see Fig. 7.5):

- modulus margin (ΔM)
- delay margin ($\Delta\tau$)
- phase margin ($\Delta\phi$)
- gain margin (ΔG)

[3]See [5] for the case of open-loop unstable systems.

Below are the definitions of the modulus margin and delay margin which will be used in the robust control design (for the definition of the gain and phase margin, see any classical control text):

Modulus Margin (ΔM)

The *modulus margin* (ΔM) is defined as the radius of the circle centred in $[-1, j0]$ and tangent to the Nyquist plot of $H_{OL}(z^{-1})$.

From the definition of the vector connecting the critical point $[-1, j0]$ with the Nyquist plot of $H_{OL}(z^{-1})$ (see Eq. (7.47)), it results that

$$\Delta M = |1 + H_{OL}(e^{-j\omega})|_{\min} = (|S_{yp}(e^{-j\omega})|_{\max})^{-1} = (\|S_{yp}\|_{\infty})^{-1} \tag{7.48}$$

As a consequence, the reduction (or minimization) of $|S_{yp}(e^{-j\omega})|_{\max}$ will imply the increase (or maximization) of the modulus margin ΔM.

In other terms the *modulus margin* ΔM is equal to the inverse of the maximum modulus of the output sensitivity function $S_{yp}(z^{-1})$ (i.e., the inverse of the H_{∞} norm of $S_{yp}(z^{-1})$). If the modulus of $S_{yp}(z^{-1})$ is expressed in dB, one has the following relationship:

$$|S_{yp}(e^{-j\omega})|_{\max}(\text{dB}) = (\Delta M)^{-1}(\text{dB}) = -\Delta M(\text{dB}) \tag{7.49}$$

The *modulus margin* is very important because

- It defines the maximum admissible value for the modulus of the output sensitivity function.
- It gives a bound for the characteristics of the nonlinear and time-varying elements tolerated in the closed-loop system (it corresponds to the circle criterion for the stability of nonlinear systems) [6].

Delay Margin ($\Delta\tau$)

For a certain frequency the phase lag introduced by a pure time delay τ is:

$$\angle\phi(\omega) = \omega\tau$$

If the Nyquist plot crosses the unit circle only once, one can therefore convert the phase margin in a delay margin, i.e., to compute the additional delay which will lead to instability. It results that:

$$\Delta\tau = \frac{\Delta\phi}{\omega_{cr}} \tag{7.50}$$

where ω_{cr} is the crossover frequency (where the Nyquist plot intersects the unit circle) and $\Delta\phi$ is the phase margin. If the Nyquist plot intersects the unit circle at several frequencies ω_{cr}^{i} (see Fig. 7.5), characterized by the associated phase margins $\Delta\phi_i$, the phase margin is defined as:

$$\Delta\phi = \min_{i} \Delta\phi_i \tag{7.51}$$

and the delay margin is defined by

$$\Delta\tau = \min_i \frac{\Delta\phi_i}{\omega_{cr}^i} \tag{7.52}$$

Remark This situation appears systematically for systems with pure time delays or with multiple vibration modes.

Typical values of the robustness margins for a *robust* controller design are

- Modulus margin: $\Delta M \geq 0.5(-6\ \text{dB})[\min : 0.4(-8\ \text{dB})]$
- Delay margin: $\Delta\tau \geq\ T_s[\min : 0.75T_s]$

where T_s is the sampling period.
Important remarks:

1. A modulus margin $\Delta M \geq 0.5$ implies that $\Delta G \geq 2(6\ \text{dB})$ and $\Delta\phi > 29°$. The converse is not generally true. *Systems with satisfactory gain and phase margins may have a very small modulus margin.*
2. Phase margin can be misleading according to Eq. (7.50). *A good phase margin may lead to a very small tolerated additional delay if ω_{cr} is high.*

The modulus margin is an intrinsic measure of the stability margin and will be subsequently used together with the delay margin for the design of robust controllers (instead of the phase and gain margin).

7.2.5 Model Uncertainties and Robust Stability

Figure 7.6 illustrates the effect of uncertainties or of the variations of the parameters of the nominal model on the Nyquist plots of the open-loop transfer function. In general the Nyquist plot corresponding to the nominal model lies inside a *tube* corresponding to the possible (or accepted) tolerances of parameter variations (or uncertainties) of the plant model.

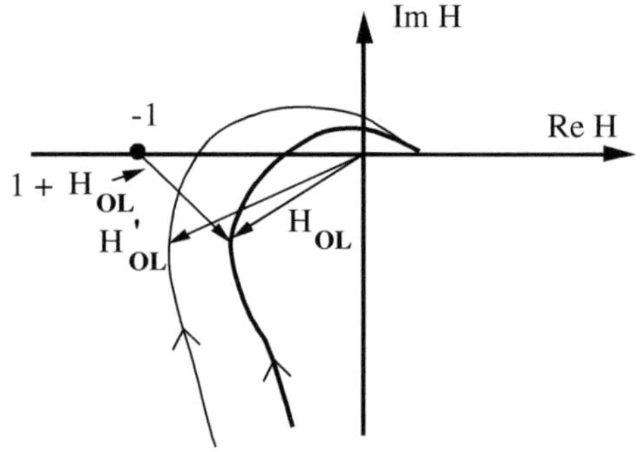

Fig. 7.6 Nyquist plot of the nominal model and perturbed model

We will consider an open-loop transfer function $H'_{OL}(z^{-1})$ which differs from the nominal one. For simplicity one assumes that the nominal transfer function $H_{OL}(z^{-1})$ as well as $H'_{OL}(z^{-1})$ are both stable (the general assumption is that both have the same number of unstable poles, see [7, 8]).

In order to assure the stability of the closed-loop system for an open loop transfer function $H'_{OL}(z^{-1})$ which differs from the nominal one $H_{OL}(z^{-1})$, the Nyquist plot of $H'_{OL}(z^{-1})$ should leave the critical point $[-1, j0]$ on the left side when traversed in the sense of the growing frequencies. Looking at Fig. 7.6 one can see that a sufficient condition for this, is that at each frequency the distance between $H'_{OL}(z^{-1})$ and $H_{OL}(z^{-1})$ be less than the distance between the nominal open-loop transfer function and the critical point. This is expressed by:

$$|H'_{OL}(z^{-1}) - H_{OL}(z^{-1})| < |1 + H_{OL}(z^{-1})| = |S_{yp}^{-1}(z^{-1})| = \left|\frac{P(z^{-1})}{A(z^{-1})S(z^{-1})}\right| \tag{7.53}$$

In other terms, the curve $|S_{yp}(e^{-j\omega})|^{-1}$ in dB (which is obtained by symmetry from $|S_{yp}(e^{-j\omega})|$) will give at each frequency a sufficient condition for the modulus of the tolerated discrepancy between the real open-loop transfer function and the nominal open-loop transfer function in order to guarantee stability.

In general, this tolerance is high in low frequencies and is low at the frequency (or frequencies) where $|S_{yp}(e^{-j\omega})|$ reaches its maximum ($=\Delta M^{-1}$). Therefore, low modulus margin will imply small tolerance to parameter uncertainties in a specified frequency region.

The relationship (7.53) expresses a robustness condition in terms of the variations of the open-loop transfer function (controller + plant). It is interesting to express this in terms of the variations of the plant model. One way to do this, is to observe that (7.53) can be rewritten as:

$$\begin{aligned}\left|\frac{B'(z^{-1})R(z^{-1})}{A'(z^{-1})S(z^{-1})} - \frac{B(z^{-1})R(z^{-1})}{A(z^{-1})S(z^{-1})}\right| &= \left|\frac{R(z^{-1})}{S(z^{-1})}\right| \cdot \left|\frac{B'(z^{-1})}{A'(z^{-1})} - \frac{B(z^{-1})}{A(z^{-1})}\right| \\ &< \left|\frac{P(z^{-1})}{A(z^{-1})S(z^{-1})}\right|\end{aligned} \tag{7.54}$$

Multiplying both sides of Eq. (7.54) by $|\frac{S(z^{-1})}{R(z^{-1})}|$ one gets

$$\left|\frac{B'(z^{-1})}{A'(z^{-1})} - \frac{B(z^{-1})}{A(z^{-1})}\right| \le \left|\frac{P(z^{-1})}{A(z^{-1})R(z^{-1})}\right| = \left|S_{up}^{-1}(z^{-1})\right| \tag{7.55}$$

The left hand side of Eq. (7.55) expresses in fact an *additive* uncertainty for the nominal plant model. The inverse of the modulus of the input sensitivity function will give a sufficient condition for the tolerated *additive* variations (or uncertainties) of the nominal plant model in order to guarantee stability. Large values of the modulus

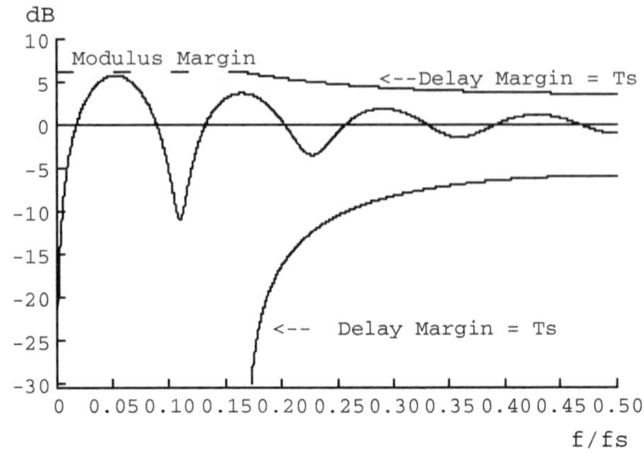

Fig. 7.7 Templates on the output sensitivity function for the modulus margin $\Delta M = 0.5$ and the delay margin $\Delta\tau = T_s$

of the input sensitivity function in certain frequency range will imply low tolerance to uncertainties in this frequency range. It will also mean that at these frequencies high activity of the input will result under the effect of disturbances.

7.2.6 Templates for the Sensitivity Functions

Robustness margins and performance specifications in the frequency domain translates easily in templates for the various sensitivity functions [2, 5]. Figure 7.7 gives the basic template for S_{yp} for assuring the modulus margin constraint ($\Delta M \geqslant 0.5$) and the delay margin ($\Delta\tau \geqslant T_s$). The template on the delay margin is an approximation (for more details see [2]). Violation of the lower or upper template does not necessarily imply violation of the delay margin (which any way can be effectively computed).

To this template, performance specification in terms of imposed attenuation and bound on the "waterbed" effect can be added (see the example in Sect. 7.3).

Templates on the modulus of the input sensitivity function S_{up} are also considered. In particular it is expected that S_{up} is low outside the frequency band of operation of the controller. Low values of the modulus of the input sensitivity functions imply a good robustness with respect to additive model uncertainties. Figure 7.8 gives an example of template on the input sensitivity function. More details can be found on the example given in Sect. 7.3.

7.2.7 Properties of the Sensitivity Functions

7.2.7.1 Output Sensitivity Function

Using an RS controller, the output sensitivity function is given by:

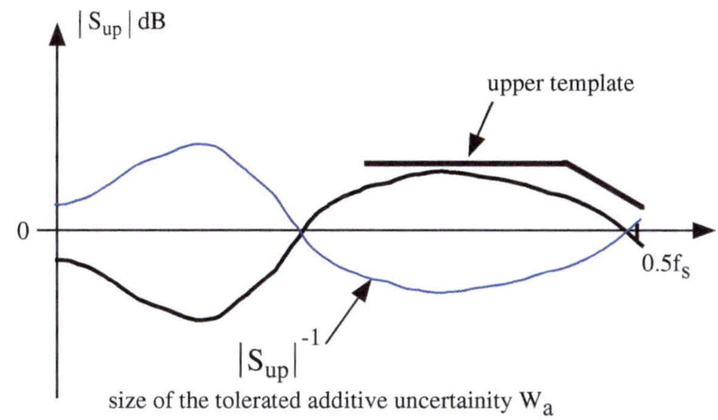

Fig. 7.8 An example of desired template for the input sensitivity function

$$S_{yp}(z^{-1}) = \frac{A(z^{-1})S(z^{-1})}{A(z^{-1})S(z^{-1}) + z^{-d}B(z^{-1})R(z^{-1})} \tag{7.56}$$

where

$$R(z^{-1}) = H_R(z^{-1})R'(z^{-1}) \tag{7.57}$$
$$S(z^{-1}) = H_S(z^{-1})S'(z^{-1}) \tag{7.58}$$

and

$$A(z^{-1})S(z^{-1}) + z^{-d}B(z^{-1})R(z^{-1}) = P_D(z^{-1}) \cdot P_F(z^{-1}) = P(z^{-1}) \tag{7.59}$$

In Eqs. (7.57) and (7.58), $H_R(z^{-1})$ and $H_S(z^{-1})$ correspond to the pre-specified parts of $R(z^{-1})$ and $S(z^{-1})$ respectively. $S'(z^{-1})$ and $R'(z^{-1})$ are the solutions of Eq. (7.16) where $P(z^{-1})$ represents the desired closed-loop poles in pole placement control strategy. The polynomial $P(z^{-1})$ is factorized in order to emphasize the dominant poles defined by $P_D(z^{-1})$ and the auxiliary poles defined by $P_F(z^{-1})$.

Property 1

The modulus of the output sensitivity function at a certain frequency gives the amplification or the attenuation of the disturbance.

At the frequencies where $|S_{yp}(\omega)| = 1(0\text{ dB})$, there is no attenuation nor amplification of the disturbance (operation in open-loop). At the frequencies where $|S_{yp}(\omega)| < 1(0\text{ dB})$, the disturbance is attenuated. At the frequencies where $|S_{yp}(\omega)| > 1(0\text{ dB})$, the disturbance is amplified.

Property 2 (The Bode Integral)

The closed-loop being asymptotically stable, the integral of the logarithm of the modulus of the output sensitivity function from 0 *to* $0.5f_S$ *is equal to* 0 *for the case of stable open-loop systems*[4]:

---

[4]See [9] for a proof. In the case of unstable open-loop systems but stable in closed-loop, this integral is positive.

$$\int_0^{0.5f_S} \log |S_{yp}(e^{-j2\pi f/f_S)}|df = 0$$

In other terms, the sum of the areas between the curve of the modulus of the output sensitivity function and the 0 dB axis taken with their sign is null. **As a consequence, the attenuation of disturbances in a certain frequency region implies necessarily the amplification of disturbances in other frequency regions.**

Property 3

The inverse of the maximum of the modulus of the sensitivity function corresponds to the modulus margin ΔM.

$$\Delta M = (|S_{yp}(e^{-j\omega})|_{\max})^{-1} \tag{7.60}$$

From the Properties 2 and 3, it results that the increase of the attenuation band or of the attenuation in a certain frequency band will in general imply an increase of $|S_{yp}(e^{-j\omega})|_{\max}$ and therefore a decrease of the modulus margin (and therefore less robustness).

Figure 7.9 shows the output sensitivity function for a closed-loop system, corresponding to a plant model $A(z^{-1}) = 1 - 0.7z^{-1}$, $B(z^{-1}) = 0.3z^{-1}$, $d = 2$. The controller has been designed using the pole placement. The desired closed-loop poles correspond to the discretization of a second order system with natural frequency $\omega_0 = 0.1f_s$ rad/s and damping $\zeta = 0.8$. The system being subject to a tonal disturbance located at $0.15f_s$ or at $0.151f_S$, a double internal model corresponding to these frequencies has been introduced in the controller fixed part H_S. In the first case a damping $\zeta = 0.3$ has been considered leading to an attenuation of 8 dB and in

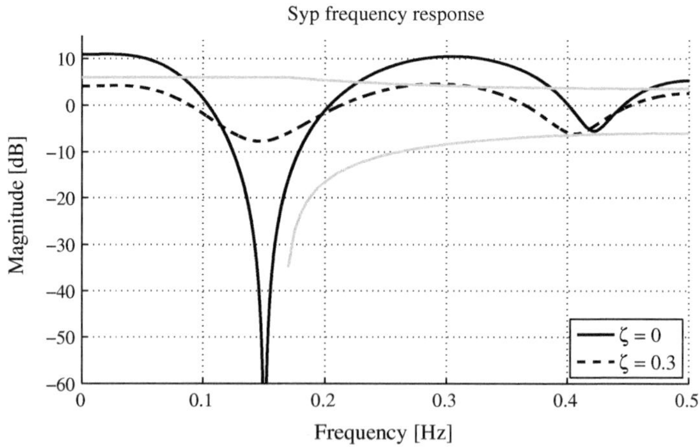

Fig. 7.9 Modulus of the output sensitivity functions for a double internal model with 0 and 0.3 damping

the second case full rejection of the disturbances have been considered using internal models with $\zeta = 0$ leading to an attenuation over 60 dB.[5]

One can clearly see that the increase of attenuation in a certain frequency region implies necessarily a stronger amplification of the disturbances outside the attenuation band. This is a direct consequence of Property 2. A similar phenomenon occurs if for a given attenuation the attenuation band is expanded.

7.2.8 *Input Sensitivity Function*

The input sensitivity function is extremely important in the design of the linear controller. The modulus of the input sensitivity function should be low at high frequencies in order to assure a good robustness of the system with respect to additive unstructured uncertainties located in the high-frequency region.[6]

The expression of the input sensitivity function using a RS controller with R and S given by (7.57) and (7.58) is

$$S_{up}(z^{-1}) = -\frac{A(z^{-1})H_R(z^{-1})R'(z^{-1})}{A(z^{-1})H_S(z^{-1})S'(z^{-1}) + q^{-d}B(z^{-1})H_R(z^{-1})R'(z^{-1})} \tag{7.61}$$

Property 1

The effect of the output disturbances upon the input is cancelled (i.e., $S_{up} = 0$) at the frequencies where

$$A(e^{-j\omega})H_R(e^{-j\omega})R'(e^{-j\omega}) = 0 \tag{7.62}$$

At these frequencies $S_{yp} = 1$ (open-loop operation). The pre-specified values assuring $S_{up} = 0$ at certain frequencies are of the same form as those used to make $S_{yp} = 1$.

Figure 7.10 illustrates the effect upon S_{up} of a pre-specified $H_R(z^{-1})$ of the form:

$$H_R(z^{-1}) = 1 + \alpha z^{-1} \; ; \; 0 < \alpha \leq 1$$

For $\alpha = 1$, one has $S_{up} = 0$ at $0.5f_s$. Using $0 < \alpha < 1$ allows to reduce more or less the input sensitivity function around $0.5f_s$.[7] This structure of $H_R(z^{-1})$ is systematically used for reducing the magnitude of the input sensitivity function in the high-frequency region.

---

[5]The structure of the H_S is $H_s = (1 + \alpha_1 q^{-1} + \alpha_2 q^{-2})(1 + \alpha_1' q^{-1} + \alpha_2' q^{-2})$.

[6]This is indeed true even in adaptive control since the uncertainties in the high-frequency region are not in general handled by the adaptive controller.

[7]The input sensitivity function correspond to the system considered previously which includes in the controller an internal model with zero damping located at $0.15f_s$.

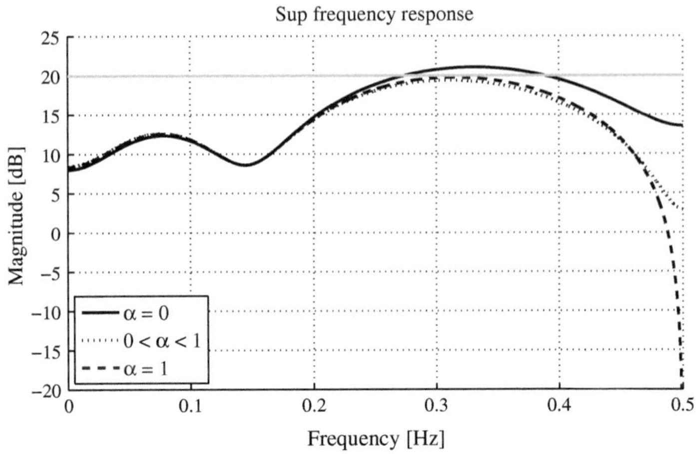

Fig. 7.10 Effect of $H_R(z^{-1}) = 1+\alpha z^{-1}, 0 < \alpha \leq 1$ upon the input sensitivity function for various values of parameter α

Property 2

At the frequencies where:

$$A(e^{-j\omega})H_S(e^{-j\omega})S'(e^{-j\omega}) = 0$$

which corresponds to perfect rejection of the output disturbances ($S_{yp} = 0$ at these frequencies), one has

$$\left|S_{up}(e^{-j\omega})\right| = \left|\frac{A(e^{-j\omega})}{B(e^{-j\omega})}\right| \tag{7.63}$$

i.e., the modulus of the input sensitivity function is equal to the inverse of the gain of the plant at this frequency.

This implies that perfect rejection of disturbances (or more generally attenuation of disturbances) should be done only in the frequency regions, where the gain of the system is large enough. If the gain is too low, $|S_{yp}|$ will be very large at these frequencies. Therefore, the robustness with respect to additive plant model uncertainties will be reduced, and the stress on the actuator will become important [10]. This also indicates that problems will occur if B has complex zeros close to the unit circle (stable or unstable). At these frequencies, rejection of disturbances should be avoided.

7.2.9 *Shaping the Sensitivity Functions for Active Vibration Control*

Two sensitivity functions are of particular interest in active vibration control:

- Output sensitivity function (the transfer function between the disturbance $p(t)$ and the output of the system $y(t)$):

$$S_{yp}(z^{-1}) = \frac{A(z^{-1})S(z^{-1})}{P(z^{-1})} \tag{7.64}$$

- Input sensitivity function (the transfer function between the disturbance $p(t)$ and the input of the system $u(t)$):

$$S_{up}(z^{-1}) = -\frac{A(z^{-1})R(z^{-1})}{P(z^{-1})} \tag{7.65}$$

In active vibration control they have to be shaped for performance and robustness purposes. The first tool for shaping the sensitivity functions, once the "performance" choices have been done (damping of some complex poles, introduction of the internal model of the disturbance, opening the loop at certain frequencies), is the introduction of the auxiliary poles.

The introduction of auxiliary asymptotically stable real poles $P_F(z^{-1})$ will cause in general a decrease of the modulus of the sensitivity function in the domain of attenuation of $1/P_F(z^{-1})$.

From Eqs. (7.56) and (7.59), one can see that the term $1/P_D(z^{-1})P_F(z^{-1})$ will introduce a stronger attenuation in the frequency domain than the term $1/P_D(z^{-1})$ if the auxiliary poles $P_F(z^{-1})$ are real (aperiodic) and asymptotically stable; however, since $S'(z^{-1})$ depends upon the poles through Eq. (7.16), one cannot guarantee this property for all the values of $P_F(z^{-1})$.

The auxiliary poles are generally chosen as high-frequency real poles under the form:

$$P_F(z^{-1}) = (1 - p_1 z^{-1})^{n_F} \ ; \ 0.05 \leq p_1 \leq 0.5$$

where:

$$n_F \leq n_p - n_D \ ; \ n_p = (\deg P)_{\max} \ ; \ n_D = \deg P_D$$

The effect of the introduction of the auxiliary poles is illustrated in Fig. 7.11, for the same system considered previously with a controller including an internal model with 0 damping at $0.05 f_S$. One observes that the introduction of 5 auxiliary real poles located at 0.5 "squeezes" the modulus of the output sensitivity function around 0 dB axis in the high-frequency range.

Note that in many applications the introduction of high-frequency auxiliary poles allows to satisfy the requirements for robustness margins.

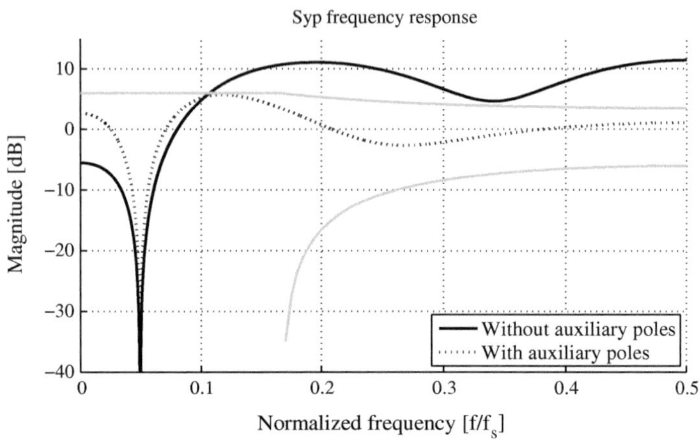

Fig. 7.11 Effect of auxiliary poles on the output sensitivity function

Simultaneous introduction of a fixed part H_{S_i} *and of a pair of auxiliary poles* P_{F_i} *in the form*

$$\frac{H_{S_i}(z^{-1})}{P_{F_i}(z^{-1})} = \frac{1 + \beta_1 z^{-1} + \beta_2 z^{-2}}{1 + \alpha_1 z^{-1} + \alpha_2 z^{-2}} \tag{7.66}$$

resulting from the discretization of the continuous-time band-stop filter (BSF):

$$F(s) = \frac{s^2 + 2\zeta_{num}\omega_0 s + \omega_0^2}{s^2 + 2\zeta_{den}\omega_0 s + \omega_0^2} \tag{7.67}$$

using the bilinear transformation[8]

$$s = \frac{2}{T_s}\frac{1 - z^{-1}}{1 + z^{-1}} \tag{7.68}$$

introduces an attenuation (a "hole") at the normalized discretized frequency

$$\omega_{disc} = 2\arctan(\frac{\omega_0 T_s}{2}) \tag{7.69}$$

as a function of the ratio $\zeta_{num}/\zeta_{den} < 1$. *The attenuation at* ω_{disc} *is given by*

$$M_t = 20\log(\frac{\zeta_{num}}{\zeta_{den}}) \ ; \ (\zeta_{num} < \zeta_{den}) \tag{7.70}$$

[8]The bilinear transformation assures a better approximation of a continuous-time model by a discrete-time model in the frequency domain than the replacement of differentiation by a difference, i.e., $s = (1 - z^{-1})T_s$ (see [6]).

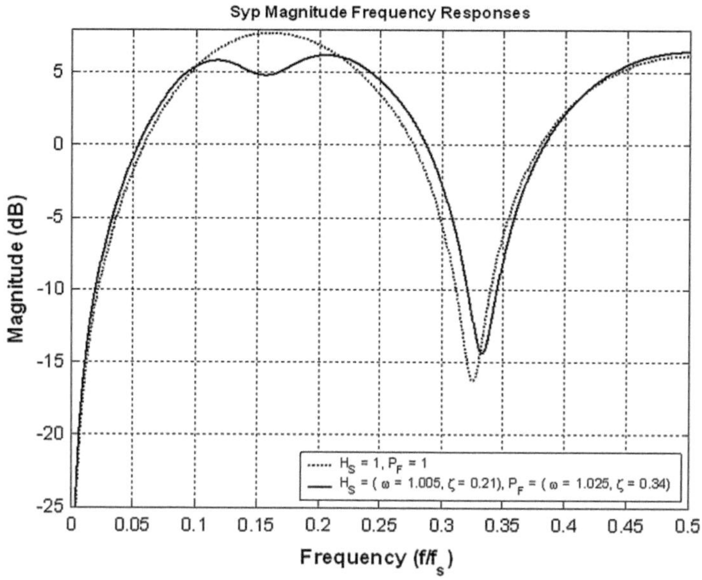

Fig. 7.12 Effects of a resonant filter H_{S_i}/P_{F_i} on the output sensitivity functions

The effect upon the frequency characteristics of S_{yp} at frequencies $f << f_{disc}$ and $f >> f_{disc}$ is negligible.

Figure 7.12 illustrates the effect of the simultaneous introduction of a fixed part H_S and a pair of poles in P, corresponding to the discretization of a resonant filter of the form of (7.67). One observes its weak effect on the frequency characteristics of S_{yp}, far from the resonance frequency of the filter.

This pole-zero filter (band-stop filter) is essential for an accurate shaping of the modulus of the sensitivity functions in the various frequency regions in order to satisfy the constraints. It allows to reduce the interaction between the tuning in different regions.

Design of the Band-Stop Filter H_{S_i}/P_{F_i}

The computation of the coefficients of H_{S_i} and P_{F_i} is done in the following way:

Specifications:

- Central normalized frequency f_{disc} ($\omega_{disc} = 2\pi f_{disc}$)
- Desired attenuation at frequency f_{disc} : M_t (dB)
- Minimum accepted damping for auxiliary poles

$$P_{F_i} : (\zeta_{den})_{min} \geq 0.3$$

Step I: Design of the continuous-time filter

$$\omega_0 = \frac{2}{T_s}\tan(\frac{\omega_{disc}}{2}) \quad 0 \leq \omega_{disc} \leq \pi \quad \zeta_{num} = 10^{M_t/20}\zeta_{den}$$

Step II: Design of the discrete-time filter using the bilinear transformation (7.68). Using (7.68) one gets

$$F(z^{-1}) = \frac{a_{z0} + a_{z1}z^{-1} + a_{z2}z^{-2}}{a_{z0} + a_{z1}z^{-1} + a_{z2}z^{-2}} = \gamma \frac{1 + \beta_1 z^{-1} + \beta_2 z^{-2}}{1 + \alpha_1 z^{-1} + \alpha_2 z^{-2}} \tag{7.71}$$

which will be effectively implemented as[9]

$$F(z^{-1}) = \frac{H_S(z^{-1})}{P_i(z^{-1})} = \frac{1 + \beta_1 z^{-1} + \beta_2 z^{-2}}{1 + \alpha_1 z^{-1} + \alpha_2 z^{-2}}$$

where the coefficients are given by[10]

$$\begin{aligned}
b_{z0} &= \frac{4}{T_s^2} + 4\frac{\zeta_{num}\omega_0}{T_s} + \omega_0^2 \; ; \; b_{z1} = 2\omega_0^2 - \frac{8}{T_s^2} \\
b_{z2} &= \frac{4}{T_s^2} - 4\frac{\zeta_{num}\omega_0}{T_s} + \omega_0^2 \\
a_{z0} &= \frac{4}{T_s^2} + 4\frac{\zeta_{den}\omega_0}{T_s} + \omega_0^2 \; ; \; a_{z1} = 2\omega_0^2 - \frac{8}{T_s^2} \\
a_{z2} &= \frac{4}{T_s^2} - 4\frac{\zeta_{den}\omega_0}{T_s} + \omega_0^2
\end{aligned} \tag{7.72}$$

$$\begin{aligned}
\gamma &= \frac{b_{z0}}{a_{z0}} \\
\beta_1 &= \frac{b_{z1}}{b_{z0}} \; ; \; \beta_2 = \frac{b_{z2}}{b_{z0}} \\
\alpha_1 &= \frac{a_{z1}}{a_{z0}} \; ; \; \alpha_2 = \frac{a_{z2}}{a_{z0}}
\end{aligned} \tag{7.73}$$

Remark For frequencies below 0.17 f_s the design can be done with a very good precision directly in discrete-time. In this case, $\omega_0 = \omega_{0den} = \omega_{0num}$ and the damping of the discrete-time filters H_{S_i} and P_{F_i} is computed as a function of the attenuation directly using Eq. (7.70).

Remark While H_S is effectively implemented in the controller, P_F is only used indirectly. P_F will be introduced in (7.17) and its effect will be reflected in the coefficients of R and S obtained as solutions of Eq. (7.59).

If the S polynomial contains the internal model of a sinusoidal disturbance, i.e., $S = S'D_p$ and D_p is a second-order polynomial with zero damping and a resonance frequency ω, the modulus of the output sensitivity function will be zero at this

[9]The factor γ has no effect on the final result (coefficients of R and S). It is possible, however, to implement the filter without normalizing the numerator coefficients.

[10]These filters can be computed using the functions *filter22.sci* (Scilab) and *filter22.m* (MATLAB®) to be downloaded from the book website.

frequency, which means total rejection of a sinusoidal disturbance. Without any shaping of the sensitivity function, there will be a "waterbed effect" in the vicinity of this frequency; however, if the objective is to introduce just a certain amount of attenuation, we should consider introduction of the "band-stop" filters which have zeros and poles. The numerator will be implemented in the "S" polynomial while the poles will be added to the desired closed-loop poles. In this case the waterbed effect will be less important.

For n narrow-band disturbances, n band-stop filters will be used

$$\frac{S_{BSF}(z^{-1})}{P_{BSF}(z^{-1})} = \frac{\prod_{i=1}^{n} S_{BSF_i}(z^{-1})}{\prod_{i=1}^{n} P_{BSF_i}(z^{-1})}. \tag{7.74}$$

A similar procedure can be used for shaping the input sensitivity function (H_S in Eq. (7.66) is replaced H_R).

7.3 Real-Time Example: Narrow-Band Disturbance Attenuation on the Active Vibration Control System Using an Inertial Actuator

This section illustrates the methodology used for the attenuation of narrow-band disturbances through an example. The active vibration control system with inertial actuator described in Sect. 2.2 will be used as a test bench. An open-loop identification for this system has been done in Sect. 6.2. The sampling frequency is $f_s = 800$ Hz.

One sinusoidal disturbance at 70 Hz is applied to the system. The disturbance is filtered by the primary path and its effects are measured by the residual force transducer. The objective is to strongly attenuate the effect of this disturbance on the residual force. The internal model principle together with the shaping of the sensitivity functions will be used for the design of a linear robust controller.

The specifications are as follows:

- the controller should eliminate the disturbance at 70 Hz (at least 40 dB attenuation).
- the maximum allowed amplification of the output sensitivity function is 6 dB (i.e., the modulus margin will be $\Delta M \geqslant 0.5$).
- a delay margin of at least one sampling period should be achieved.
- the gain of the controller has to be zero at 0 Hz (since the system has a double differentiator behaviour).
- the gain of the controller should be zero at $0.5 f_s$ where the system has low gain and uncertainties exist.
- the effect of disturbances on the control input should be attenuated above 100 Hz in order to improve robustness with respect to unmodelled dynamics ($S_{up}(e^{j\omega}) < -40$ dB, $\forall \omega \in [100, 400$ Hz$]$).

The steps for the design of the linear controller are

1. include all (stable) secondary path poles in the closed-loop characteristic polynomial;
2. design the fixed part of the controller denominator in order to cancel the 70 Hz disturbance (IMP)
$$H_S(q^{-1}) = 1 + a_1 q^{-1} + q^{-2}, \tag{7.75}$$
where $a_1 = -2\cos(2\pi f/f_S)$, $f = 70$ Hz. The modulus of the resulting output sensitivity function is shown in Fig. 7.13 (curve IMP). As one can see the maximum of the modulus of the output sensitivity function is larger than 6 dB;
3. open the loop at 0 Hz and at 400 Hz by setting the fixed part of the controller numerator as
$$H_R = (1 + q^{-1}) \cdot (1 - q^{-1}) = 1 - q^{-2}. \tag{7.76}$$
The resulting output sensitivity function is shown also in Fig. 7.13 (curve IMP + Hr). As it can be seen, it has an unacceptable value around 250 Hz (violation of the delay margin constraint);
4. to improve robustness two complex conjugate poles have been added to the characteristic polynomial, one at 65 Hz and the second at 75 Hz, both of them with 0.2 damping factor. The resulting output sensitivity function (curve IMP + Hr + aux. poles) has the desired characteristics; however, as one can see in Fig. 7.14 (curve IMP + Hr + aux. poles), the modulus of the input sensitivity function is higher than −40 dB between 100 and 400 Hz;

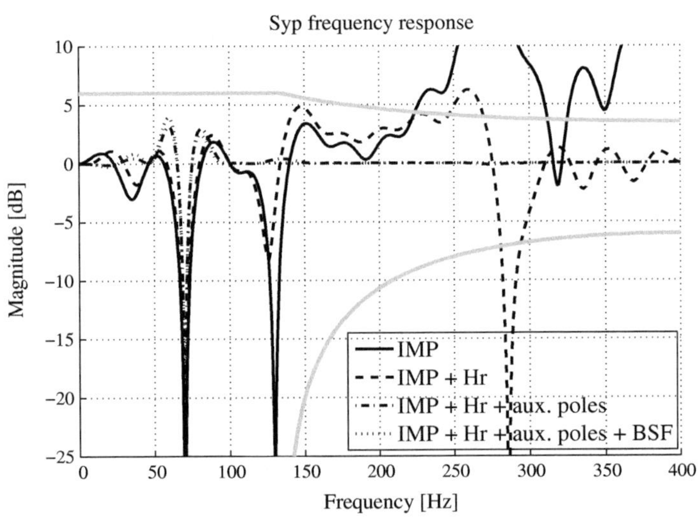

Fig. 7.13 Output sensitivity functions for the various controllers (*grey lines* represent the templates for modulus and delay margins)

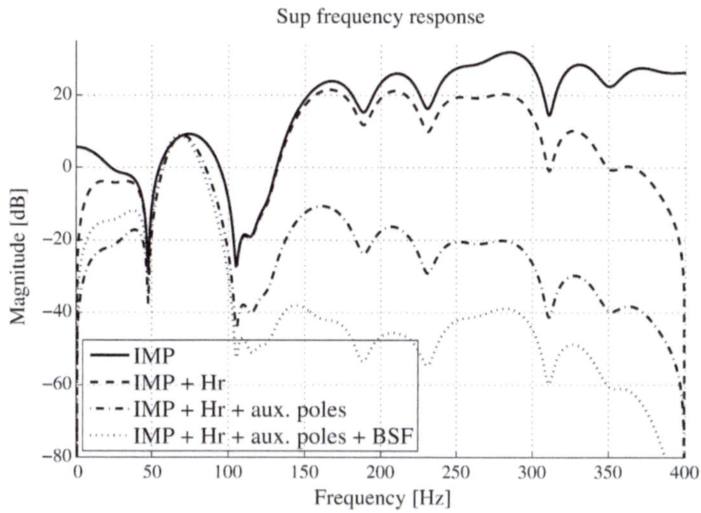

Fig. 7.14 Input sensitivity functions for the various controllers

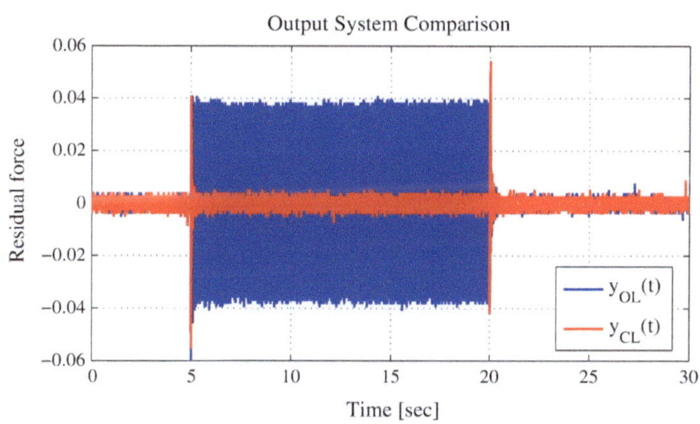

Fig. 7.15 Time response results for a 70 Hz disturbance in open-loop and in closed-loop

5. add band-stop filters (BSF) on the S_{up} sensitivity function: one at 160 Hz, the other at 210 Hz, with -20 and -15 dB attenuation respectively. Both have 0.9 damping factor for the denominator. One can see that this has the desired effect on the input sensitivity functions and no effects on the output sensitivity function.

The resulting modulus margin is 0.637 and the resulting delay margin is $2.012 \cdot T_s$. The final controller satisfies the desired specifications both in terms of performance and robustness.

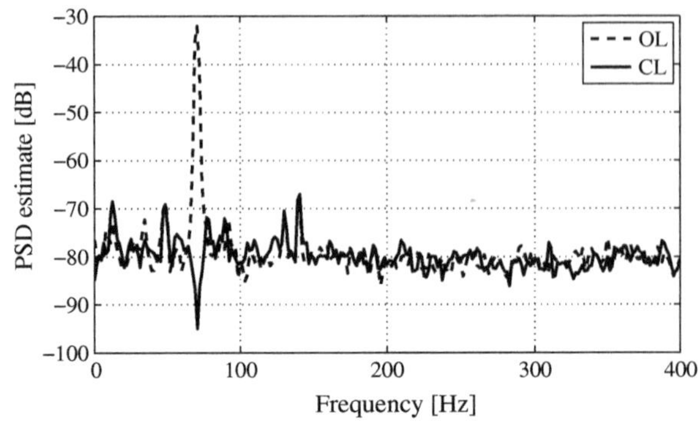

Fig. 7.16 PSD of the open-loop disturbance (*dashed line*) and effective attenuation (*solid line*) for 70 Hz in closed-loop

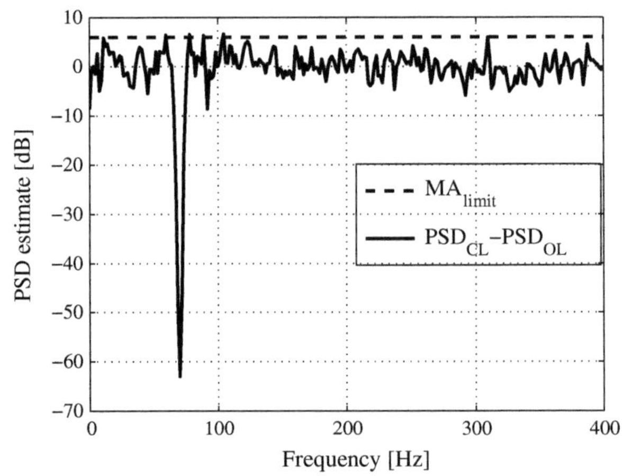

Fig. 7.17 Effective residual attenuation/amplification PSD estimates computed as the difference between the open-loop PSD and the closed-loop PSD

Real-Time Results

Time domain results in open-loop ($y_{OL}(t)$) and in closed-loop ($y_{CL}(t)$) are shown in Fig. 7.15. A frequency domain analysis has been done and is shown in Figs. 7.16 and 7.17. It can be seen that the controller achieves all the desired specifications. Under the effect of the controller the residual force is almost at the level of the system's noise.

7.4 Pole Placement with Sensitivity Function Shaping by Convex Optimisation

In [11] it was shown that the problem of shaping the sensitivity functions in the context of pole placement can be formulated as a convex optimisation problem, and routines for convex optimisation can be used (available in the toolbox OPTREG[11]).

[11]To be downloaded from the book website.

We will present this method which will be used in the context of active damping. This method takes first advantage of the Youla–Kučera parametrization. It is assumed that

- the fixed parts of the controller H_R and H_S have been defined (in order to achieve certain performances);
- a "central" stabilizing controller is already designed;
- the templates for the output and input sensitivity functions have been defined (in order to obtain the required robustness margins and performance specifications).

One considers the Youla–Kučera parmetrization for the controller as follows:

$$R = H_R(R_0 + AH_S Q) \tag{7.77}$$

$$S = H_S(S_0 - z^{-d} BH_R Q) \tag{7.78}$$

where the fixed parts of the controller (H_R, H_S), and (A, B) are polynomials in z^{-1} ($z^{-d}B/A$ is the nominal model) and Q is a rational transfer function proper and asymptotically stable.[12]

The central controller R_0/S_0 ($Q = 0$) can be obtained by solving the following Bezout equation for R_0 and S_0:

$$AH_S S_0 + z^{-d} BH_R R_0 = P_D, \tag{7.79}$$

where P_D is an asymptotically stable polynomial defined by the designer and which contains the desired dominant poles for the closed-loop system. Expressing Q as a ratio of proper transfer functions in z^{-1} such as

$$Q(z^{-1}) = \frac{B_Q(z^{-1})}{A_Q(z^{-1})} \tag{7.80}$$

one gets

$$\frac{R}{S} = \frac{H_R(R_0 A_Q + AH_S B_Q)}{H_S(S_0 A_Q - z^{-d} BH_R B_Q)}. \tag{7.81}$$

The poles of the closed-loop system will be given by

$$P = AS + z^{-d} BR = P_D A_Q,$$

where the zeros of P_D are the fixed poles of the closed-loop (defined by the central controller) and the zeros of A_Q are the additional poles which will be introduced by the optimization procedure. The output and input sensitivity functions can be written as:

$$S_{yp} = \frac{AS}{AS + z^{-d} BR} = \frac{AH_S}{P_D}\left(S_0 - B_{nom} H_R \frac{B_Q}{A_Q}\right); \tag{7.82}$$

---

[12] This particular YK parametrization allows to preserve the fixed parts H_R and H_S in the resulting controller given in Eq. (7.4).

$$S_{up} = \frac{AH_R}{P_D}\left(R_0 + A_{nom}H_S\frac{B_Q}{A_Q}\right). \tag{7.83}$$

As shown in the above Eqs. (7.82) and (7.83), the sensitivity functions can obviously be expressed in the form $T_1 + T_2\frac{\beta}{\alpha}$.

Imposing a certain frequency-dependent limit W (template) on the modulus of the sensitivity functions (attenuation band, modulus margin, delay margin, restrictions on the input sensitivity function) leads then to a condition of the form

$$\left|T_1 \arg z + T_2 \arg z \frac{\beta' \arg z}{\alpha' \arg z}\right| \leq |W \arg z| \quad \forall |z| = 1 \tag{7.84}$$

Condition (7.84) is equivalent to the condition

$$\left\|\bar{T}_1 + \bar{T}_2\frac{\beta'}{\alpha'}\right\|_\infty < 1 \tag{7.85}$$

Thus, Eq. (7.84) implies the existence of α and β such that by setting $\bar{T}_1 = W^{-1}T_1$ and $\bar{T}_2 = W^{-1}T_2$ one obtains

$$\left|W^{-1}T_1\alpha + W^{-1}T_2\beta\right| \leq Re\{\alpha\} \tag{7.86}$$

and this is obviously a convex condition on α and β. Details can be found in [11, 12].

For point-wise testing of the conditions a frequency gridding is carried out (e.g. 32 points between $f = 0$ and $f = 0.5f_s$).

For the optimization procedures the polynomials A_Q et B_Q will take the form (Ritz method):

$$A_Q(x_a) = 1 + \sum_{k=1}^{N} x_{ak}\alpha_k\ ; \tag{7.87}$$

$$B_Q(x_b) = x_{b_0} + \sum_{k=1}^{N} x_{bk}\beta_k, \tag{7.88}$$

where α_k, β_k are stable polynomials (affine in x_{ak} et x_{bk}) and N is the order of the parametrization (i.e., the number of points on the sensitivity functions where the constraints have to be verified). The parameters to be optimized are x_{ak} et x_{bk}.

For the discrete-time cases α_k and β_k can be chosen as

$$\alpha_k = \beta_k = \left(\frac{z_0 - z^{-1}}{1 - z_0 z^{-1}}\right)^k,$$

where z_0 is the time constant of the parametrization (which can be adjusted).

Using the parametrization and the constraints indicated above an (RS) controller with desired properties can be obtained by convex optimization. For more details on the optimization procedure see [13, 14].

The MATLAB® toolbox Optreg provides the appropriate routines for specifying the constraints and finding the optimal controller. The method will be used in Chap. 10 for active damping.

7.5 Concluding Remarks

- The design of polynomial RS controllers for active vibration control systems has been discussed in this chapter.
- The design of the controller requires the knowledge of the plant model (the secondary path in active vibration control).
- Asymptotic rejection of tonal disturbances can be achieved using the Internal Model Principle (it requires the knowledge of the frequency of the disturbance).
- The Youla–Kučera parametrization of the controller provides a separation between disturbance compensation and feedback stabilization.
- Robustness is not an intrinsic property of a control strategy. It results from an appropriate choice of some control objectives related to the sensitivity functions.
- Two sensitivity functions are of major interest: the output sensitivity function and the input sensitivity function.
- *Modulus margin* and *delay margin* are basic robustness indicators.
- Shaping of the sensitivity functions is a key issue in active vibration control in order to achieve desired performance and robustness objectives.
- Performance and robustness specifications translate in desired templates for the sensitivity functions.
- Pole placement combined with tools for shaping the sensitivity functions is an efficient approach for designing active vibration control systems.
- Shaping of the sensitivity functions can be conveniently achieved by the selection of the auxiliary poles and the use of band-stop filters.
- Pole placement combined with convex optimization can provide almost an automatic solution to the design problem, once the desired templates for the sensitivity functions are defined.

7.6 Notes and References

The first issue in the design of AVC systems (assuming that the plant model is known) is the translation of the performance and robustness specifications in desired templates for the sensitivity functions. Then any design method which allows to achieve the desired sensitivity functions can be used, such as Pole placement

[1, 6, 7, 15], Linear Quadratic Control [6, 7, 15], H_∞ control [8, 16], CRONE control [17–19], Generalized Predictive Control [2, 20].

The shaping of the sensitivity function can be converted in a convex optimization problem [12] and the use of this approach is detailed in [11, 13, 14].

The Bode integral constraint in the context of AVC is discussed in [21, 22].

References

1. Goodwin G, Sin K (1984) Adaptive filtering prediction and control. Prentice Hall, Englewood Cliffs
2. Landau ID, Lozano R, M'Saad M, Karimi A (2011) Adaptive control, 2nd edn. Springer, London
3. Anderson B (1998) From Youla–Kučera to identification, adaptive and nonlinear control. Automatica 34(12):1485–1506. doi:10.1016/S0005-1098(98)80002-2, http://www.sciencedirect.com/science/article/pii/S0005109898800022
4. Tsypkin Y (1997) Stochastic discrete systems with internal models. J Autom Inf Sci 29(4&5):156–161
5. Landau I, Constantinescu A, Rey D (2005) Adaptive narrow band disturbance rejection applied to an active suspension - an internal model principle approach. Automatica 41(4):563–574
6. Landau I, Zito G (2005) Digital control systems - design, identification and implementation. Springer, London
7. Astrom KJ, Wittenmark B (1984) Computer controlled systems. Theory and design. Prentice-Hall, Englewood Cliffs
8. Doyle JC, Francis BA, Tannenbaum AR (1992) Feedback control theory. Macmillan, New York
9. Sung HK, Hara S (1988) Properties of sensitivity and complementary sensitivity functions in single-input single-output digital control systems. Int J Control 48(6):2429–2439. doi:10.1080/00207178808906338
10. Landau ID, Alma M, Constantinescu A, Martinez JJ, Noë M (2011) Adaptive regulation - rejection of unknown multiple narrow band disturbances (a review on algorithms and applications). Control Eng Pract 19(10):1168–1181. doi:10.1016/j.conengprac.2011.06.005
11. Larger J, Constantinescu A (1979) Pole placement design using convex optimisation criteria for the flexible transmission benchmark. Eur J Control 5(2):193–207
12. Rantzer A, Megretski A (1981) A convex parametrization of robustly stabilizig controllers. IEEE Trans Autom Control 26:301–320
13. Langer J, Landau I (1999) Combined pole placement/sensitivity function shaping method using convex optimization criteria. Automatica 35:1111–1120
14. Langer J (1998) Synthèse de régulateurs numériques robustes. Application aux structures souples. Ph.D. thesis, INP Grenoble, France
15. Franklin GF, Powell JD, Workman ML (1998) Digital control of dynamic systems, vol 3. Addison-Wesley, Menlo Park
16. Zhou K, Doyle J (1998) Essentials of robust control. Prentice-Hall International, Upper Saddle River
17. Oustaloup A, Mathieu B, Lanusse P (1995) The CRONE control of resonant plants: application to a flexible transmission. Eur J Control 1(2):113–121. doi:10.1016/S0947-3580(95)70014-0, http://www.sciencedirect.com/science/article/pii/S0947358095700140
18. Lanusse P, Poinot T, Cois O, Oustaloup A, Trigeassou J (2004) Restricted-complexity controller with CRONE control-system design and closed-loop tuning. Eur J Control 10(3):242–251. doi:10.3166/ejc.10.242-251, http://www.sciencedirect.com/science/article/pii/S0947358004703647

19. Oustaloup A, Cois O, Lanusse P, Melchior P, Moreau X, Sabatier J (2006) The CRONE aproach: theoretical developments and major applications. In: IFAC Proceedings 2nd IFAC workshop on fractional differentiation and its applications, vol 39(11), pp 324–354. doi:10.3182/20060719-3-PT-4902.00059, http://www.sciencedirect.com/science/article/pii/S1474667015365228
20. Camacho EF, Bordons C (2007) Model predictive control. Springer, London
21. Hong J, Bernstein DS (1998) Bode integral constraints, collocation, and spillover in active noise and vibration control. IEEE Trans Control Syst Technol 6(1):111–120
22. Chen X, Jiang T, Tomizuka M (2015) Pseudo Youla–Kučera parameterization with control of the waterbed effect for local loop shaping. Automatica 62:177–183

Chapter 8
Identification in Closed-Loop Operation

8.1 Introduction

There are two reasons for considering identification in closed-loop operation in the context of active vibration control systems:

- obtaining improved system models for controller redesign; and
- retuning of the controller without opening the loop.

The objective of identification in closed-loop is to obtain a plant model describing as precisely as possible the behaviour of the real closed-loop system for a given controller. In other words, the objective of system identification in closed-loop is to search for a plant model that in feedback with the controller operating on the true plant will lead to a closed-loop transfer function (sensitivity function) that is as close as possible to that of the real closed-loop system. If the performance of the closed-loop system is not satisfactory, it is expected that this model identified in closed-loop will allow the redesign of the controller in order to improve the performance of the real-time control system.

It has been shown in [1, 2], as well as in many other references, that identification in closed-loop, provided that appropriate identification algorithms are used, leads in general to better models for controller design.

In order to understand the potential of the identification in closed-loop as well as the difficulties which can be encountered, let us consider the case of the plant model identification in closed-loop where the external excitation is added to the controller output (see Fig. 8.1a). Figure 8.1b shows an equivalent scheme that emphasizes the transfer function between the external excitation r_u and the plant input u, as well as the effect of the measurement noise upon the plant input. Assume that the external excitation is a PRBS that has almost constant frequency spectrum from 0 to $0.5f_s$.

One observes that the effective plant input corresponds to the external excitation filtered by the output sensitivity function S_{yp} (see Sect. 7.1), whose magnitude has a maximum in the frequency regions close to the critical point $[-1, j0]$ (see Sect. 7.2.4). Therefore the frequency spectrum of the effective input applied to the

I.D. Landau et al., *Adaptive and Robust Active Vibration Control*,
Advances in Industrial Control, DOI 10.1007/978-3-319-41450-8_8

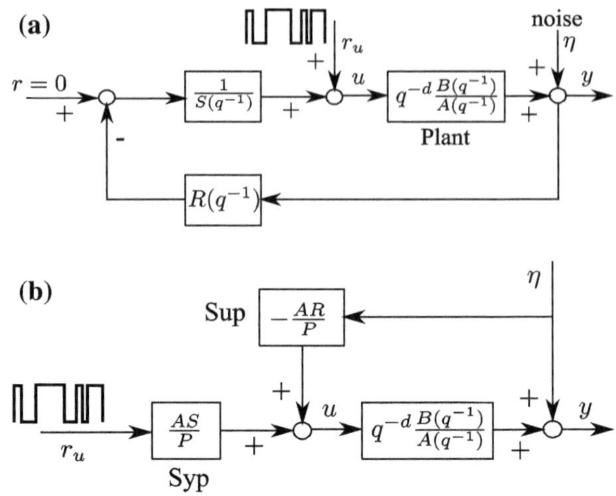

Fig. 8.1 Identification in closed-loop: **a** excitation added to the control output, **b** equivalent representation

plant will be enhanced in these frequency regions. As a consequence, the quality of the identified model in these critical regions for stability and performance will be improved. Unfortunately, in the meantime, the feedback introduces a correlation between the measurement noise and the plant input. This leads to an important bias on the estimated parameters if one would like to identify the plant model with open-loop techniques.

Therefore, for a good identification in closed-loop operation one needs identification methods *that take advantage of the "improved" characteristics of the effective excitation signal applied to the plant input but which are not affected by the noise in the context of feedback*. An efficient solution for this problem is provided by the "closed-loop output error" method (CLOE) that will be presented next.

8.2 Closed-Loop Output Error Identification Methods

The Principle

The principle of closed-loop output error identification algorithms is illustrated in Fig. 8.2. The upper part represents the true closed-loop system and the lower part represents an adjustable predictor of the closed-loop. This closed-loop predictor uses a controller identical to the one used on the real-time system.

The prediction error between the output of the real-time closed-loop system and the closed-loop predictor (closed-loop output error) is a measure of the difference between the true plant model and the estimated one. This error can be used to adapt the estimated plant model such that the closed-loop prediction error is minimized (in the sense of a certain criterion). In other words, the objective of the identification

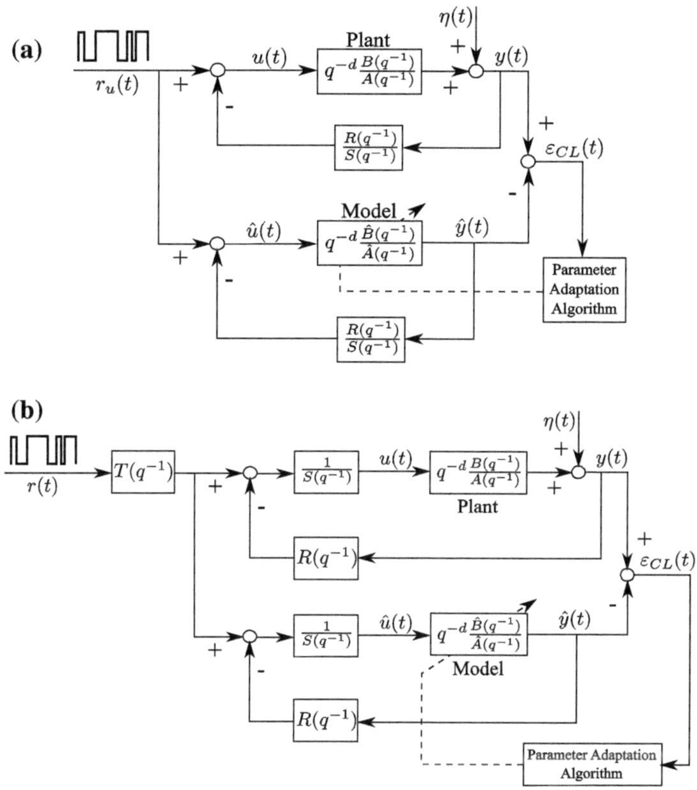

Fig. 8.2 Closed-loop output error identification method. **a** Excitation superposed to control output. **b** Excitation superposed to the reference

in closed-loop is to find the best plant model which minimizes the prediction error between the measured output of the true closed-loop system and the predicted closed-loop output. The use of these methods requires the knowledge of the controller.

As it can be seen from Fig. 8.2, the minimization of the closed-loop prediction error will minimize the difference between real and estimated sensitivity functions. For the case of the excitation added to the controller output, the difference between

$$S_{yv} = \frac{q^{-d}BS}{AS + q^{-d}BR} \tag{8.1}$$

and

$$\hat{S}_{yv} = \frac{q^{-d}\hat{B}S}{\hat{A}S + q^{-d}\hat{B}R} \tag{8.2}$$

will be minimized, where $\hat{A}$ and $\hat{B}$ are the estimates of the A and B polynomials.[1]

For the case of the excitation added to the reference, with $T = R$, the difference between

$$S_{yr} = \frac{q^{-d}BR}{AS + q^{-d}BR} \tag{8.3}$$

and

$$\hat{S}_{yr} = \frac{q^{-d}\hat{B}R}{\hat{A}S + q^{-d}\hat{B}R} \tag{8.4}$$

will be minimized. Since $|S_{yr} - \hat{S}_{yr}| = |S_{yp} - \hat{S}_{yp}|$, the difference between the true and the estimated output sensitivity function will also be minimized.

In the context of active vibration control, we will be in general interested to get a model which allows a better estimation of the output sensitivity function. Therefore, often, the configuration of Fig. 8.2b will be used with $T = R$.[2]

The Algorithms

$$G(q^{-1}) = \frac{q^{-d}B(q^{-1})}{A(q^{-1})}, \tag{8.5}$$

where

$$B(q^{-1}) = b_1 q^{-1} + \cdots + b_{n_B} q^{-n_B} = q^{-1}B^*(q^{-1}) \tag{8.6}$$

$$A(q^{-1}) = 1 + a_1 q^{-1} + \cdots + a_{n_A} q^{-n_A} = 1 + q^{-1}A^*(q^{-1}) \tag{8.7}$$

The plant is operated in closed-loop with an RST digital controller (without lack of generality). The output of the plant operating in closed-loop is given by (see Fig. 8.2a):

$$y(t+1) = -A^*y(t) + B^*u(t-d) + A\eta(t+1) = \theta^T\varphi(t) + A\eta(t+1), \tag{8.8}$$

where $u(t)$ is the plant input, $y(t)$ is the plant output, $\eta(t)$ is the output noise and:

$$\theta^T = [a_1 \ldots, a_{n_A}, b_1 \ldots, b_{n_B}] \tag{8.9}$$

$$\varphi^T(t) = [-y(t) \ldots, -y(t-n_A+1), u(t-d) \ldots, u(t-n_B+1-d)] \tag{8.10}$$

$$u(t) = -\frac{R}{S}y(t) + r_u, \tag{8.11}$$

where r_u is the external excitation added to the output of the controller (r_u is equal to $\frac{T}{S}r$ if the external excitation is applied on the reference as in Fig. 8.2b).

[1] In this case, S_{yv} corresponds to the transfer function between $r_u(t)$ and $y(t)$.

[2] This is equivalent to sending the excitation to the input of the filter R in Fig. 8.2b.

For a fixed value of the estimated parameters, the predictor of the closed-loop (the design system) can be expressed as:

$$\hat{y}(t+1) = -\hat{A}^*\hat{y}(t) + \hat{B}^*\hat{u}(t-d) = \hat{\theta}^T\phi(t), \tag{8.12}$$

where

$$\hat{\theta}^T = [\hat{a}_1 \ldots, \hat{a}_{n_A}, \hat{b}_1 \ldots, \hat{b}_{n_B}] \tag{8.13}$$

$$\phi^T(t) = [-\hat{y}(t) \ldots, -\hat{y}(t-n_A+1), \hat{u}(t-d) \ldots, \hat{u}(t-n_B+1-d)] \tag{8.14}$$

$$\hat{u}(t) = -\frac{R}{S}\hat{y}(t) + r_u \tag{8.15}$$

The closed-loop prediction (output) error is defined as:

$$\varepsilon_{CL}(t+1) = y(t+1) - \hat{y}(t+1) \tag{8.16}$$

It clearly results from Fig. 8.2a that for constant values of the estimated parameters, the predictor regressor vector $\phi(t)$ depends only upon the external excitation. Therefore under the assumption that the external excitation (r or r_u) and the stochastic noise η are independent, $\phi(t)$ and $\eta(t)$ are not correlated (as well as $\phi(t)$ and $\varepsilon_{CL}(t+1)$), the scheme has the structure of an output error prediction.

If known fixed parts should be included in the estimated plant model, the equation of the predictor for the closed-loop has to be modified in order to preserve the input/output behaviour. See for details Sect. 8.2.4 and [3].

For all the methods, the parameter adaptation algorithm (PAA) has the general form

$$\hat{\Theta}(t+1) = \hat{\Theta}(t) + F(t)\Phi(t)\nu(t+1) \tag{8.17}$$

$$F(t+1)^{-1} = \lambda_1(t)F(t)^{-1} + \lambda_2(t)\Phi(t)\Phi^T(t) \tag{8.18}$$

$$0 < \lambda_1(t) \leq 1\,;\, 0 \leq \lambda_2(t) < 2\,;$$

$$F(0) > 0\,;\, F(t)^{-1} > \alpha F^{-1}(0)\,;\, 0 < \alpha < \infty \tag{8.19}$$

$$F(t+1) = \frac{1}{\lambda_1(t)}\left[F(t) - \frac{F(t)\Phi(t)\Phi^T(t)F(t)}{\frac{\lambda_1(t)}{\lambda_2(t)} + \Phi^T(t)F(t)\Phi(t)}\right] \tag{8.20}$$

$$\nu(t+1) = \frac{\nu^\circ(t+1)}{1+\Phi^T(t)F(t)\Phi(t)}, \tag{8.21}$$

where $\nu^\circ(t+1) = f_1(\hat{\Theta}(t), \hat{\Theta}(t-1), \ldots, y(t+1), \nu(t), \nu(t-1), \ldots)$ is the *a priori* adaptation error, $\nu(t+1) = f_2(\hat{\Theta}(t+1), \hat{\Theta}(t), \ldots, y(t+1), \nu(t), \nu(t-1), \ldots)$ is the *a posteriori* adaptation error and $\Phi(t)$ is the observation vector.

For each recursive identification algorithm Θ, Φ, and $\nu^\circ(t+1)$ will have specific expressions. Note that the sequences $\lambda_1(t)$ and $\lambda_2(t)$ allow to define the time profile of the adaptation gain $F(t)$. For convergence analysis in the stochastic environment, it is assumed that a PAA with decreasing adaptation gain is used (i.e., $\lambda_1(t) \equiv 1$, $\lambda_2(t) = \lambda_2 > 0$).

The fundamental differences with respect to the open-loop output error identification algorithm come from the structure of the adjustable predictor and of the observation vector.

8.2.1 The Closed-Loop Output Error Algorithm

Replacing now the fixed predictor of the closed-loop given in (8.12) by an adjustable predictor, one gets:

- *a priori predicted output*:

$$\hat{y}^{\circ}(t+1) = \hat{y}(t+1|\hat{\theta}(t)) = \hat{\theta}^T(t)\phi(t); \tag{8.22}$$

- *a posteriori predicted output*:

$$\hat{y}(t+1) = \hat{y}(t+1|\hat{\theta}(t+1)) = \hat{\theta}^T(t+1)\phi(t); \tag{8.23}$$

- *a priori* prediction error as:

$$\varepsilon_{CL}^{\circ}(t+1) = y(t+1) - \hat{y}^{\circ}(t+1); \tag{8.24}$$

- *a posteriori* prediction error as:

$$\varepsilon_{CL}(t+1) = y(t+1) - \hat{y}(t+1). \tag{8.25}$$

The equation for the *a posteriori* prediction error becomes in the deterministic environment (no noise, see [4] for details):

$$\varepsilon_{CL}(t+1) = \frac{S}{P}[\theta - \hat{\theta}(t+1)]^T\phi(t) \tag{8.26}$$

The rules given in Chap. 4 suggest a PAA with:

$$\begin{aligned} \hat{\Theta}(t) &= \hat{\theta}(t) \\ \Phi(t) &= \phi(t) \\ \nu^{\circ}(t+1) &= \varepsilon_{CL}^{\circ}(t+1) \end{aligned}$$

This is termed the *Closed-Loop Output Error* (CLOE) algorithm [1, 2, 4]. It can be shown (see [2, 4] that in both deterministic and stochastic environment the sufficient condition for stability and unbiased asymptotic convergence is:

$$H'(z^{-1}) = \frac{S(z^{-1})}{P(z^{-1})} - \frac{\lambda_2}{2} \tag{8.27}$$

should be strictly positive real (where $\max_t \lambda_2(t) \leq \lambda_2 < 2$).

To relax this condition, the following two solutions have been proposed.

8.2.2 Filtered and Adaptive Filtered Closed-Loop Output Error Algorithms (F-CLOE, AF-CLOE)

Equation (8.26) for $\hat{\theta} = constant$ can also be rewritten as:

$$\varepsilon_{CL}(t+1) = \frac{S}{P} \cdot \frac{\hat{P}}{S}[\theta - \hat{\theta}]\frac{S}{\hat{P}}\phi(t) = \frac{\hat{P}}{P}[\theta - \hat{\theta}]\phi_f(t), \tag{8.28}$$

where

$$\phi_f(t) = \frac{S}{\hat{P}}\phi(t) \tag{8.29}$$

$$\hat{P} = \hat{A}S + q^{-d}\hat{B}R \tag{8.30}$$

In Eq. (8.30), $\hat{P}$ is an estimation of the true closed-loop poles based on an initial estimation of the plant model (for example using an open-loop experiment). This formulation leads to the *Filtered Closed-Loop Output Error* (F-CLOE) algorithm [2] which uses the same adjustable predictor as CLOE (see Eqs. (8.22) and (8.23)) and the PAA with:

$$\hat{\Theta}(t) = \hat{\theta}(t)$$
$$\Phi(t) = \phi_f(t)$$
$$\nu^{\circ}(t+1) = \varepsilon^{\circ}_{CL}(t+1)$$

It can be shown that by neglecting the non-commutativity of time-varying operators (an exact algorithm can however be derived), under the sufficient condition that:

$$H'(z^{-1}) = \frac{\hat{P}(z^{-1})}{P(z^{-1})} - \frac{\lambda_2}{2} \tag{8.31}$$

is strictly positive real, both asymptotic stabilities in deterministic environment and asymptotic unbiasedness in a stochastic environment are assured [2].

One can further relax the condition of Eq. (8.31) by filtering $\phi(t)$ through a time-varying filter $S/\hat{P}(t)$, where $\hat{P}(t)$ corresponds to the current estimate of the closed-loop given by: $\hat{P}(t) = \hat{A}(t)S + q^{-d}\hat{B}(t)R$, where $\hat{A}(t)$ and $\hat{B}(t)$ are the current estimates of the A and B polynomials (the AF-CLOE algorithm).

8.2.3 Extended Closed-Loop Output Error Algorithm (X-CLOE)

For the case where the noise model is $\eta(t+1) = \frac{C}{A}e(t+1)$, where $e(t+1)$ is a zero mean gaussian white noise and $C(q^{-1}) = 1 + q^{-1}C^*(q^{-1})$ is an asymptotically stable polynomial, an extended output error prediction model can be defined:

$$\hat{y}(t+1) = -\hat{A}^*\hat{y}(t) + \hat{B}^*\hat{u}(t-d) + \hat{H}^*\frac{\varepsilon_{CL}(t)}{S}$$
$$= \hat{\theta}^T\phi(t) + \hat{H}^*\frac{\varepsilon_{CL}(t)}{S} = \hat{\theta}_e^T\phi_e(t) \tag{8.32}$$

Equation (8.8) for the plant output becomes in this case:

$$y(t+1) = \theta^T\phi(t) + H^*\frac{\varepsilon_{CL}(t)}{S} - C^*\varepsilon_{CL}(t) + Ce(t+1) \tag{8.33}$$
$$= \theta_e^T\phi_e(t) - C^*\varepsilon_{CL}(t) + Ce(t+1), \tag{8.34}$$

where

$$H^* = h_1 + h_2q^{-1} + \cdots + h_{n_H}q^{-n_H+1} = C^*S - A^*S - q^{-d}B^*R, \tag{8.35}$$
$$H = 1 + q^{-1}H^* = 1 + CS - P, \tag{8.36}$$
$$\theta_e^T = [\theta^T, h_1, \ldots, h_{n_H}], \tag{8.37}$$
$$\hat{\theta}_e^T = [\hat{\theta}^T, \hat{h}_1, \ldots, \hat{h}_{n_H}], \tag{8.38}$$
$$\phi_e^T(t) = [\phi^T(t), \varepsilon_{CLf}(t), \ldots, \varepsilon_{CLf}(t-n_H+1)], \tag{8.39}$$
$$\varepsilon_{CLf}(t) = \frac{1}{S}\varepsilon_{CL}(t). \tag{8.40}$$

Subtracting (8.32) from (8.34), one obtains the following expression for the closed-loop prediction error (for details see [5]):

$$\varepsilon_{CL}(t+1) = \frac{1}{C}[\theta_e - \hat{\theta}_e]^T\phi_e(t) + e(t+1). \tag{8.41}$$

Equation (8.41) clearly shows that for $\hat{\theta}_e = \theta_e$ the closed-loop prediction error tends asymptotically towards $e(t+1)$.

Replacing the fixed predictor (8.32) with an adjustable one, a recursive identification algorithm (X-CLOE) can be obtained by using a PAA with:

$$\hat{\Theta}(t) = \hat{\theta}_e(t)$$
$$\Phi(t) = \phi_e(t)$$
$$\nu^\circ(t+1) = \varepsilon_{CL}^\circ(t+1) = y(t+1) - \hat{\theta}_e^T(t)\phi_e(t)$$

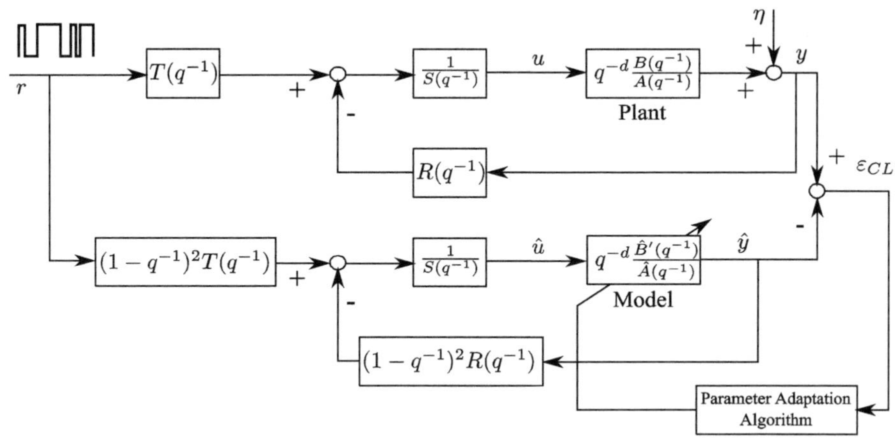

Fig. 8.3 Taking into account the double differentiator behaviour for identification in closed-loop

The analysis in the deterministic case ($C = 1, e = 0$) using the theorem given in Chap. 4 shows that global asymptotic stability is assured without any positive real condition (since the *a posteriori* closed-loop prediction error equation in this case is $\varepsilon_{CL} = [\theta_e - \hat{\theta}_e(t+1)]^T \phi_e(t))$.

Asymptotic unbiased estimates in a stochastic environment can be obtained under the sufficient condition [2, 5] that:

$$H'(z^{-1}) = \frac{1}{C(z^{-1})} - \frac{\lambda_2}{2} \tag{8.42}$$

is strictly positive real (where $\max_t \lambda_2(t) \leq \lambda_2 < 2$).

8.2.4 Taking into Account Known Fixed Parts in the Model

In the context of active vibration control systems, like for identification in open-loop operation, it is wise to take into account that the secondary path has a known double differentiator behaviour. This will require a modification of the controller used in the closed-loop predictor. To take into account the double differentiator behaviour when external excitation is superposed to the input of the controller (at the input of the filter R) one should modify the CLOE configuration as shown in Fig. 8.3.[3]

[3]The external excitation effect is equivalently obtained by filtering the signal through R and adding it to the output of the filter R in the upper part of the Fig. 8.3. Using algorithms from the CLID toolbox, both T and R should be modified according to the Fig. 8.3.

8.2.5 Properties of the Estimated Model

It is very important to asses the properties of the estimated model in the frequency domain. This will allow to know in what frequency region the approximation of the true plant will be best (it is expected that this should be particularly true in the critical regions for design). Nevertheless, the properties of the estimated models will depend on the point where the external excitation is applied. There are several options. When the excitation is superposed to the output of the controller (like in Fig. 8.2a), the properties of the estimated model in the frequency domain (bias distribution) result from ([2]):

$$\hat{\theta}^* = \arg\min_{\hat{\theta}\in\mathscr{D}} \int_{-\pi}^{\pi} |S_{yp}|^2[|G-\hat{G}|^2|\hat{S}_{yp}|^2\phi_{r_u}(\omega)+\phi_\eta(\omega)]\mathrm{d}\omega, \tag{8.43}$$

where $\phi_{r_u}(\omega)$ and $\phi_\eta(\omega)$ are the power spectral densities of the excitation and the measurement noise, respectively. This expression shows that

- The estimation of the plant model parameters is unbiased when G is in the model set[4];
- The bias distribution is not affected by the spectrum of the noise (which is the case when using the (filtered) open-loop identification methods [2]);
- The approximation of the true model is not only weighted by the sensitivity function but is further weighted by the estimated output sensitivity function; and
- Quality of the estimated model is enhanced in the critical region for design.[5]

By contrast the bias distribution in the frequency domain for the open-loop output error is given by:

$$\hat{\theta}^* = \arg\min_{\hat{\theta}\in\mathscr{D}} \int_{-\pi}^{\pi} [|G-\hat{G}|^2|\phi_{r_u}(\omega)+\phi_\eta(\omega)]\mathrm{d}\omega \tag{8.44}$$

As one can see, the basic difference is that in open-loop identification using output error algorithm one has an equal weight for all the frequencies. The comparison between (8.43) and (8.44), explains why identification in closed-loop may provide better models for design.

When the external excitation signal is superposed to the input of the controller, with $T=R$, the asymptotic bias distribution is given by:

$$\hat{\theta}^* = \arg\min_{\hat{\theta}\in\mathscr{D}} \int_{-\pi}^{\pi} |S_{yp}|^2[|G-\hat{G}|^2|\hat{S}_{up}|^2\phi_{r_u}(\omega)+\phi_\eta(\omega)]\mathrm{d}\omega, \tag{8.45}$$

[4]Both true plant model and estimated plant model have the same orders.

[5]Recall that the maximum of the output sensitivity function corresponds to the minimum distance with respect to the Nyquist point.

where $\hat{S}_{up} = -\hat{A}R/\hat{P}$ is the estimated input sensitivity function.

For more details see [2, 6].

8.2.6 Validation of Models Identified in Closed-Loop Operation

As in open-loop identification, it is the model validation that will tell us on one hand if the identified model is acceptable and on the other hand it will allow us to select the best model among the models provided by various identification methods.

The objective of the model validation in closed-loop operation is to find what plant model combined with the current controller provides the best prediction of the behaviour of the closed-loop system. The results of model validation in closed-loop will depend upon the controller used.

Four validation procedures can be defined:

1. Statistical validation tests on the closed-loop output error (uncorrelation test between $\varepsilon_{CL}(t+1)$ and $\hat{y}(t)$).
2. Closeness of the computed and identified poles of the closed-loop system.
3. Closeness of the computed and identified sensitivity functions of the closed-loop system.
4. Time response validation (comparison of time responses of the real closed-loop system and of the closed-loop predictor).

Statistical Validation

The statistical validation follows the same principles as for open-loop model identification; however, in this case one considers the residual prediction error between the output of the plant operating in closed-loop and the output of the closed-loop predictor. An uncorrelation test will be used.

Using the schemes shown in Fig. 8.2b (or Fig. 8.3), where the predictor is given by Eq. (8.12) through (8.15), one computes with the identified values of the parameters:

- The cross correlations between the residual closed-loop output error $\varepsilon_{CL}(t+1)$ and the predicted output $\hat{y}(t)$);
- The covariance of the residual closed-loop output error.

This type of test is motivated on one hand by the fact that uncorrelation between the predicted output and the residual closed-loop prediction error leads to unbiased parameter estimates and on the other hand this uncorrelation implies the uncorrelation between the closed-loop output error and the external excitation. This means that the residual prediction error does not contain any information which depends upon the external excitation and therefore all the correlations between the external excitation and the output of the closed-loop system are captured by the closed-loop predictor.

One computes:

$$R_\varepsilon(0) = \frac{1}{N}\sum_{t=1}^{N}\varepsilon_{CL}^2(t) \tag{8.46}$$

$$R_{\hat{y}}(0) = \frac{1}{N}\sum_{t=1}^{N}\hat{y}^2(t) \tag{8.47}$$

$$R_{\varepsilon\hat{y}}(i) = \frac{1}{N}\sum_{t=1}^{N}\varepsilon_{CL}(t)\hat{y}(t-i) \qquad i = 1, 2, \ldots, n_A \tag{8.48}$$

$$RN_{\varepsilon\hat{y}}(i) = \frac{R_{\varepsilon\hat{y}}(i)}{[R_{\hat{y}}(0)\cdot R_\varepsilon(0)]^{1/2}} \tag{8.49}$$

As a confidence test, one can use the criterion

$$|RN(i)| \leq \frac{2.17}{\sqrt{N}}, \tag{8.50}$$

where N is the number of data (see also Sect. 5.5), as well as the practical criterion $|RN(i)| \leq 0.15$.

In many practical situations, one either has a previous plant model identified in open-loop or several identification algorithms are used on the data collected in closed-loop. Then a comparative validation has to be done and useful comparison indicators are provided by $R_\varepsilon(0)$ and max $|RN_{\varepsilon\hat{y}}|$ for each model (however other comparison criteria can be considered).

Pole Closeness Validation

If the model identified in closed-loop in feedback with the controller used during identification allows constructing a good predictor for the real system, this implies that the poles of the closed-loop system and of the closed-loop predictor are close (assuming that a persistent excitation has been applied for identification). As a consequence, the closeness of the closed-loop predictor poles (which can be computed) and those of the real closed-loop system (which can be identified by an open-loop type identification between the external excitation and the output) will give an indication of the quality of the identified model.

The closeness of the two sets of poles can be judged by a visual examination of the poles chart but a quantification of the closeness can be done (see next).

Sensitivity Functions Closeness Validation

From the same arguments as above it results that if the identified model is good, the sensitivity functions of the closed-loop predictor (which can be computed) are close to the sensitivity functions of the real system (which can be identified by an open-loop type identification between the external excitation and the output).

To some extent the closeness of the sensitivity functions can be assessed by visual inspection. Moreover it is possible to quantify rigorously the distance between two transfer functions by computing the Vinnicombe distance (see Appendix A).

Extensive simulations and a large number of experimental results have shown that the statistical tests and the poles or sensitivity functions closeness give coherent results and allow a clear comparison between several models ([1]).

Time Domain Validation

For the validation in the time domain, the time responses of the closed-loop system and of the closed-loop predictor are compared. Unfortunately in practice it is in general not easy to compare accurately several models using this technique. In fact a good validation by poles or sensitivity functions closeness will imply a good superposition of the time domain responses while the reciprocal is not always true.

8.3 A Real-Time Example: Identification in Closed-Loop and Controller Redesign for the Active Control System Using an Inertial Actuator

A first controller for this system has been designed in Sect. 7.3 using a plant model identified in open-loop and it has been tested in real time. The objective in this section is to illustrate the procedure for identification in closed-loop operation. For carrying the identification in closed-loop operation the controller designed on the basis of the open-loop identified model will be used. The identification experiment is done in the absence of the narrow-band output disturbance.

In this example the objective of the identification in closed-loop will be to heavily weight the differences between the estimated model and the true model in the frequency regions close to the Nyquist point. This is achieved by adding the excitation signal to the control signal (see Sect. 8.2.5).

To take into account the double differentiator behaviour of the secondary path model, the solution indicated in Fig. 8.3 has been used, i.e., the double differentiator has been added to the polynomials $T(q^{-1}) = S(q^{-1})$ and $R(q^{-1})$.

Before running the identification algorithms, the input and output signals have been centred. The orders of the model used for identification in closed-loop operation are the same as those of the model identified in open-loop ($n_B = 23$ and $n_A = 22$). The final order for the secondary path numerator after adding the known fixed part will be $n_B = 25$.

A parameter adaptation algorithm with decreasing gain has been used for all the identification methods. The best results in terms of validation have been obtained using the X-CLOE method. The uncorrelation validation test result for the closed-loop identification is shown in Fig. 8.4. It can be seen that the model is valid. The loss function is 7.7×10^{-5} an it is very small compared to the measured output.

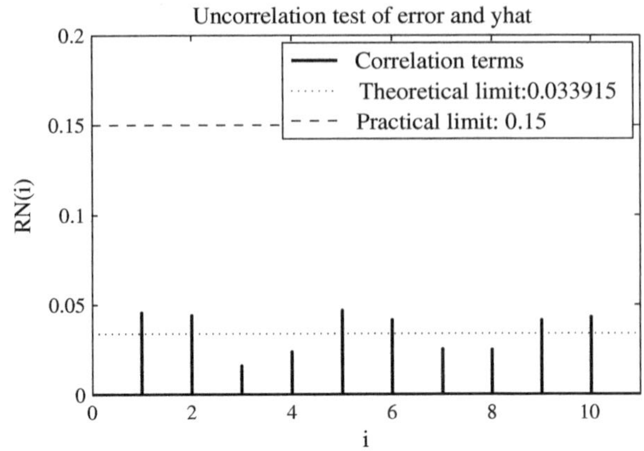

Fig. 8.4 Uncorrelation test for the model identified in closed-loop operation with XCLOE

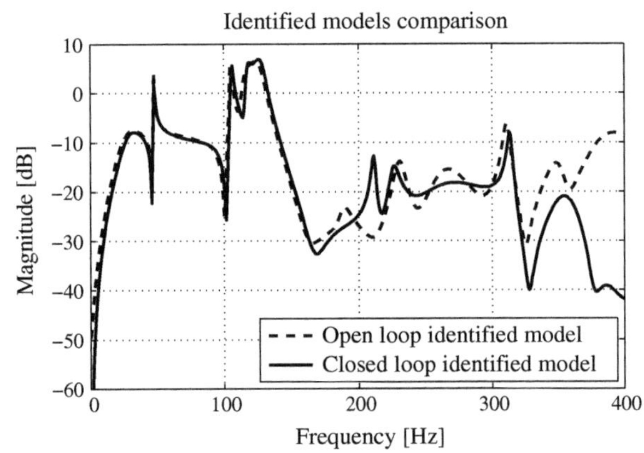

Fig. 8.5 Comparison between the frequency characteristics of the secondary path models identified in open-loop and in closed-loop operation

A comparison with an open-loop identification of the closed-loop has also been accomplished to validate the model. The open-loop model does not pass the uncorrelation test on the closed-loop data; the loss function for the open-loop identified model on the closed-loop data is 1.3×10^{-3} (much higher than for the model identified in closed-loop). One can conclude already that the model identified in closed-loop operation is better than the model identified in open-loop operation. A Bode magnitude comparison between the open-loop identified model from Sect. 6.2 and the closed-loop identified model in the presence of the controller designed in Sect. 7.3 is shown in Fig. 8.5. It can be observed that the two models are very close in the frequency region of interest (50–95 Hz). Note that the differences between the two transfer functions appear in the frequency region over 150 Hz, where the magnitude of the input sensitivity function is very low (see Fig. 7.14) and therefore there will be a little impact on performances.

Further comparison between the two models requires an estimation of the closed-loop transfer function. The closed-loop between excitation and measurement has

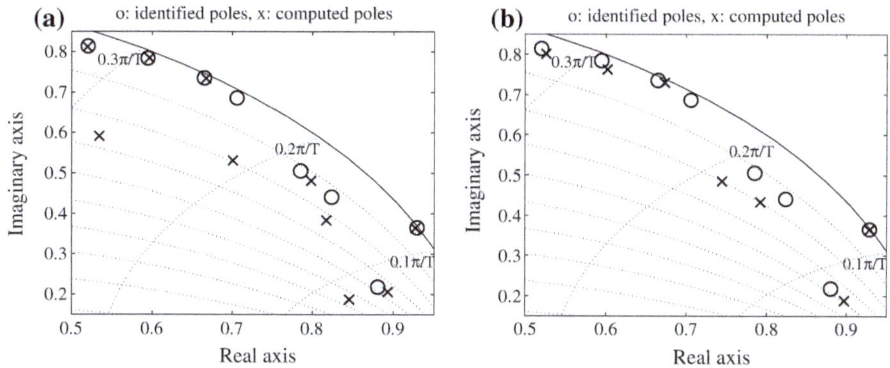

Fig. 8.6 Closed-loop poles closeness comparison using the model identified in closed-loop operation (**a**) and the open-loop identified model (**b**)

been identified as an input/output model using XOLOE method. The identified model of the closed-loop passed the whiteness test (i.e., it is a valid model). This allows to compare the identified closed-loop poles with the calculated closed-loop poles using the two models identified in open and in closed-loop operation. The pole closeness between the poles of the identified closed-loop model and the poles computed with the open-loop identified model and with the model identified in closed-loop are shown in Fig. 8.6. The model identified in closed-loop gives a slightly better result.

Using the same specifications and controller design steps as described in Sect. 7.3, a new controller has been obtained on the basis of the model identified in closed-loop operation. The controller has been tested using the same procedure as before. Time domain results in open-loop and in closed-loop are shown in Fig. 8.7. Frequency

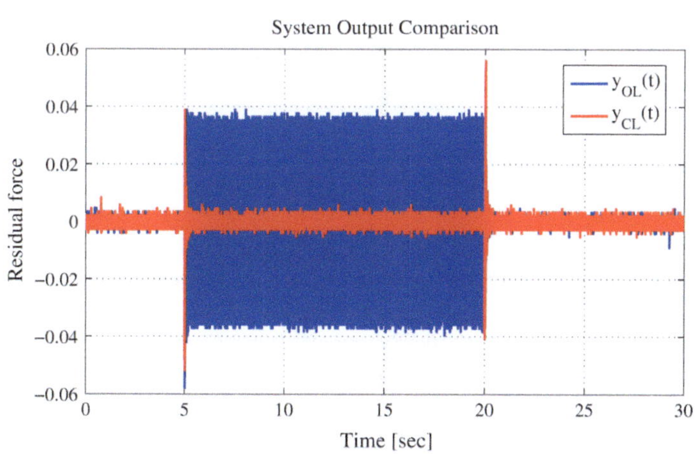

Fig. 8.7 The response results for a 70 Hz disturbance in open-loop and in closed-loop with the redesigned controller

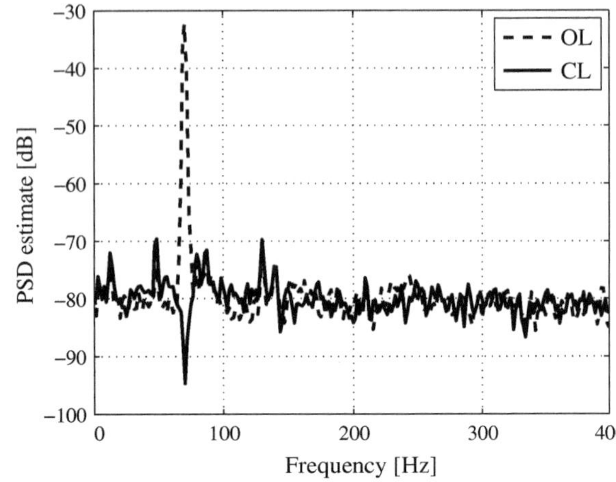

Fig. 8.8 PSD of the residual force in open-loop (*dashed line*) and closed-loop (*solid line*) for 70 Hz disturbance using the redesigned controller

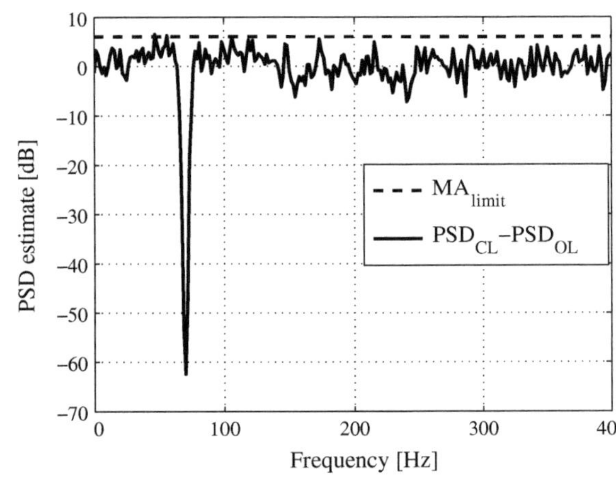

Fig. 8.9 Effective residual attenuation/amplification PSD estimates computed as the difference between the open-loop PSD and the closed-loop PSD using the redesigned controller

domain analysis has also been done and the results are shown in Figs. 8.8 and 8.9. It can be seen that the controller effectively reduces the disturbance and the residual force is at the level of the system's noise. These figures have to be compared with Figs. 7.15, 7.16, and 7.17.

The global attenuation of the newly designed controller is 49 dB, while for the first controller it was 48.4 dB. As for the first controller, the maximum amplification does not exceed the 6 dB limit (dashed line in Figs. 7.17 and 8.9). The disturbance attenuation is of 62.4 dB for the new controller and 63 dB for the initial one. The differences are negligible taking also into account that they were obtained on the basis of a single trial (one realization of a stochastic process).[6] One can conclude that in this particularvadjust case, already the quality of the model identified in open-

[6]It was not possible to conduct a sufficiently large number of measurements for this example.

loop was sufficient to get a good controller. Therefore, the initial controller based on the open-loop identified model will be used in Sect. 9.4 to design a reduced order controller.

8.4 Concluding Remarks

- Plant model identification in closed-loop operation provides efficient tools either for improving open-loop identified models or for redesign and retuning of existing controllers.
- The objective of identification in closed-loop operation is to obtain, for a given controller, a plant model allowing the best description of the behaviour of the closed-loop system.
- Identification in closed-loop is based on the use of an adaptive predictor for the closed-loop which is re-parameterized in terms of the plant model to be identified.
- The estimated parameters minimize asymptotically a criterion in terms of the closed-loop prediction error.
- As for the case of identification in open-loop, there is no single algorithm which gives the best results in all the situations.
- Comparative validation of the identified models is crucial for the selection of the best identified model.
- In addition to the statistical validation test, the pole closeness between the true closed-loop poles (obtained through identification of the closed-loop) and the computed ones, based on the identified model is a very useful validation tool.

8.5 Notes and References

Plant model identification in closed-loop operation has been considered for a long time as a very difficult problem. See [7] for a survey.

It is the work done on the topics of "Identification for Control" and "Iterative Identification and Controller Re-design" which contributed to put the problem of identification in closed-loop operation in an appropriate context. See [8–13] for details.

The original reference for the closed-loop output error is [4]. Further details and comparative evaluations can be found in [1, 2, 6, 14].

References

1. Landau I, Karimi A (1997) Recursive algorithms for identification in closed loop. A unified approach and evaluation. Automatica 33(8):1499–1523
2. Landau ID, Lozano R, M'Saad M, Karimi A (2011) Adaptive control, 2nd edn. Springer, London
3. Landau I, Zito G (2005) Digital control systems - Design, identification and implementation. Springer, London
4. Landau ID, Karimi A (1997) An output error recursive algorithm for unbiased identification in closed loop. Automatica 33(5):933–938. doi:10.1016/S0005-1098(96)00223-3
5. Landau ID, Karimi A (1996) An extended output error recursive algorithm for identification in closed loop. In: Proceedings of the 35th IEEE conference on decision and control, vol 2, pp 1405–1410. doi:10.1109/CDC.1996.572708
6. Karimi A, Landau I (1998) Comparison of the closed loop identification methods in terms of the bias distribution. Syst Control Lett 4:159–167
7. Soderstrom T, Stoica P (1989) System identification. Prentice Hall, New York
8. Gevers M (1993) Towards a joint design of identification and control. In: Trentelman HL, Willems JC (eds) Essays on control: perspectives in the theory and its applications. Birkhäuser, Boston, pp 111–152
9. Gevers, M (1995) Identification for control. In: Prepr. IFAC Symposium ACASP 95. Budapest, Hungary
10. Van den Hof P, Schrama R (1993) An indirect method for transfer function estimation from closed loop data. Automatica 29(6):1751–1770
11. Van den Hof P, Schrama R (1995) Identification and control—closed-loop issues. Automatica 31(12):1751–1770
12. Zang Z, Bitmead RR, Gevers M (1991) Iterative weighted model refinement and control robustness enhancement. In: Proceedings of the 30th IEEE-CDC. Brighton, UK
13. Zang Z, Bitmead RR, Gevers M (1995) Iterative weighted least-squares identification and weighted LQG control design. Automatica 31(11):1577–1594
14. Landau ID, Karimi A (1999) A recursive algorithm for ARMAX model identification in closed loop. IEEE Trans Autom Control 44(4):840–843

Chapter 9
Reduction of the Controller Complexity

9.1 Introduction

The complexity (order of the polynomials R and S) of the controllers designed on the basis of identified models depends upon

- the complexity of the identified model;
- the performance specifications; and
- the robustness constraints.

The controller will have a minimum complexity equal to that of the plant model but as a consequence of performance specifications and robustness constraints, this complexity increases (often up to the double of the size of the model, in terms of number of parameters, and in certain cases even more). In many applications, the necessity of reducing the controller complexity results from constraints on the computational resources in real time (reduction of the number of additions and multiplications).

Therefore one should ask the question: can we obtain a simpler controller with almost the same performance and robustness properties as the nominal one (design based on the plant model)?

Consider the system shown in Fig. 9.1 where the plant model transfer function is given by

$$G(z^{-1}) = \frac{z^{-d}B(z^{-1})}{A(z^{-1})} \tag{9.1}$$

and the nominal controller is given by:

$$K(z^{-1}) = \frac{R(z^{-1})}{S(z^{-1})} \tag{9.2}$$

I.D. Landau et al., *Adaptive and Robust Active Vibration Control*,
Advances in Industrial Control, DOI 10.1007/978-3-319-41450-8_9

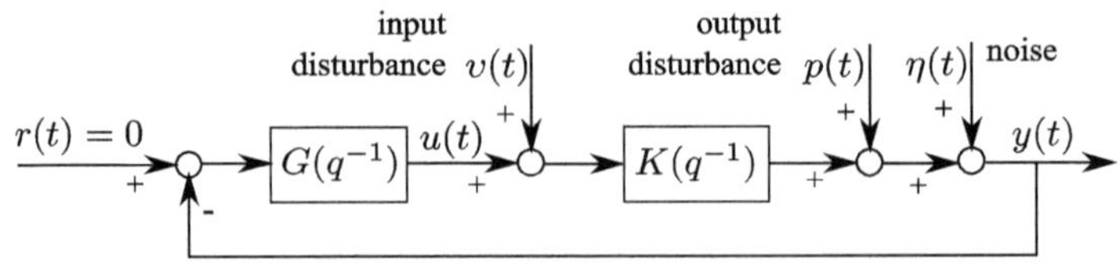

Fig. 9.1 The true closed-loop system

where:

$$R(z^{-1}) = r_0 + r_1 z^{-1} + \cdots + r_{n_R} z^{-n_R} \tag{9.3}$$

$$S(z^{-1}) = 1 + s_1 z^{-1} + \cdots + s_{n_S} z^{-n_S} = 1 + z^{-1} S^*(z^{-1}) \tag{9.4}$$

Different sensitivity functions have been defined in Sect. 7.1 for the system given in Fig. 9.1.

The system given in Fig. 9.1 will be denoted the "true closed-loop system". Throughout this chapter, feedback systems which will use either an estimation of G (denoted $\hat{G}$) or a reduced order estimation of K (denoted $\hat{K}$) will be considered. The corresponding sensitivity functions will be denoted as follows:

- S_{xy}—Sensitivity function of the true closed-loop system (K, G).
- $\hat{S}_{xy}$—Sensitivity function of the nominal simulated closed-loop system (nominal controller K + estimated plant model $\hat{G}$).
- $\hat{\hat{S}}_{xy}$—Sensitivity function of the simulated closed-loop system using a reduced order controller (reduced order controller $\hat{K}$ + estimated plant model $\hat{G}$).

Similar notations are used for $P(z^{-1})$, $\hat{P}(z^{-1})$ when using K and $\hat{G}$, $\hat{\hat{P}}(z^{-1})$ when using $\hat{K}$ and $\hat{G}$.

The specific objective will be to reduce the orders n_R and n_S of controller polynomials R and S.

The basic rule for developing procedures for controller complexity reduction is to search for controllers of reduced orders which preserve the properties of the closed-loop as much as possible. A direct simplification of the controller transfer function by traditional techniques (cancellation of poles and zeros which are close, approximations in the frequency domain, balanced truncation, etc.) without taking into account the properties of the closed-loop leads in general to unsatisfactory results (see [1, 2]).

Two approaches can be considered for the controller complexity reduction

1. **Indirect Approach**

This approach is implemented in three steps

a. Reduction of the complexity of the model used for design, trying to preserve the essential characteristics of the model in the critical frequency regions for design.

b. Design of the controller on the basis of the reduced model.
c. Test of the resulting controller on the nominal model.

2. **Direct Approach**

Search for a reduced order approximation of the nominal controller which preserves the properties of the closed-loop.

The indirect approach has a number of drawbacks

- Does not guarantee the complexity of the resulting controller (since the robustness specifications will be more severe when using reduced order models).
- The errors resulting from model reduction will propagate in the design of the controller.

The direct approach seems the most appropriate for the reduction of the controller complexity since the approximation is done in the last stage of the design and the resulting performance can be easily evaluated. A combination of the two approaches is also possible (see Chap. 10), i.e., the resulting controller obtained by the indirect approach, after it has been tested on the nominal plant model is further reduced through the direct approach.

9.2 Criteria for Direct Controller Reduction

Two criteria can be considered for direct reduction of the controller complexity

- *Closed-loop input matching* (CLIM). In this case, one would like that the control generated in closed-loop by the reduced order controller be as close as possible to the control generated in closed-loop by the nominal controller.
- *Closed-loop output matching* (CLOM). In this case, one would like that the closed-loop output obtained with the reduced order controller be as close as possible to the closed-loop output obtained with the nominal controller.

These two criteria are illustrated in Fig. 9.2, where the nominal controller is denoted by K and is given in (9.2) and the reduced controller is denoted by $\hat{K}$ and is given by

$$\hat{K}(z^{-1}) = \frac{\hat{R}(z^{-1})}{\hat{S}(z^{-1})} \tag{9.5}$$

where:

$$\hat{R}(z^{-1}) = r_0 + r_1 z^{-1} + \cdots + r_{n_R} z^{-n_R} \tag{9.6}$$

$$\hat{S}(z^{-1}) = 1 + s_1 z^{-1} + \cdots + s_{n_S} z^{-n_S} = 1 + z^{-1}\hat{S}^*(z^{-1}) \tag{9.7}$$

The *closed-loop input matching* is equivalent to minimizing the following norm:

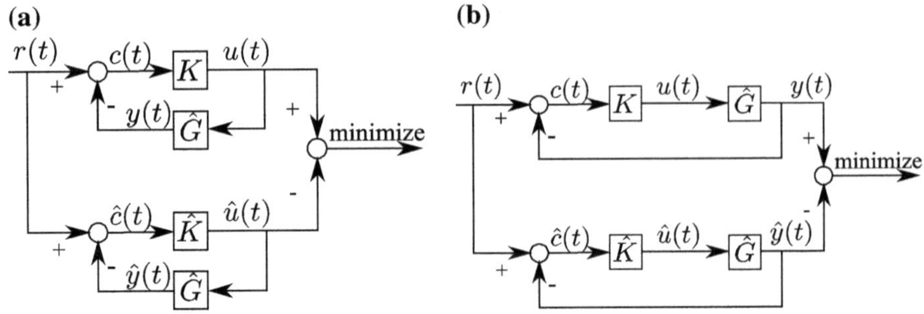

Fig. 9.2 Criteria for controller complexity reduction. **a** Input matching. **b** Output matching

$$\|\hat{S}_{up} - \hat{\hat{S}}_{up}\| = \left\| \frac{K}{1+K\hat{G}} - \frac{\hat{K}}{1+\hat{K}\hat{G}} \right\| \tag{9.8}$$

where $\hat{S}_{up}$ is the input sensitivity function of the nominal simulated closed-loop and $\hat{\hat{S}}_{up}$ is the input sensitivity function when using the reduced order controller. Therefore the optimal reduced order controller will be given by

$$\hat{K}^* = \arg\min_{\hat{K}} \|\hat{S}_{up} - \hat{\hat{S}}_{up}\| = \arg\min_{\hat{K}} \|\hat{S}_{yp}(K - \hat{K})\hat{\hat{S}}_{yp}\| \tag{9.9}$$

As it can be seen, the difference between the two controllers is heavily weighted by the output sensitivity function. The maximum of its modulus corresponds to the critical region for design. Therefore, the reduced order controller will very well approximate the nominal controller in this critical frequency region for design.

If we now consider preservation of performance in tracking using the *closed-loop output matching*, the reduced order controller should minimize the following norm:

$$\|\hat{S}_{yr} - \hat{\hat{S}}_{yr}\| = \left\| \frac{K\hat{G}}{1+K\hat{G}} - \frac{\hat{K}\hat{G}}{1+\hat{K}\hat{G}} \right\| \tag{9.10}$$

To preserve the performance for output disturbance rejection, the reduced order controller should minimize

$$\|\hat{S}_{yp} - \hat{\hat{S}}_{yp}\| = \left\| \frac{1}{1+K\hat{G}} - \frac{1}{1+\hat{K}\hat{G}} \right\| \tag{9.11}$$

Fortunately, these two norms are equal and the reduced order controller can be obtained using the following expression:

$$\hat{K}^* = \arg\min_{\hat{K}} \|\hat{S}_{yp} - \hat{\hat{S}}_{yp}\| = \arg\min_{\hat{K}} \|\hat{S}_{yp}(K - \hat{K})\hat{\hat{S}}_{yv}\| \tag{9.12}$$

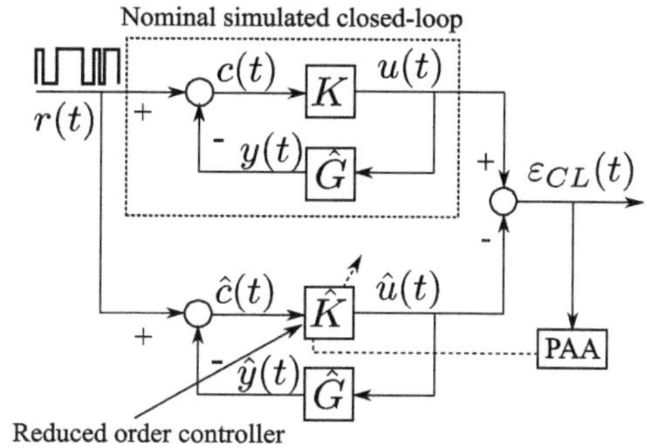

Fig. 9.3 Estimation of reduced order controllers by the closed-loop input matching (CLIM) method. Use of simulated data

Equations (9.9) and (9.12) show that a weighted norm of $K - \hat{K}$ should be minimized.

For closed-loop input matching (Fig. 9.2a) one tries to find a reduced order controller which will minimize the difference between the input sensitivity function of the nominal simulated system and the input sensitivity function of the simulated system using a reduced order controller. This is equivalent to the search for a reduced controller which minimizes the error between the two loops (in the sense of a certain criterion) for a white noise type excitation (like PRBS).

For the tracking of the nominal output (Fig. 9.2b) the principle remains the same, except that in this case one tries to minimize the difference between the nominal complementary sensitivity function (7.8) and the reduced order complementary sensitivity function computed with $\hat{K}$ and $\hat{G}$.

It can be seen immediately that in both cases the problem of finding a reduced order controller can be formulated as an identification in closed-loop (see Chap. 8) where the plant model is replaced by the reduced order controller to be estimated and the controller is replaced by the available estimated model of the plant (dual problem).

The reduction procedures and the validation techniques for reduced order controllers to be presented next are available in the MATLAB® toolbox REDUC® [3] (to be downloaded from the book website) or in the stand alone software iReg which includes a module for controller complexity reduction.[1]

9.3 Estimation of Reduced Order Controllers by Identification in Closed-Loop

9.3.1 Closed-Loop Input Matching (CLIM)

The principle of closed-loop input matching approach is illustrated in Fig. 9.3.

[1] See the website: http://tudor-bogdan.airimitoaie.name/ireg.html.

The upper part represents the simulated nominal closed-loop system. It is made up of the nominal controller (K) and the best identified plant model ($\hat{G}$). This model should assure the best closeness behaviour of the true closed-loop system and the nominal simulated one. Identification of this plant model in closed-loop can be considered if the nominal controller can be implemented.

The lower part is made up of the estimated reduced order controller ($\hat{K}$) in feedback connection with the plant model ($\hat{G}$) used in the nominal simulated system. The parameter adaptation algorithm (PAA) will try to find the best reduced order controller which will minimize the closed-loop input error. The closed-loop input error is the difference between the plant input generated by the nominal simulated closed-loop system and the plant input generated by the simulated closed-loop using the reduced order controller.

The output of the nominal controller is given by

$$u(t+1) = -S^*(q^{-1})u(t) + R(q^{-1})c(t+1) = \theta^T \psi(t) \tag{9.13}$$

where

$$c(t+1) = r(t+1) - y(t+1) \tag{9.14}$$

$$y(t+1) = -\hat{A}^* y(t) + \hat{B}^* u(t-d) \tag{9.15}$$

$$\psi^T(t) = [-u(t), \ldots, -u(t-n_S+1), c(t+1), \ldots, c(t-n_R+1)] \tag{9.16}$$

$$\theta^T = [s_1, \ldots, s_{n_S}, r_0, \ldots, r_{n_R}] \tag{9.17}$$

To implement and analyze the algorithm, we need respectively the *a priori* (based on $\hat{\theta}(t)$) and the *a posteriori* (based on $\hat{\theta}(t+1)$) predicted outputs of the estimated reduced order controller (of orders $n_{\hat{S}}$ and $n_{\hat{R}}$) which are given by (see the lower part of Fig. 9.3).

a priori:

$$\begin{aligned}\hat{u}^\circ(t+1) &= \hat{u}(t+1|\hat{\theta}(t)) = -\hat{S}^*(t, q^{-1})\,\hat{u}(t) + \hat{R}(t, q^{-1})\,\hat{c}(t+1)\\ &= \hat{\theta}^T(t)\phi(t)\end{aligned} \tag{9.18}$$

a posteriori:

$$\hat{u}(t+1) = \hat{\theta}^T(t+1)\,\phi(t) \tag{9.19}$$

where

$$\hat{\theta}^T(t) = [\hat{s}_1(t), \ldots, \hat{s}_{n_{\hat{S}}}(t), \hat{r}_0(t), \ldots, \hat{r}_{n_{\hat{R}}}(t)] \tag{9.20}$$

$$\phi^T(t) = [-\hat{u}(t), \ldots, -\hat{u}(t-n_{\hat{S}}+1), \hat{c}(t+1), \ldots, \hat{c}(t-n_{\hat{R}}+1)] \tag{9.21}$$

$$\hat{c}(t+1) = r(t+1) - \hat{y}(t+1) = r(t+1) + \hat{A}^*\hat{y}(t) - \hat{B}^*\hat{u}(t-d) \tag{9.22}$$

The closed-loop input error is given by

a priori:

$$\varepsilon^{\circ}_{CL}(t+1) = u(t+1) - \hat{u}^{\circ}(t+1) \tag{9.23}$$

a posteriori:

$$\varepsilon_{CL}(t+1) = u(t+1) - \hat{u}(t+1) \tag{9.24}$$

The equation governing the *a posteriori* prediction error becomes (see [4, 5] for details)

$$\varepsilon_{CL}(t+1) = \frac{\hat{A}}{P}[\theta - \hat{\theta}(t+1)]^T \phi(t) \tag{9.25}$$

and the parameter adaptation algorithm will be given by

$$\hat{\theta}(t+1) = \hat{\theta}(t) + F(t)\Phi(t)\varepsilon_{CL}(t+1) \tag{9.26}$$

$$F^{-1}(t+1) = \lambda_1(t)F^{-1}(t) + \lambda_2(t)\Phi(t)\Phi^T(t) \tag{9.27}$$

$$0 < \lambda_1(t) \leq 1;\ 0 \leq \lambda_2(t) < 2; \quad F(0) > 0$$

$$\varepsilon_{CL}(t+1) = \frac{\varepsilon^{\circ}_{CL}(t+1)}{1 + \Phi^T(t)F(t)\Phi(t)} = \frac{u(t+1) - \hat{u}^{\circ}(t+1)}{1 + \Phi^T(t)F(t)\Phi(t)} \tag{9.28}$$

As we can see from (9.28), the *a posteriori* closed-loop input error $\varepsilon_{CL}(t+1)$ can be expressed in terms of the *a priori* (measurable) closed-loop input error $\varepsilon^{\circ}_{CL}(t+1)$. Therefore, the right-hand side of (9.26) will depend only on measurable quantities at $t+1$.

Specific algorithms will be obtained by an appropriate choice of the observation vector $\Phi(t)$ as follows:

- CLIM: $\Phi(t) = \phi(t)$
- F-CLIM: $\Phi(t) = \frac{\hat{A}(q^{-1})}{\hat{P}(q^{-1})}\phi(t)$

where

$$\hat{P}(q^{-1}) = \hat{A}(q^{-1})S(q^{-1}) + q^{-d}\hat{B}(q^{-1})R(q^{-1}). \tag{9.29}$$

The introduction of the filtering of ϕ is motivated by the elimination of a positive realness sufficient condition for stability and convergence which, in the case of the CLIM algorithm, depends on $\hat{A}/\hat{P}$. A detailed analysis of the properties of these algorithms can be found in [5].

The properties of the estimated controller in the frequency domain results from the following expression (bias distribution) [5]:

$$\hat{\theta}^* = \arg\min_{\hat{\theta}\in\mathscr{D}} \int_{-\pi}^{\pi} |\hat{S}_{yp}|^2 \left[|K - \hat{K}|^2 |\hat{\hat{S}}_{yp}|^2 \phi_r(\omega) + \phi_\eta(\omega)\right] \mathrm{d}\omega \tag{9.30}$$

where $\phi_r(\omega)$ is the excitation spectrum and $\phi_\eta(\omega)$ is the measurement noise spectrum (it does not have effect upon the minimization of $|K - \hat{K}|$).

Estimation of reduced order controllers is also possible using real-time data (if the prototype of the nominal controller can be implemented on the real system) [5].

9.3.2 Closed-Loop Output Matching (CLOM)

The principle of this method is illustrated in Fig. 9.4. Despite that, the point where the external excitation is applied and the output variable is different with respect to Fig. 9.2b, the transfer function between $r(t)$ and $u(t)$ in Fig. 9.4 is the same as the transfer function between $r(t)$ and $y(t)$ in Fig. 9.2b. This means that in the absence of disturbances (it is the case in simulation) $u(t)$ generated by the upper part of the scheme given in Fig. 9.4 is equal to y(t) generated in Fig. 9.2b. This allows one to use for closed-loop output matching the CLIM (or F-CLIM) algorithm. For effective implementation of the algorithm, the only changes occur in Eqs. (9.13) and (9.18), where $c(t)$ is replaced by:

$$x(t) = \hat{G}(r(t) - u(t)) \tag{9.31}$$

and $\hat{c}(t)$ is replaced by:

$$\hat{x}(t) = \hat{G}(r(t) - \hat{u}(t)) \tag{9.32}$$

One should note that the order of the blocks in the upper part of Fig. 9.4 can be interchanged (like the upper part of Fig. 9.2b) without affecting the operation of the algorithm.

9.3.3 Taking into Account the Fixed Parts of the Nominal Controller

It is often required that the reduced order controller contains some of the fixed filters incorporated in the nominal controller (for example, model of the disturbance, opening of the loop at $0.5 f_S$ or at other frequency). In order to do this, one first

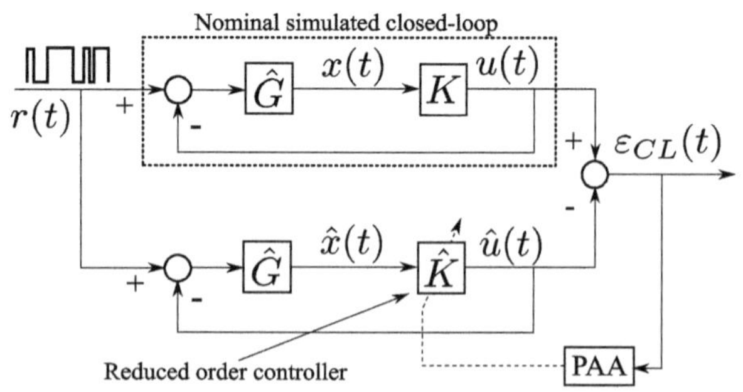

Fig. 9.4 Estimation of reduced order controllers by the closed-loop output matching (CLOM) method. Use of simulated data

factorizes the nominal controller under the form $K = K_F K'$, where K_F represents all the fixed parts that one would like to be also incorporated in the reduced order controller. The reduced order controller is factorized as $\hat{K} = K_F \hat{K}'$.

One replaces in the CLIM algorithm the input $\hat{c}$ of the controller $\hat{K}$ by the input to the controller $\hat{K}'$, denoted $\hat{c}'$, where $\hat{c}'$ is given by

$$\hat{c}'(t) = K_F(q^{-1})\hat{c}(t) \tag{9.33}$$

and in $\Phi(t)$, $\hat{c}(t)$ is replaced by $\hat{c}'(t)$. In the CLOM algorithm, one replaces $\hat{x}$ by $\hat{x}'$ given by

$$\hat{x}'(t) = K_F(q^{-1})\hat{G}(q^{-1})(r(t) - \hat{u}(t)). \tag{9.34}$$

9.3.3.1 Validation of Reduced Order Controllers

Once a reduced order controller has been estimated, it should be validated before considering its implementation on the real system.

It is assumed that the nominal controller stabilizes the nominal plant model (used for controller reduction). One implicitly assumes that the model uncertainties have been taken into account in the design of the nominal controller. The reduced order controller should satisfy the following conditions:

- It stabilizes the nominal plant model.
- The reduced sensitivity functions (computed with the reduced order controller) are close to the nominal sensitivity functions in the critical frequency regions for performance and robustness. In particular, the output and input sensitivity functions should be examined.
- The *generalized stability margin* (see Appendix A) of the system using the reduced order controller should be close to the *generalized stability margin* of the nominal closed-loop. This condition is expressed as

$$|b(K, \hat{G}) - b(\hat{K}\hat{G})| < \varepsilon; \varepsilon > 0 \tag{9.35}$$

where $b(K, \hat{G})$ and $b(\hat{K}\hat{G})$ are the *generalized stability margins* corresponding to the nominal controller and to the reduced order controller respectively and ε is a small positive number. The closeness of the two stability margins allows maintaining the robustness properties of the initial design.

The proximity of the nominal and reduced sensitivity functions can be judged by visual examination of their frequency characteristics. There is however the possibility to make a numerical evaluation of this proximity by computing the *Vinnicombe distance* (ν gap) between these transfer functions (see Appendix A). The *Vinnicombe distance* allows with one number (between 0 and 1), to make a first evaluation of the proximity of the reduced and nominal sensitivity functions.

9.4 Real-Time Example: Reduction of Controller Complexity

In Sect. 7.3, a controller based on the open-loop identified model has been designed for the active vibration control system using an inertial actuator (see Sect. 2.2) and tested experimentally. It was shown in Sect. 8.3 that the controller designed on the basis of the model identified in open-loop provides similar performance to that of the controller designed on the basis of the model identified in closed-loop. Therefore in this section, the reduction of the complexity of the controller designed on the basis of the model identified in open-loop (which achieves the specifications) will be considered.

The criterion given in Eq. (9.8) will be considered, which corresponds to CLIM with external excitation added to the input of the controller. The model of the plant identified in closed-loop operation has been used. The excitation used was a PRBS with the following characteristics: N = 11 (number of cells) and p = 2 (clock frequency divider). The fixed parts of the controller have been preserved (internal model of the disturbance, opening the loop at $0.5 f_S$ and at 0 Hz).

Table 9.1 presents a summary of the controller order reduction results for various values of n_R and n_S. The first column represents the controller number (the controller with number 00 represents the initial nominal controller). The orders of the reduced controllers are indicated in columns n_R and n_S. The next column gives the Vinnicombe gap (Vg) between the initial controller and the reduced order controller. Similarly, the Vinnicombe gaps for the input and output sensitivity functions are also given in columns 5 and 6, respectively. A Vg of 0 indicates perfect matching while a Vg of 1 indicates very important differences between the two transfer functions. The *generalized stability margin* (see Appendix A) is given in column 7. For robustness reasons, it should be close to the value obtained for the nominal controller. The maximum of the output sensitivity function and the frequency in Hz for which it is obtained are given in columns 8 and 9, respectively. Finally, the stability of the closed-loop is indicated in the last column (1 represents a stable closed-loop, 0—unstable).

Only the first 12 reduced order controllers are shown in the table.[2] For experimental evaluation, controller 11 has been considered ($n_R = 19, n_S = 22$).

The output and input sensitivity functions obtained with the nominal and reduced order controllers are shown in Figs. 9.5 and 9.6, respectively. As it can be observed, the differences are very small within the frequency region of interest (except for the input sensitivity function at the 50 Hz—but this does not affect nor the robustness nor the performance). In Fig. 9.7 the transfer functions of the two controllers are shown.

It is important to remind that the comparison of the Bode characteristics of the two controllers does not guarantees that the reduced order controller stabilizes the system or that it assures good performances. It is the comparison of the sensitivity functions and the stability test which gives the right answers.

---

[2]These results have been obtained using the software iREG. Similar results are obtained with the *compcon.m* function from the toolbox REDUC.

Table 9.1 Summary of the controller order reduction results

No.	n_R	n_S	$Vg(\frac{R}{S})$	$Vg(S_{up})$	$Vg(S_{yp})$	St-margin	$\max(S_{yp})$	$[f_{\max}]$	stable
00	29	32	0	0	0	0.3297	3.92	[60.0147]	1
01	29	32	0	0	0	0.3297	3.92	[60.0147]	1
02	28	31	0.001	0.003	0	0.3297	3.92	[60.0147]	1
03	27	30	0.0101	0.0284	0.0031	0.3296	3.8742	[60.0147]	1
04	26	29	0.0095	0.0282	0.0035	0.3306	3.8958	[60.0147]	1
05	25	28	0.0096	0.0327	0.004	0.3286	3.8958	[60.0147]	1
06	24	27	0.0103	1	0.0017	0.3263	3.9329	[60.0147]	1
07	23	26	0.0154	0.0498	0.0041	0.3213	3.9459	[60.0147]	1
08	22	25	0.0153	0.0545	0.0048	0.3232	3.9548	[60.0147]	1
09	21	24	0.0159	0.0514	0.0045	0.3232	3.9406	[60.0147]	1
10	20	23	0.0253	0.0972	0.0109	0.3268	3.9676	[60.0147]	1
11	19	22	0.0604	0.2645	0.0328	0.3089	3.9345	[59.3959]	1
12	18	21	1	1	1	0	3.7477	[59.3959]	0

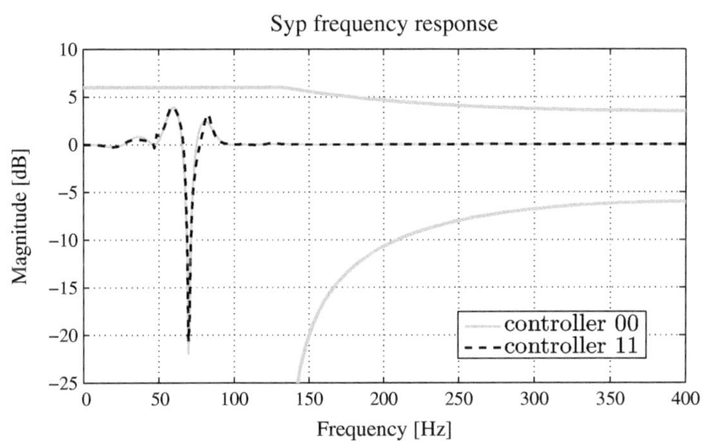

Fig. 9.5 Output sensitivity functions for initial and reduced order controllers

Finally, the controller has been tested in real time in the presence of a 70 Hz sinusoidal disturbance. Time domain results in open and in closed-loop operation are shown in Fig. 9.8. The difference between the two power spectral densities for open-loop and closed-loop is shown in Fig. 9.9.[3]

For the reduced order controller, the following results have been obtained: (1) the global attenuation is 48.2 dB (instead of 48.4 dB for the nominal controller), the

[3]Figures 9.8 and 9.9 should be compared with Figs. 7.15 and 7.17.

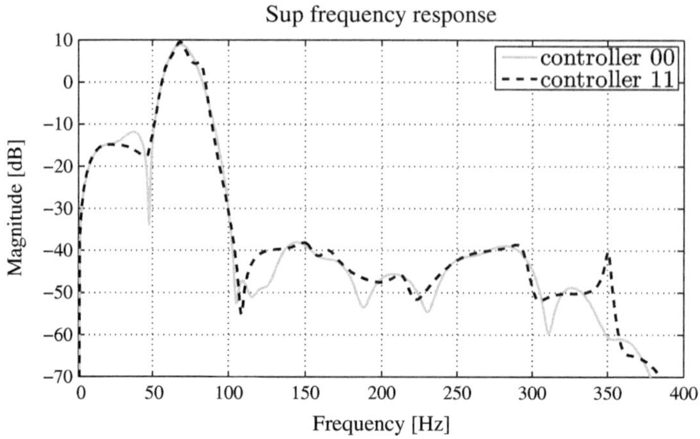

Fig. 9.6 Input sensitivity functions for initial and reduced order controllers

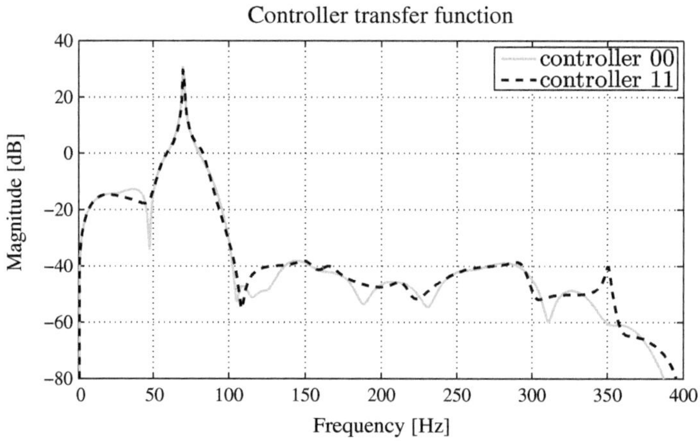

Fig. 9.7 Controller transfer function comparison between initial and reduced order controller

disturbance attenuation is 56.4 dB (instead of 62.4 dB but still much more than the required attenuation) and the maximum amplification is 7.5 dB (instead of maximum 6 dB specified). A small reduction in performance with respect to the initial nonreduced controller is observed but the number of parameters has been reduced from 62 to 44. These results presented above have been obtained using a single trial.

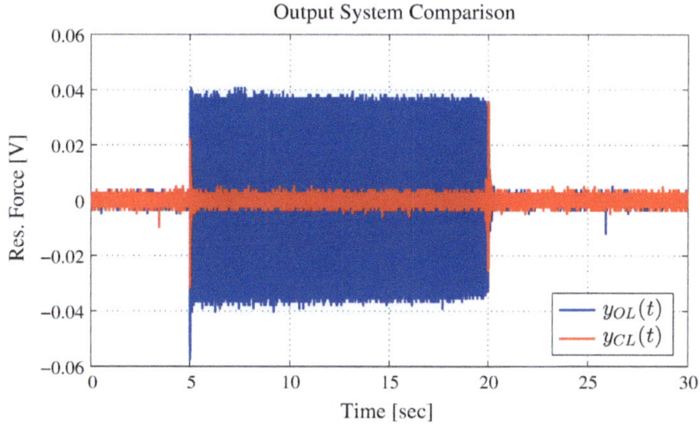

Fig. 9.8 Time response results for a 70 Hz sinusoidal disturbance in open- and in closed-loop operation using the reduced order controller

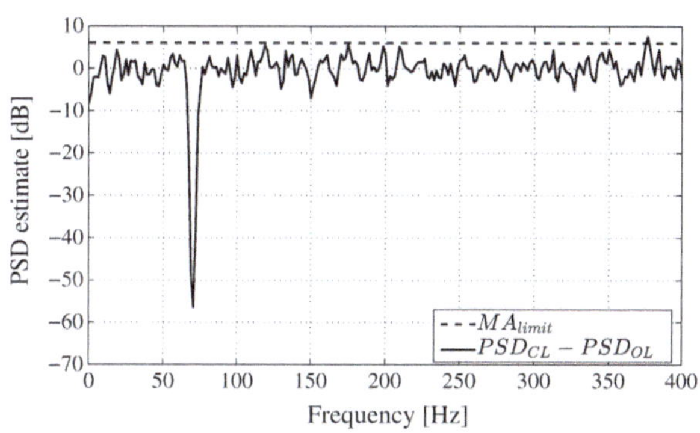

Fig. 9.9 Effective residual attenuation/amplification PSD estimates computed as the difference between the open-loop PSD and the closed-loop PSD (reduced order controller)

9.5 Concluding Remarks

- The objective of controller reduction is to find a controller of reduced complexity such that the characteristics of the closed-loop using the reduced order controller are as close as possible to the characteristics of the closed-loop using the nominal controller.
- Two specific objectives have been considered
  - closed-loop input matching (CLIM); and
  - closed-loop output matching (CLOM).
- The CLOM (CLIM) objective corresponds to the estimation of a reduced order controller such that the error between the output (the control input) of the closed-loop using the reduced order controller and the output (the control input) of the

closed-loop using the nominal controller be minimized in the sense of a certain criterion.
- Controller reduction can be viewed as a dual problem with respect to plant model identification in closed-loop (similar algorithms will be used).
- The reduced order controllers should be validated before their effective use.
- Techniques for validation of the reduced order controllers have been provided in this chapter.

9.6 Notes and References

The problem of controller reduction is clearly presented in [1, 2]. See also [6].

The basic references for the algorithms discussed in this chapter (analysis and evaluation) are [5, 7, 8]. A unified view of identification in closed-loop and controller reduction can be found in [8].

References

1. Anderson B, Liu Y (1989) Controller reduction: concepts and approaches. IEEE Trans Autom Control 34(8):802–812
2. Anderson B (1993) Controller design: moving from theory to practice. IEEE Control Mag 13(4):16–25
3. Adaptech (1999) Adaptech, 4 rue de la Tour de l'Eau, St. Martin dHères, France: REDUC® Controller order reduction by closed-loop identification (Toolbox for MATLAB®)
4. Landau ID, Karimi A (1997) An output error recursive algorithm for unbiased identification in closed loop. Automatica 33(5):933–938. doi:10.1016/S0005-1098(96)00223-3
5. Landau I, Karimi A, Constantinescu A (2001) Direct controller order reduction by identification in closed loop. Automatica 37:1689–1702
6. Zhou K, Doyle J (1998) Essentials of robust control. Prentice-Hall International, Upper Saddle River
7. Constantinescu A, Landau I (2003) Controller order reduction by identification in closed-loop applied to a benchmark problem. Eur J Control 9(1):84–99. doi:10.3166/ejc.9.84-99
8. Landau I, Karimi A (2002) A unified approach to model estimation and controller reduction (duality and coherence). Eur J Control 8(6):561–572

Part III
Active Damping

Chapter 10
Active Damping

10.1 Introduction

As indicated in the introduction of the book, Sect. 1.3, passive dampers despite that they provide a good attenuation over a wide band of frequencies, they always have a significant resonance peak at a certain frequency within the frequency range of operation. To correct this situation an active vibration isolation (control) has to be considered. The test bench described in Sect. 2.1, belongs to this category. Such a system has a primary path through which the disturbances are attenuated in certain frequency ranges and amplified around the resonance of the system. The secondary path is expected to correct the behaviour of the primary path in the frequency region where the primary path shows a significant resonance (amplification of the vibrations in this zone) through the appropriate use of feedback control. The use of the feedback should attenuate the effect of the resonance of the primary path without deteriorating the attenuation provided by the primary path at other frequencies. This means that the "water bed" effect due to the Bode integral should be carefully managed by shaping the sensitivity functions. Recall also that active damping consists of damping a resonance mode without changing its frequency.

The methodology of designing active damping systems will be illustrated by considering the active suspension described in Sect. 2.1.

The first step of the design consists of defining the control specifications. Roughly, the control objective is illustrated in Fig. 10.1 where the PSD (power spectral density) of the residual force is represented (thin line). We would like to attenuate the resonance but it is the meantime that the tolerated amplification at other frequencies with respect to the open-loop characteristics should be very low.[1] The desired template for the PSD corresponds to the curve in thick grey line is shown in Fig. 10.1. The final objective of the design will be to find the lowest complexity controller which allows matching the performance specifications.

[1]As a consequence of the Bode integral, the level of attenuation imposed is related to the level of tolerated amplification at other frequencies.

I.D. Landau et al., *Adaptive and Robust Active Vibration Control*,
Advances in Industrial Control, DOI 10.1007/978-3-319-41450-8_10

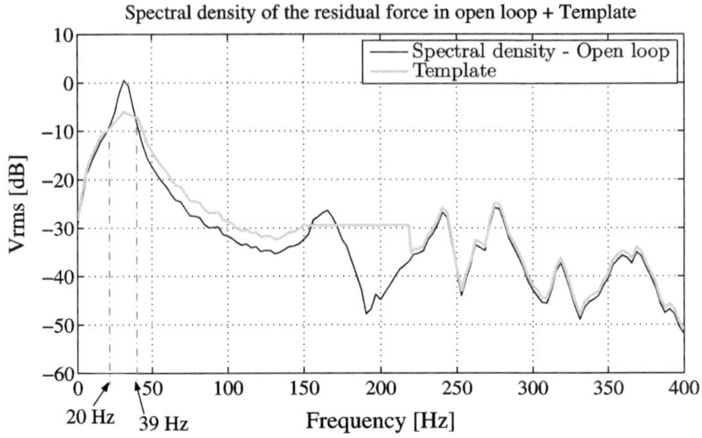

Fig. 10.1 Template imposed on the spectral density of the residual force

Once the performance specifications are formulated, the methodology of design is illustrated in Fig. 10.2. It comprises a number of steps:

- Open-loop identification of the secondary path (one needs a model of the secondary path for controller design).
- Design of a robust controller allowing to match the performance specifications (the design uses the model identified in open-loop operation).
- Implementation and test.
- Identification of the secondary path model in closed-loop operation (an improved model is expected).
- Redesign (retuning) of the controller based on the model identified in closed-loop operation.
- Implementation and validation of the new controller.
- Controller order reduction preserving the stability and performance of the system.
- Implementation and validation of the reduced order controller.

It may happen in practice that one stops after the test of the controller designed on the basis of the model of the secondary path identified in open-loop operation; however, once the implementation of the controller is done it is easy to do an identification in closed loop and the procedure can go further. The complexity controller reduction may not be necessary in some cases if there are no constraints on the computer power or on the cost of the control.

10.2 Performance Specifications

In active damping the desired performances are specified in the frequency domain. A template for the expected power spectral density (PSD) of the residual force or acceleration has to be defined. For the active suspension described in Sect. 2.1, the desired template is shown in Fig. 10.1 and the details are given below:

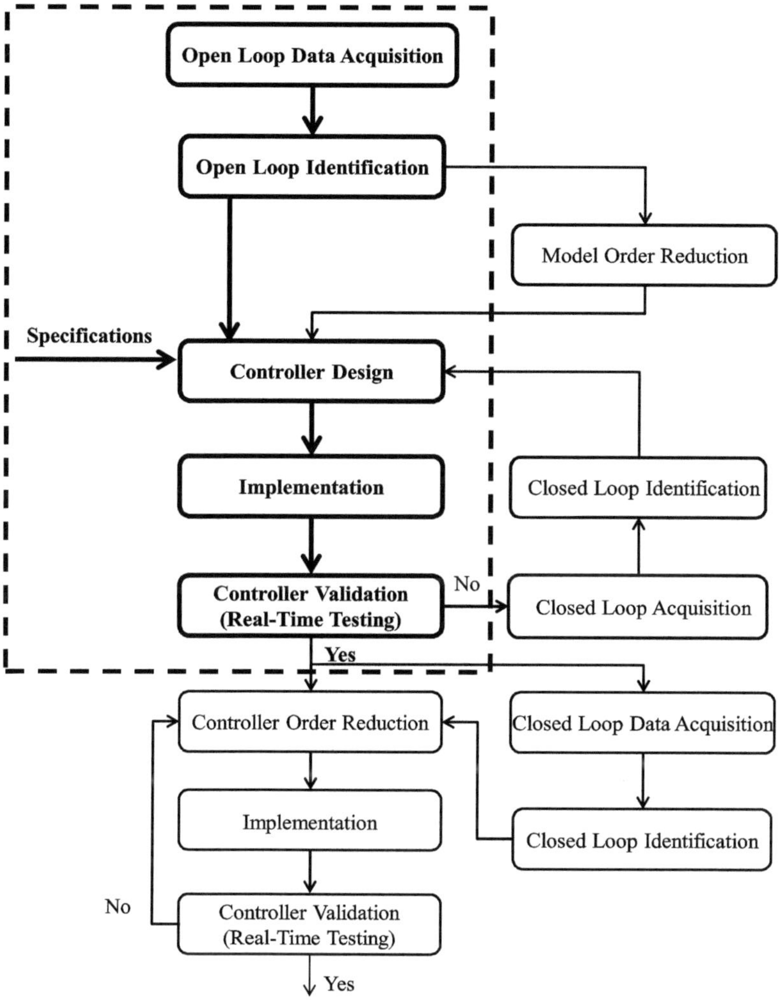

Fig. 10.2 Design methodology

- for frequencies below 20 Hz, maximum amplification with respect to the open loop: 1 dB;
- at 20 Hz, 0 dB amplification;
- at 31.25 Hz (the resonance) an attenuation of at least 6.6 dB;
- at 39 Hz, maximum 0 dB amplification;
- between 39 and 150 Hz maximum 3 dB amplification with respect to the open-loop PSD;
- between 150 and 220 Hz amplification/attenuation below −30 dB with respect to the value of the open-loop PSD at the resonance; and

- from 220 Hz above, maximum amplification of 1 dB with respect to the open-loop PSD.

In addition, as for any feedback control systems, robust specifications should be considered:

- modulus margin ≥ -6 dB;
- delay margin ≥ 1.25 ms (one sampling period);
- $S_{up} < 10$ dB, between 0 and 35 Hz; $S_{up} < 0$ dB, between 40 and 150 Hz; $S_{up} < -20$ dB, between 150 and 220 Hz and < -30 dB above 220 Hz; and
- opening the loop at $0.5\ f_s$.

The reduction of the magnitude of S_{up} is related to the robustness with respect to additive uncertainties and the fact that the system has low gain in high frequencies (robustness requires low level control action at the frequencies where the system has no gain—see Sect. 7.2). Opening the loop at $0.5\ f_s$ will lower drastically the gain of the controller at high frequencies close to $0.5\ f_s$.

One of the steps in the design procedure is to transform the objectives shown in Fig. 10.1 and detailed above in specifications for the design of the feedback system. The active damping can be interpreted as an additional attenuation/amplification of the disturbance (vibration) acting upon the system. In other terms the difference between the PSD of the residual force in open-loop operation and the desired PSD will give the desired attenuation and the tolerated amplification for the feedback loop around the secondary path. The attenuation/amplification introduced by a feedback system is characterized by the frequency domain behaviour of the output sensitivity function S_{yp}. Therefore the difference between the open-loop PSD of the residual acceleration (force) and the desired PSD will generate a *desired template* for the modulus of the output sensitivity function to be achieved. Figure 10.3 shows the

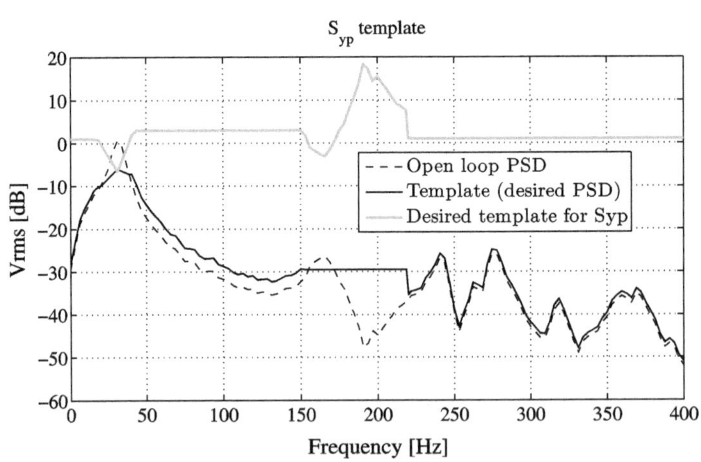

Fig. 10.3 Desired template for the output sensitivity function S_{yp} (without the robustness constraints)

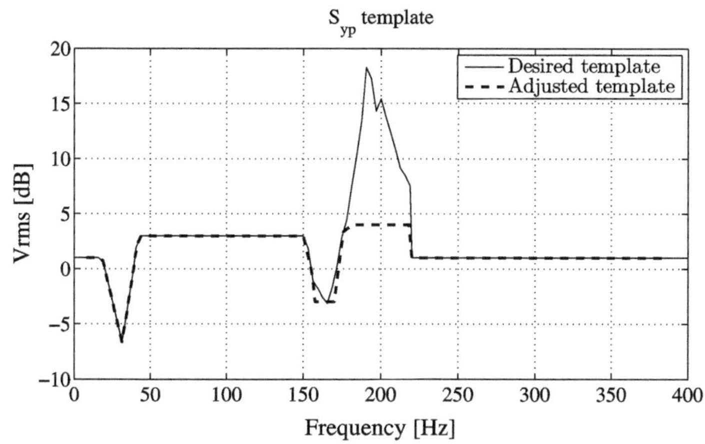

Fig. 10.4 Desired template for the output sensitivity function S_{yp} and adjusted template taking into account the robustness constraints

open-loop PSD, the desired PSD when active damping operates and their difference which constitutes a first template for the desired output sensitivity function.

Nevertheless, this template has to take into account also the robustness constraints imposed in terms of modulus margin and delay margin. Modulus margin imposes a maximum of 6 dB and this maximum decreases in high frequencies as a consequence of the constraints on the delay margin. Figure 10.4 shows the desired template as well as the adjusted one which takes into account the modulus and the delay margins. Figure 10.5 shows the template for shaping the input sensitivity function resulting from the specifications defined earlier (nominal template).

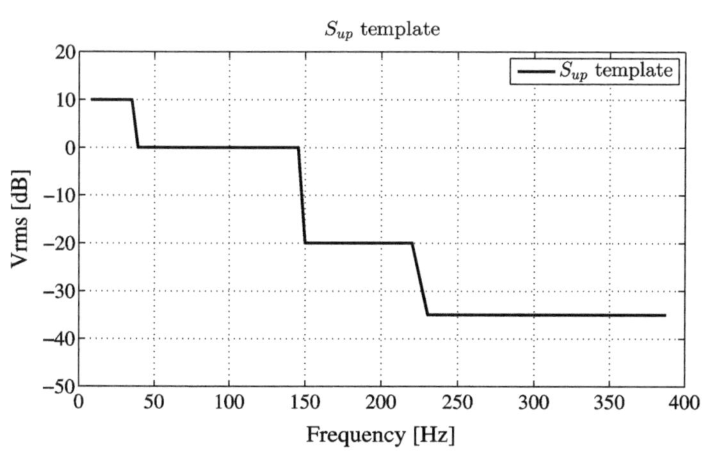

Fig. 10.5 Template for the input sensitivity function S_{up}

10.3 Controller Design by Shaping the Sensitivity Functions Using Convex Optimization

The convex optimization procedure for controller design has been presented in Sect. 7.4. Since the objective is also to obtain a low complexity controller, a first step which was considered in this approach was to use a reduced order secondary path model taking into account that according to the control objective, the control will not have to be effective in high frequencies. One of the most commonly used and efficient methods for model reduction is balancing. Because in the case of the active suspension we are interested in specific frequency intervals, the approach considered for the model reduction is the frequency-weighted balancing method which is suitable when a certain frequency range is of interest. Given the nominal full order model G and the input and output weighting matrices W_i and W_o, the objective is to find a stable and minimum-phase lower order model G_r such that the weighted error

$$\| W_o(G - G_r)W_i \|_\infty \tag{10.1}$$

is as small as possible.

The identified model of the secondary path has been presented in Sect. 6.1.1. A reduced order model with $n_A = 8$, $n_B = 11$, $d = 0$ has been obtained by using the "balanced truncation" technique in which the low frequencies have been appropriately weighted. The parameters of the reduced order model are given in Table 10.1. The frequency characteristics of the nominal and reduced order models are shown in Fig. 10.6.

Nevertheless, once the design is done on the reduced order model, the resulting controller has to be tested on the full order model before implementation. After a

Table 10.1 Parameters of the reduced order model

Coeff.	A	Coeff.	B
a_0	1.0000	b_0	0.0000
a_1	−2.1350	b_1	0.1650
a_2	2.1584	b_2	−1.0776
a_3	−2.2888	b_3	3.6137
a_4	2.2041	b_4	−8.1978
a_5	−1.8433	b_5	15.4346
a_6	1.4035	b_6	−19.4427
a_7	−0.2795	b_7	14.2604
a_8	−0.2057	b_8	−10.8390
a_9	–	b_9	11.9027
a_{10}	–	b_{10}	−7.2010
a_{11}	–	b_{11}	1.3816

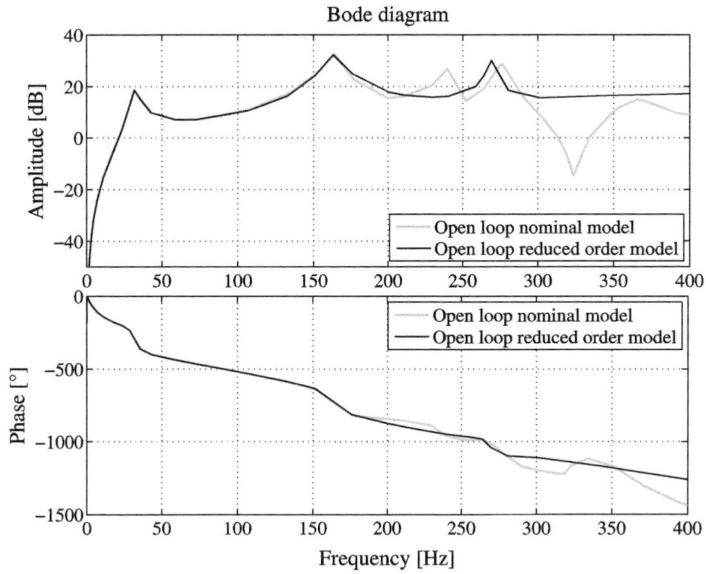

Fig. 10.6 Bode diagram (amplitude and phase) of the open-loop nominal and reduced order models

trial it was found that the basic templates have to be modified in certain frequency regions in order that the controller designed on the reduced order model matches the original templates when used with the nominal full order model.

For initializing the optimization procedure for controller design, a pair of poles at the resonance frequency $f = 31.939$ Hz with a damping $\xi = 0.8$, and a fixed real pole corresponding to the lowest frequency pole of the system (located at the intersection of the 5.73 Hz curve with the real axis) have been assigned. The region of optimization for the poles has been considered to be a circle with a radius 0.99. A fixed part in the controller $H_R = 1 + q^{-1}$ is introduced in order to open the loop at 0.5 f_s.

For convenience, the controller designed will be denoted OLBC (**O**pen **L**oop **B**ased **C**ontroller—controller designed using the open-loop identified model). The parameters of the resulting OLBC controller ($n_R = 27$, $n_S = 30$) are given in Table 10.2.

In Fig. 10.7, the achieved sensitivity functions with the full nominal model are shown. Clearly, the controller allows matching the specifications. The achieved modulus margin is -2.775 dB and the achieved delay margin is 4.1 T_s ($T_s = 1.25$ ms).

The performance on the real system is shown in Fig. 10.8. As it can be seen the specifications are satisfied.

Nevertheless, the full design procedure will be illustrated since in certain cases:

- the results obtained with the controller designed on the basis of the open-loop model may not necessarily be fully satisfactory; and
- the complexity of the controller has to be reduced.

Table 10.2 Parameters of the controller based on the reduced order open-loop identified model (OLBC)

Coeff.	R	Coeff.	S	Coeff.	R	Coeff.	S
r_0	0.0162	s_0	1.0000	r_{16}	0.0071	s_{16}	−0.1070
r_1	−0.0515	s_1	−5.1406	r_{17}	−0.0111	s_{17}	0.1031
r_2	0.0695	s_2	11.9134	r_{18}	−0.0068	s_{18}	−0.0384
r_3	−0.0255	s_3	−15.9616	r_{19}	0.0263	s_{19}	0.1284
r_4	−0.0666	s_4	12.7194	r_{20}	−0.0198	s_{20}	−0.0601
r_5	0.1315	s_5	−4.5490	r_{21}	0.0032	s_{21}	−0.0939
r_6	−0.1245	s_6	−2.0666	r_{22}	−0.0059	s_{22}	0.0027
r_7	0.0570	s_7	3.1609	r_{23}	0.0188	s_{23}	0.1820
r_8	0.0485	s_8	0.7437	r_{24}	−0.0180	s_{24}	−0.1586
r_9	−0.1405	s_9	−6.0665	r_{25}	0.0066	s_{25}	0.0457
r_{10}	0.1456	s_{10}	8.5544	r_{26}	0.0003	s_{26}	−0.0534
r_{11}	−0.0610	s_{11}	−6.8795	r_{27}	−0.0007	s_{27}	0.1081
r_{12}	−0.0242	s_{12}	3.6997	r_{28}	–	s_{28}	−0.0901
r_{13}	0.0422	s_{13}	−1.8094	r_{29}	–	s_{29}	0.0345
r_{14}	−0.0212	s_{14}	1.0885	r_{30}	–	s_{30}	−0.0049
r_{15}	0.0051	s_{15}	−0.4045	–	–	–	–

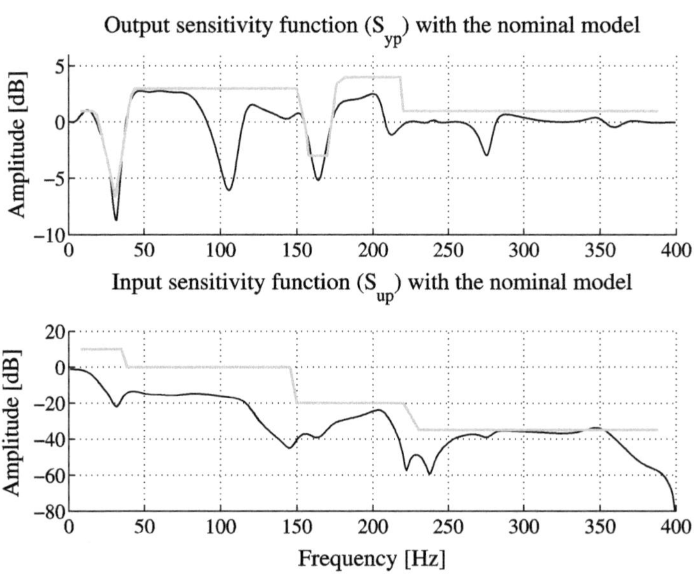

Fig. 10.7 Achieved sensitivity functions (*black*) with the OLBC controller and the nominal model

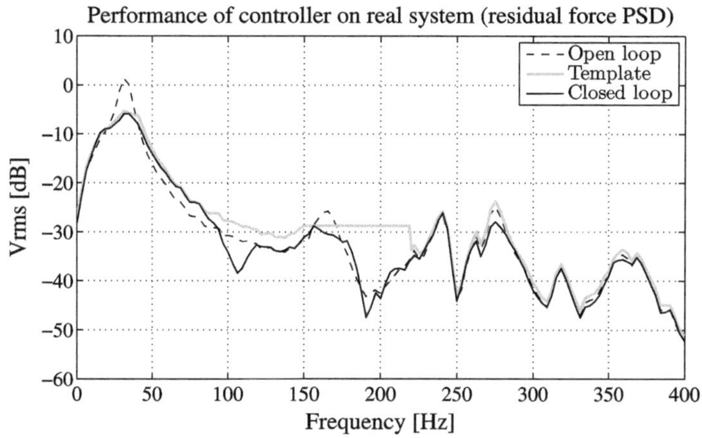

Fig. 10.8 Performance of the OLBC controller on the real system (PSD of the residual force)

10.4 Identification in Closed-Loop of the Active Suspension Using the Controller Designed on the Model Identified in Open-Loop

The methodology of identification in closed-loop operation has been presented in Chap. 8. A model with $n_A = 14$, $n_B = 16$ and $d = 0$ will be identified (same orders as for the model identified in open-loop operation).

One would like to identify a model which will minimize the error between the true output sensitivity function and the estimated sensitivity function, taking also into account that the plant model has a double differentiator. To achieve this, the excitation has been added to the input of the filter R (see Chap. 8 for details). Within this context, data acquisition was done with the same PRBS sequence as in open-loop identification (generated by 9-bit shift register and a clock frequency of $f_s/4$).

The best identified model in terms of statistical validation was the model identified with X-CLOE using a time-varying forgetting factor with $\lambda_0 = \lambda_1 = 0.95$. The parameters of this model are given in Table 10.3.

It is very important to assess if the model identified in closed loop is better than the model identified in open-loop for describing the behaviour of the closed-loop system using the OLBC controller. Figure 10.9 shows the identified poles of the closed loop (using an RELS algorithm for the closed-loop system identification considered as an input/output map from the excitation to the residual force) and the computed closed-loop poles using the open-loop identified model (OLID-M) and the OLBC controller. Figure 10.10 shows the same type of comparison but the computed closed-loop poles are calculated using the model identified in closed loop (CLID-M). Visual comparison shows clearly that the CLID-M model gives a better description of the real closed-loop system using the OLBC controller (this is obvious in the low frequency range which defines the main behaviour of the closed-loop system in terms of performance).

Table 10.3 Parameters of the model identified in closed-loop

Coeff.	A	Coeff.	B	Coeff.	A	Coeff.	B
a_0	1.0000	b_0	0.0000	a_9	0.6201	b_9	0.2716
a_1	−0.3003	b_1	−0.1556	a_{10}	−0.1095	b_{10}	1.8255
a_2	0.3504	b_2	0.1843	a_{11}	0.1593	b_{11}	1.1575
a_3	−0.6740	b_3	0.5518	a_{12}	−0.1580	b_{12}	1.3638
a_4	−0.2478	b_4	−1.4001	a_{13}	−0.0957	b_{13}	−0.8958
a_5	−0.4929	b_5	3.4935	a_{14}	−0.2030	b_{14}	1.6724
a_6	−0.3217	b_6	−0.3536	a_{15}	–	b_{15}	−1.7691
a_7	0.6157	b_7	−2.7181	a_{16}	–	b_{16}	−0.2240
a_8	0.1459	b_8	−3.0041	–	–	–	–

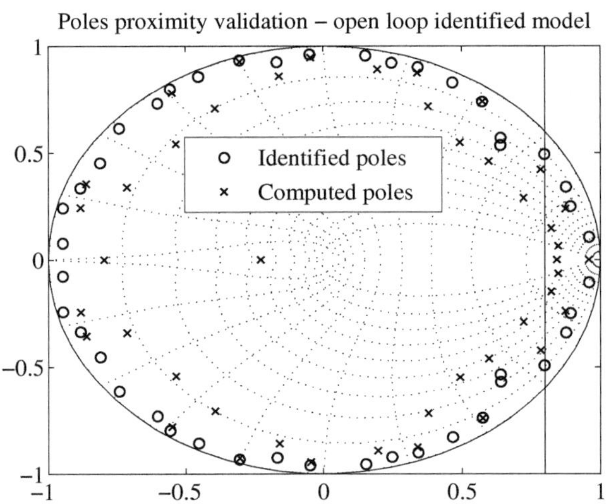

Fig. 10.9 Proximity poles validation of the full order open-loop identified model. Identified and computed closed-loop poles

This is also confirmed by the comparison of the real-time results with the simulated results obtained with the OLID-M model and the CLID-M model (see Fig. 10.11). A small improvement is observed.

10.5 Redesign of the Controller Based on the Model Identified in Closed Loop

Similar to the open-loop situation a reduced order model obtained by balanced truncation will be used. This model has the following dimensions: $n_A = 8$, $n_B = 11$, $d = 0$. The frequency characteristics of this reduced model and those of the full

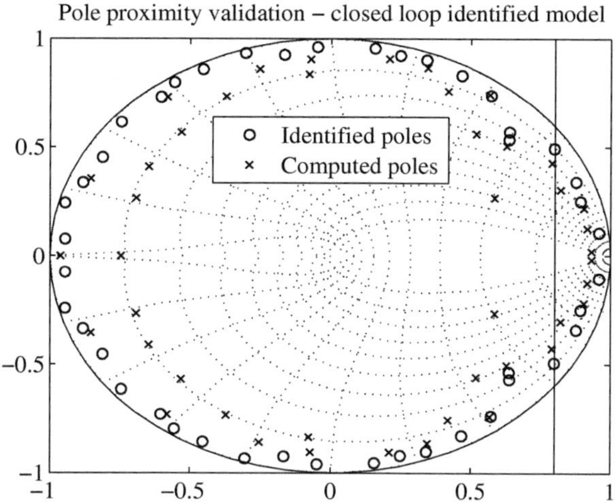

Fig. 10.10 Proximity poles validation of the full order closed-loop identified model. Identified and computed closed-loop poles

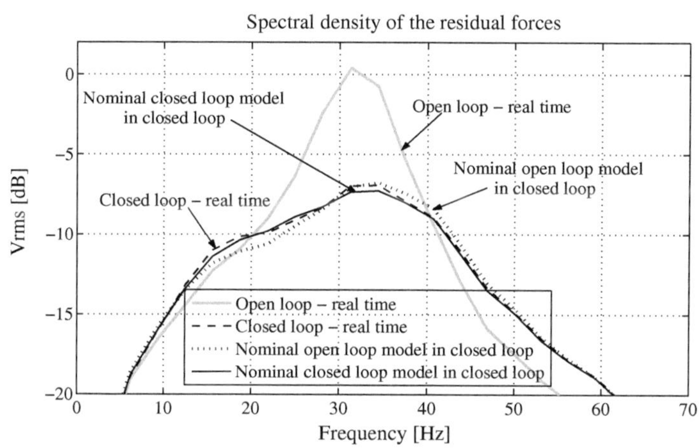

Fig. 10.11 Spectral density of the simulated and real-time closed-loop output (zoom)

order model identified in closed loop are shown in Fig. 10.12.[2] It can be observed that the reduced order model approximates very well the frequency characteristics of the nominal model identified in closed loop in the low frequency range of interest.

Applying the same design procedure based on convex optimization but now using the reduced order model obtained from the nominal model identified in closed loop a new controller (CLBC—Closed-Loop Based Controller) is obtained whose

[2]The option of identifying in closed loop a reduced order model instead of a model of nominal order followed by an order reduction using balanced truncation has provided less good results. For details see [1].

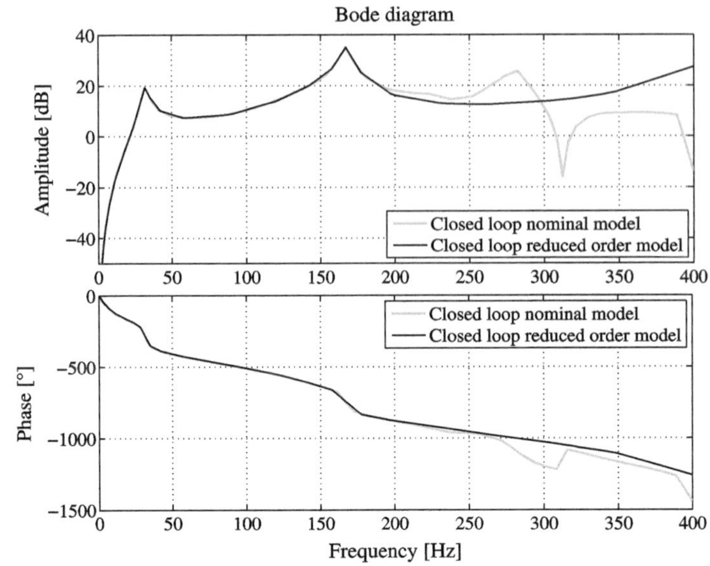

Fig. 10.12 Bode diagram (amplitude and phase) of the nominal model identified in closed-loop operation and of the corresponding reduced order model

parameters are given in Table 10.4. The sensitivity functions with the nominal CLID-M model are shown in Fig. 10.13. The robustness margins are: (1) Modulus Margin $= -3.702$ dB; (2) Delay Margin $= 1.834\, T_S$.

Figure 10.14 shows a comparison of the real-time results obtained with the OLBC controller and with the CLBC (controller based on the closed-loop identified model). The results are very close indicating that already the open-loop identified model was very good.

10.6 Controller Complexity Reduction

Once the CLBC controller is tested and the performance results are satisfactory (see Sect. 10.5), one can pass to the last step of the design methodology presented in Fig. 10.2 which is the reduction of the complexity of the controller.

The techniques for controller complexity reduction by identification in closed loop of the reduced order controller described in Chap. 9 will be used.

One aspect which is very important when reducing the complexity of a controller is that the controller reduction should be done such as to preserve as much as possible the desirable closed-loop properties. Direct simplification of the controller using standard techniques (poles–zeros cancellation within a certain radius, balanced reduction) without taking into account the closed loop behaviour produces in general unsatisfactory results [2, 3].

Table 10.4 Parameters of the controller based on the model identified in closed-loop operation (reduced order model) (CLBC)

Coeff.	R	Coeff.	S	Coeff.	R	Coeff.	S
r_0	0.0195	s_0	1.0000	r_{16}	−0.0488	s_{16}	0.8567
r_1	−0.0618	s_1	−4.5610	r_{17}	0.0446	s_{17}	−0.6306
r_2	0.1030	s_2	9.4917	r_{18}	−0.0495	s_{18}	0.3005
r_3	−0.1238	s_3	−12.4447	r_{19}	0.0437	s_{19}	−0.1080
r_4	0.1263	s_4	12.6103	r_{20}	−0.0255	s_{20}	0.0162
r_5	−0.1087	s_5	−11.5883	r_{21}	0.0078	s_{21}	0.1348
r_6	0.0581	s_6	9.8694	r_{22}	0.0055	s_{22}	−0.2960
r_7	0.0050	s_7	−7.4299	r_{23}	−0.0178	s_{23}	0.3737
r_8	−0.0389	s_8	5.3112	r_{24}	0.0254	s_{24}	−0.3835
r_9	0.0499	s_9	−4.0129	r_{25}	−0.0215	s_{25}	0.3633
r_{10}	−0.0648	s_{10}	2.9544	r_{26}	0.0102	s_{26}	−0.3058
r_{11}	0.0727	s_{11}	−2.1480	r_{27}	−0.0022	s_{27}	0.2004
r_{12}	−0.0602	s_{12}	1.9636	r_{28}	–	s_{28}	−0.0883
r_{13}	0.0511	s_{13}	−1.9125	r_{29}	–	s_{29}	0.0218
r_{14}	−0.0597	s_{14}	1.4914	r_{30}	–	s_{30}	−0.0019
r_{15}	0.0616	s_{15}	−1.0471	–	–	–	–

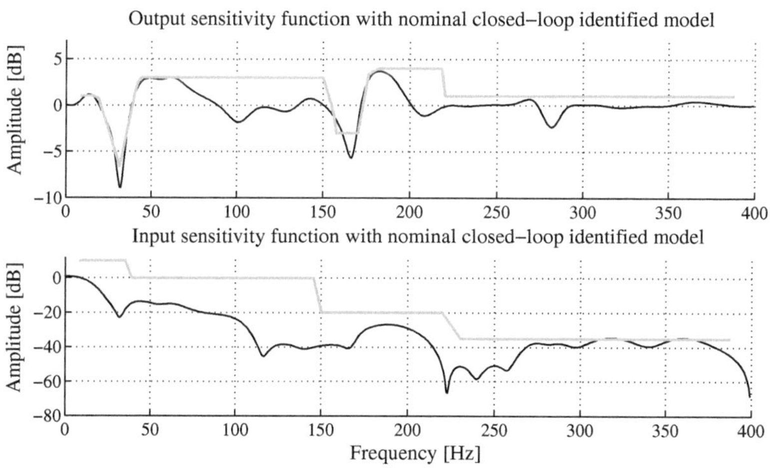

Fig. 10.13 Achieved sensitivity functions (*black thin line*) with the CLBC controller and the nominal model identified in closed-loop operation

The orders of the nominal CLBC controller to be reduced are $n_R = 27, n_S = 30$, and its coefficients have been presented in Table 10.4. The model which will be used for the reduction of the controller is the nominal closed-loop identified model

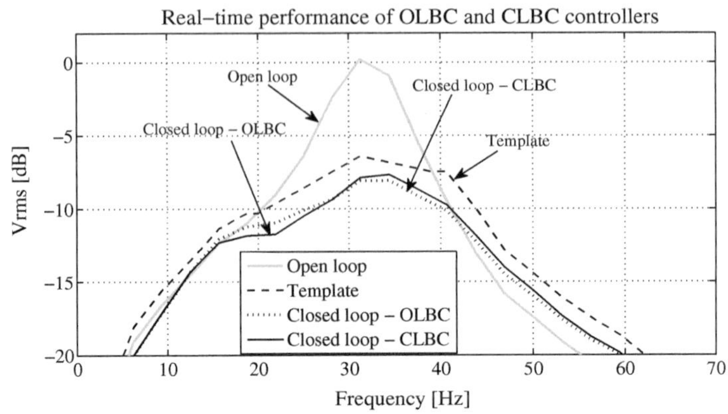

Fig. 10.14 Real-time performance of the OLBC and CLBC controllers (detail)

CLID-M (see Sect. 10.4). The parameters of the model have been given in Table 10.3 (see Sect. 10.4).

Since in active damping we are concerned with attenuation of the disturbances, the main objective for controller reduction will be to obtain an output sensitivity function for the reduced order controller as close as possible to the output sensitivity function obtained with the nominal order controller. As indicated in Chap. 9 and [4], in order to achieve this, the CLOM procedure has to be used. The reduction procedures have been run with simulated data.

A variable forgetting factor with $\lambda_1(0) = 0.95$ and $\lambda_0 = 0.9$ $(\lambda_1(t) = \lambda_0\lambda_1(t-1) + 1 - \lambda_0)$ has been used in the algorithm for the controller parameters estimation. The external input was a PRBS generated by a 9-bit shift register with a $p = 4$ frequency divider (4096 samples). In addition a fixed part $H_R = 1 + q^{-1}$ has been introduced in the reduced order controllers ($R = H_R R'$) which preserves the opening of the loop at 0.5 f_s.

10.6.1 CLOM Algorithm with Simulated Data

Two reduced order controllers have been computed: CLBC-CLOM16 with the orders $n_R = 14$, $n_S = 16$ and CLBC-CLOM5 with the orders $n_R = 4$, $n_S = 5$.

The frequency characteristics of the output and input sensitivity functions (S_{yp} and S_{up}) for the nominal controller CLBC and the two reduced order controllers CLBC-CLOM16 and CLBC-CLOM5 are shown in Figs. 10.15 and 10.16, respectively.

Note that the reduced controller CLBC-CLOM16 corresponds to the complexity of the pole placement controller with the fixed part H_R, while controller CLBC-CLOM5 has a lower complexity.

The values of the various ν-gap are summarized in Table 10.5 (the last two rows give real-time results). It can be remarked that the Vinnicombe stability margins

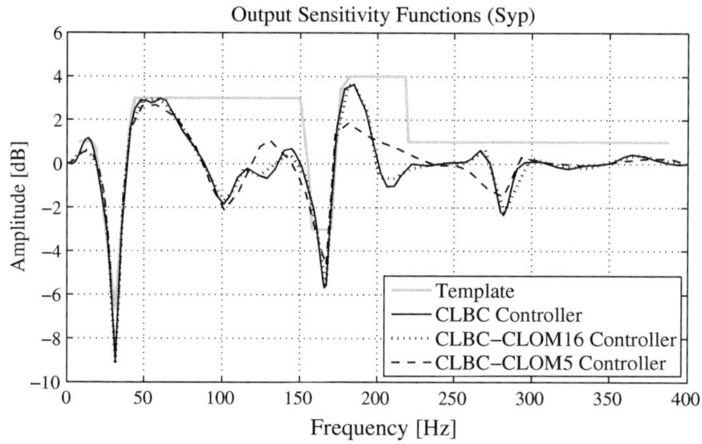

Fig. 10.15 Output sensitivity functions (controller order reduction with CLOM algorithm and simulated data)

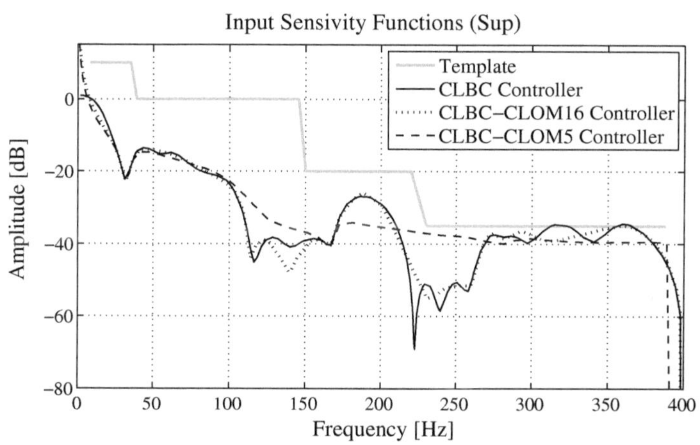

Fig. 10.16 Input sensitivity functions (controller order reduction with CLOM algorithm and simulated data)

$b(K_i, G)$ computed with the nominal model CLID-M for the various reduced order controllers are close to the stability margin obtained with the nominal controller.

The last two rows of Table 10.5 give real-time results. Row 6 gives the ν-gap between the input/output transfer function corresponding to the input sensitivity function S_{up} of the true closed-loop system constituted by the nominal designed controller with the real plant (obtained by system identification between the input r and the output y) and the input/output transfer function of the simulated closed-loop system ($\hat{\hat{S}}_{up}$) constituted by the various controllers (including the nominal one and the reduced ones obtained using simulated data) in feedback connection with the plant

Table 10.5 Comparison of the nominal and reduced order controllers (controller reduction using CLOM algorithm and simulated data)

Controller	*CLBC* $nR = 27$ $nS = 30$	*CLBC-CLOM16* $nR = 14$ $nS = 16$	*CLBC-CLOM5* $nR = 4$ $nS = 5$
$\delta_\nu(K_n, K_i)$	0	0.6577	0.6511
$\delta_\nu(S^n_{up}, S^i_{up})$	0	0.6577	0.6511
$\delta_\nu(S^n_{yp}, S^i_{yp})$	**0**	**0.0386**	**0.1308**
$b(K_i, G)$	0.0303	0.0135	0.0223
$\delta_\nu(CL(K_n), CL(K_i))$	0.2610	0.2963	0.4275
Closed-loop error variance	0.13582	0.14755	0.17405

model. This quantity is denoted by $\delta_\nu(CL(K_n), CL(K_i))$. This is a good criterion for the validation of the reduced order controllers in real time. It can be observed that the CLBC-CLOM16 controller gives results which are very close to those of the nominal CLBC controller. Row 7 gives the variance of the residual closed-loop input error between the true system and the simulated one. The results are coherent to those of row 6, showing that CLBC-CLOM16 gives performance very close to those of the nominal controller.

10.6.2 Real-Time Performance Tests for Nominal and Reduced Order Controllers

The spectral densities of the residual forces in open loop and in closed loop corresponding to the nominal controller CLBC and the reduced order ones obtained with the CLOM method (CLBC-CLOM16 and CLBC-CLOM5) are presented in Fig. 10.17.

It can be seen that the performance of reduced order controllers are very close to that of the nominal controller designed using a reduced model of the closed-loop identified model. Note also that the reduction in terms of number of parameters is significant. Very close results have been obtained using the CLIM reduction procedure (see [1, 5]).

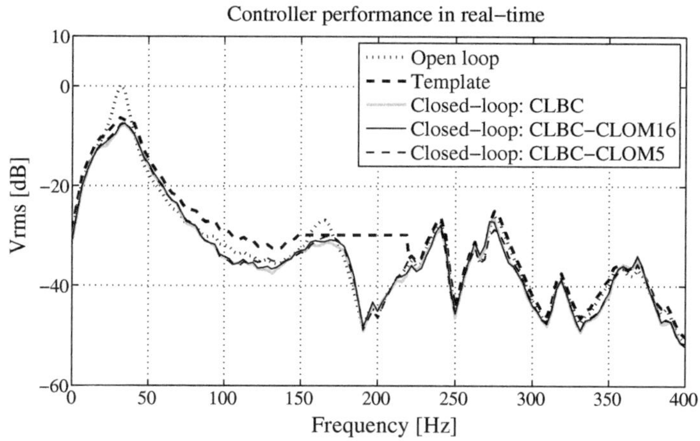

Fig. 10.17 Spectral density of the residual forces in open and closed loop for the nominal and reduced order controllers (CLOM)

10.7 Design of the Controller by Shaping the Sensitivity Function with Band-Stop Filters

The objective of this section is to provide an alternative design procedure for active damping which does not require the use of the convex optimization procedure, but uses only band-stop filters which are iteratively introduced in order to shape the sensitivity functions. This method has been introduced in Sect. 7.2.9. The frequency and damping of the poles of the open-loop identified model are given in Table 6.3.

All asymptotically stable poles will be included as initial desired closed-loop poles. Only the pole located at -0.2177 which corresponds in fact to a pair of damped oscillatory poles near $0.5f_s$ will not be included. All the poles remain unchanged in terms of damping, except the complex poles located at 31.939 Hz for which the damping imposed in closed loop will be $\xi = 0.8$ and the complex poles at 164.34 Hz for which a damping of 0.167 will be chosen. These two damped poles will help to satisfy the desired template on the output sensitivity function. 16 real auxiliary poles are assigned at 0.15 (this will not augment the size of the resulting controller).[3]

Figure 10.18 (curve "Controller 1") shows the resulting output sensitivity function S_{yp}. As it can be seen, it almost satisfies the robustness constraints on the modulus margin and delay margin (it is inside the basic template for robustness at all frequencies except around 55 Hz). Nevertheless, when compared to the specification for the output sensitivity function in Fig. 10.19 (dotted line), it can be observed that the desired disturbance attenuation is not satisfied. The input sensitivity function satisfies the specified template, see Fig. 10.20.

[3]The design using BSF has been done with iReg software which provides a convenient interactive environment.

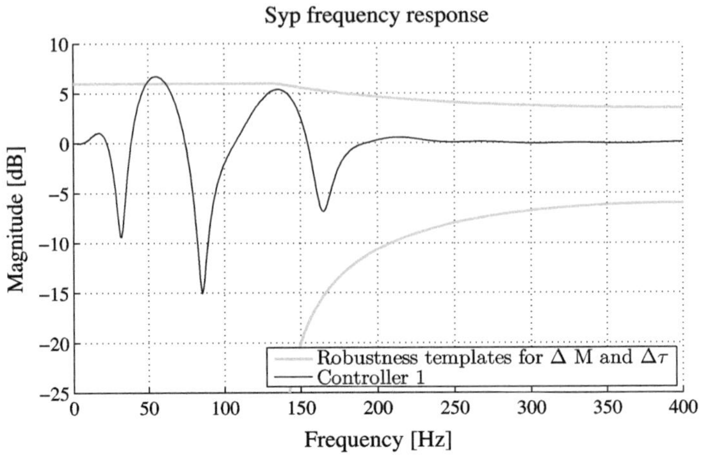

Fig. 10.18 Output sensitivity function for the Controller 1 (with modulus and delay margin templates)

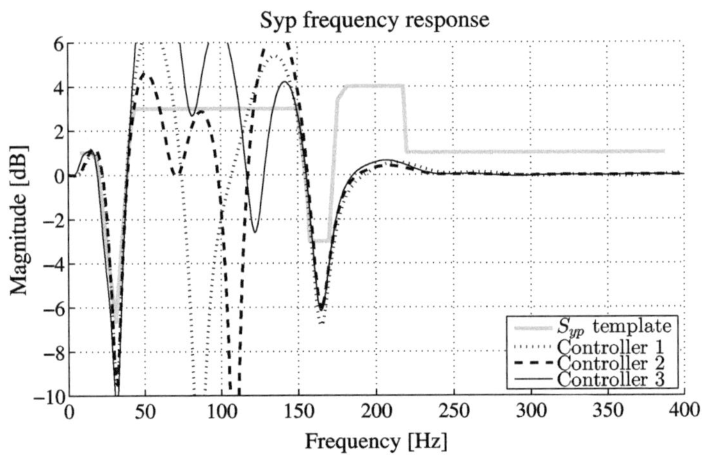

Fig. 10.19 Output sensitivity function with the three initial controllers

To have zero gain on the input sensitivity function at $0.5f_s$, one zero at -1 is added to the fixed part of the controller numerator by including into H_R the first order polynomial $(1+q^{-1})$. One more characteristic pole at 0.15 is then added (this will not increase the order of the controller but avoid to have a pole assigned to 0). The result can be seen in Figs. 10.19 and 10.20, "Controller 2" curve. One can see that the template is still violated in several frequency regions.

For shaping the output sensitivity function in the frequency region of the first attenuation mode around 30 Hz three BSF have been added at 14, 24 and 38.7 Hz, with attenuation of -2.5, -7 and -5.5 dB respectively. The resulting controller sensitivity functions are shown in Figs. 10.19 and 10.20 (curve "Controller 3"). The

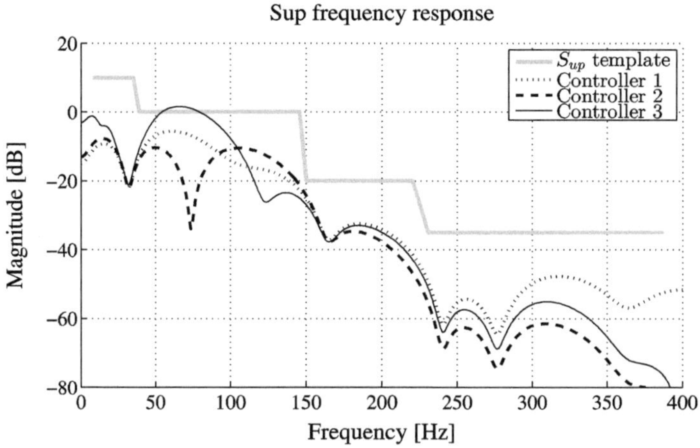

Fig. 10.20 Input sensitivity function with the three initial controllers

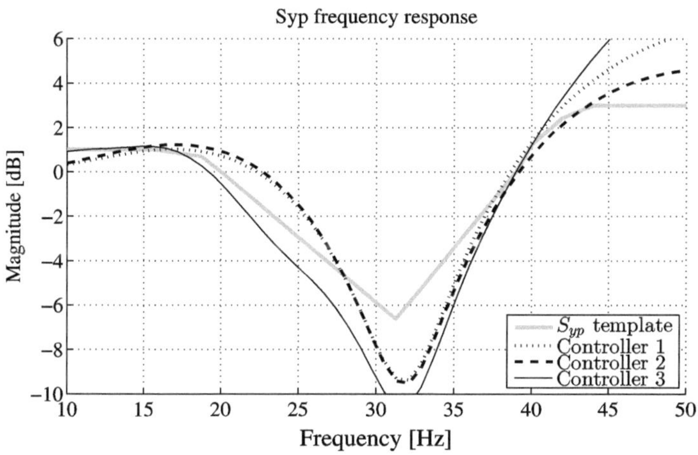

Fig. 10.21 Output sensitivity function with the three initial controllers (zoom)

result in the region of the first attenuation mode around 30 Hz can be better evaluated using Fig. 10.21, where a zoom between 10 and 50 Hz is shown. For all three BSF, the denominator damping has been chosen equal to 0.5. It can be observed that "Controller 3" satisfies the imposed template in the lower frequency region below 30 Hz.

The final design step is to improve the shape of the sensitivity functions at the other frequencies. Two additional BSF have been added for shaping the output sensitivity function and five for shaping the input sensitivity function. In addition, the initial three BSF have been slightly modified as each newly added BSF has, however, a slight influence at neighbouring frequencies. Tables 10.6 and 10.7 summarize the characteristics of the various BSF. A sensitivity functions comparison between

Table 10.6 Band-stop filters for the output sensitivity function

Controller number	Frequency (Hz)	Attenuation (dB)	Denominator damping
1	14	−9.1	0.95
2	23.5	−14.759	0.95
3	41.158	−5.2	0.5
4	69.45	−15.11	0.95
5	132.5	−14.759	0.95

Table 10.7 Band-stop filters for the input sensitivity function

Controller number	Frequency (Hz)	Attenuation (dB)	Denominator damping
1	51.5	−16	0.95
2	70.74	−14.052	0.5
3	92.6	−15.1	0.95
4	115.76	−9.1	0.5
5	313.826	−2.733	0.95

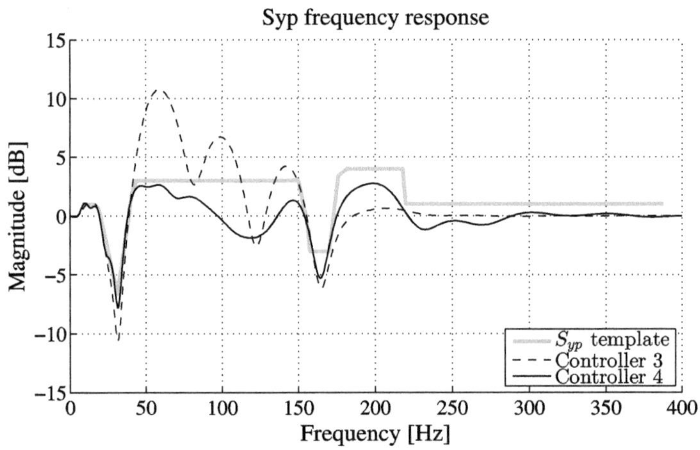

Fig. 10.22 Output sensitivity function comparison between "Controller 3" and "Controller 4"

"Controller 3" and "Controller 4" is given in Figs. 10.22 (output sensitivity functions) and 10.23 (input sensitivity functions).

Finally, Figs. 10.24 and 10.25 give a comparison of "Controller 4" and the controller designed using convex optimization (see previous sections). A zoom between 10 and 50 Hz is shown in Fig. 10.26 for comparative evaluation of the obtained characteristics around the first attenuation region. As it can be seen, both controllers satisfy the template in the low frequency region while in the high frequency region

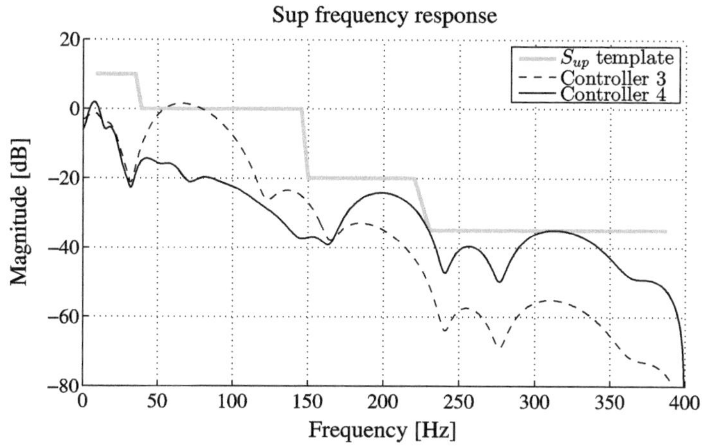

Fig. 10.23 Input sensitivity function comparison between "Controller 3" and "Controller 4"

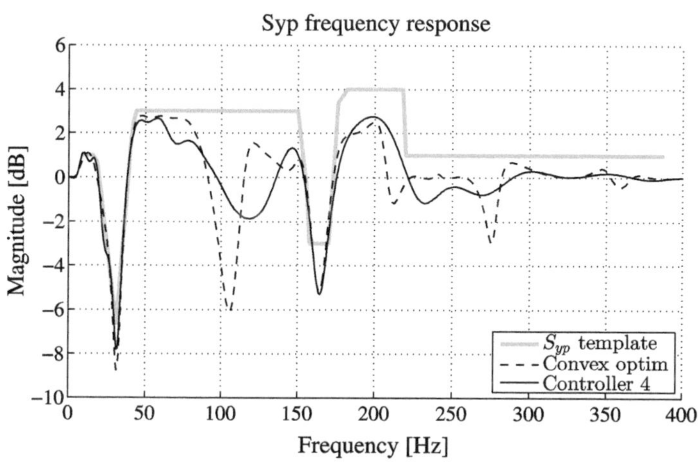

Fig. 10.24 Output sensitivity function comparison between the convex optimization controller and "Controller 4" obtained using iREG

the controller designed by convex optimization slightly exceeds the imposed template. Concerning their complexity, "Controller 4" designed using BSF filters has 71 parameters ($n_R = 34$ and $n_S = 36$) while the controller designed by convex optimization has 58 parameters ($n_R = 27$ and $n_S = 30$).

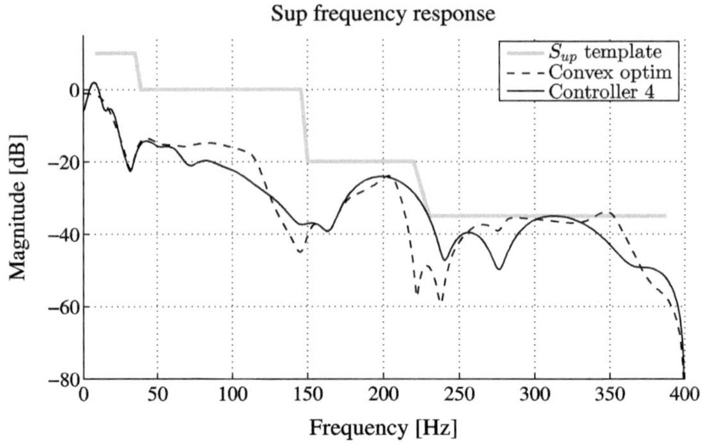

Fig. 10.25 Input sensitivity function comparison between the convex optimization controller and "Controller 4" obtained using iREG

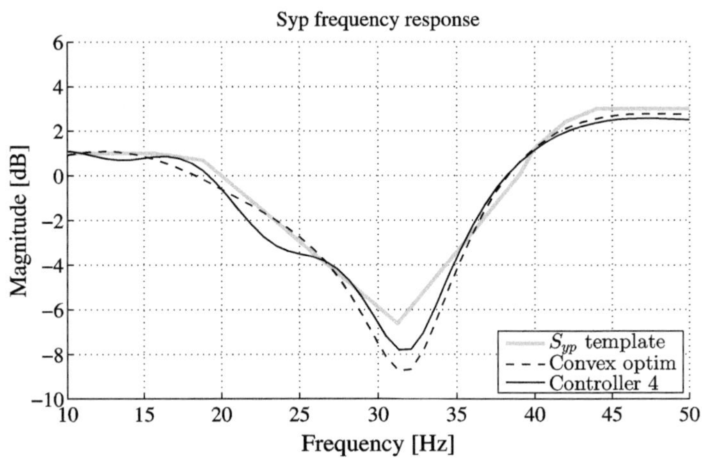

Fig. 10.26 Output sensitivity function comparison between the convex optimization controller and "Controller 4" obtained using iREG (zoom)

10.8 Concluding Remarks

- The design of active damping systems consists of the following major steps:
  - Definition of the control performance specifications in the frequency domain.
  - Design of the controller ensuring the desired performance.
  - Validation of the controller.
- Design of the controller for active damping include several steps:

- Open-loop identification of the secondary path.
- Design of the controller based on the secondary path model identified in open-loop operation.
- Implementation and validation of the controller.

• If the performance is not satisfactory, the following procedure has to be followed:

- Identification in closed-loop operation of a new model for the secondary path and validation of the identified model.
- Redesign of the controller on the basis of the model identified in closed loop.
- Implementation and validation of the controller designed on the basis of the model identified in closed-loop operation.

• The effective design of the controller requires the shaping of the sensitivity functions.
• Shaping of the sensitivity functions can be achieved using convex optimization or band-stop filters combined with poles placement.
• If constraints on the computational load exist, the final step in the design is the reduction of the controller complexity with the objective of preserving the closed-loop performance.
• The reduced order controller should be implemented and validated.

10.9 Notes and References

Active damping for disturbance attenuation and control of lightly damped structures which has different objectives use, however, similar feedback techniques [6].

Suspension bridges and cable-stayed bridges require active damping to reduce the effects of various phenomena. Active damping solutions have been proposed in [7–14]. Active tendon control of cable-stayed bridges using hydraulic suspension is considered in [11].

An important issue in active damping is the construction of physical systems allowing to achieve active damping. Use of piezoelectric devices is a very efficient approach for many applications. See [14] for a survey and [8] for a detailed modelling of this type of devices. Applications of piezoelectric devices for active damping have been reported for: (i) Large space structures [14], (ii) Wafer stepper in lithography [15] and (iii) Active tendon control of cable structures in space [16, 17].

Other references related to active damping include [18–21].

A key issue in active damping is a careful shaping of the sensitivity functions. Other techniques than those presented in this chapter can be used. H_∞ where the shaping of the sensitivity function is converted in a weighted frequency criterion minimization can be considered [22, 23]. Linear-Quadratic Control with frequency weighting can also be considered [24].

References

1. Constantinescu A (2001) Commande robuste et adaptative d'une suspension active. Thèse de doctorat, Institut National Polytechnique de Grenoble
2. Anderson B, Liu Y (1989) Controller reduction: concepts and approaches. IEEE Trans Autom Control 34(8):802–812
3. Anderson B (1993) Controller design: moving from theory to practice. IEEE Control Mag 13:16–25
4. Landau I, Karimi A (2002) A unified approach to model estimation and controller reduction (duality and coherence). Eur J Control 8(6):561–572
5. Landau I, Karimi A, Constantinescu A (2001) Direct controller order reduction by identification in closed loop. Automatica 37:1689–1702
6. Karkosch HJ, Preumont A (2002) Recent advances in active damping and vibration control. Actuator 2002, 8th International Conference on New ActuatorsBremen, Germany, pp 248–253
7. Cannon RJ, Rosenthal D (1984) Experiments in control of flexible structures with noncolocated sensors and actuators. J Guid Control Dyn 7(5):546–553
8. Preumont A (2011) Vibration control of active structures - an introduction. Springer, Berlin
9. Achkire Y, Preumont A (1996) Active tendon control of cable-stayed bridges. Earthq Eng Struct Dyn 25(6):585–597
10. Achkire Y, Bossens F, Preumont A (1998) Active damping and flutter control of cable-stayed bridges. J Wind Eng Ind Aerodyn 74–76:913–921
11. Bossens F, Preumont A (2001) Active tendon control of cable-stayed bridges: a large-scale demonstration. Earthq Eng Struct Dyn 30(7):961–979
12. Auperin M, Dumoulin C (2001) Structural control: point of view of a civil engineering company in the field of cable-supported structures. In: Casciati F, Magonette G (eds) Structural control for civil and infrastructure engineering. Singapore, pp 49–58
13. Preumont A, Dufour JP, Malekian C (2015) An investigation of the active damping of suspension bridges. Math Mech Complex Syst 3(4):385–406
14. Preumont A, Dufour JP, Malekian C (1992) Active damping by a local force feedback with piezoelectric actuators. J Guid Control Dyn 15(2):390–395
15. Jansen B (2000) Smart disk tuning and application in an ASML wafer stepper. Msc. thesis, Control Laboratory, University of Twente, Enschede, The Netherlands
16. Preumont A, Achkire Y, Bossens F (2000) Active tendon control of large trusses. AIAA J 38(3):493–498
17. Preumont A, Bossens F (2000) Active tendon control of vibration of truss structures: theory and experiments. J Intell Mat Syst Struct 2(11):91–99
18. Preumont A, Loix N (1994) Active damping of a stiff beam-like structure with acceleration feedback. Exp Mech. 34(1):23–26
19. Preumont A, Achkire Y (1997) Active damping of structures with guy cables. J Guid Control Dyn 20(2):320–326
20. Chen X, Jiang T, Tomizuka M (2015) Pseudo Youla-Kučera parameterization with control of the waterbed effect for local loop shaping. Automatica 62:177–183
21. Sievers LA, von Flotow AH (1988) Linear control design for active vibration isolation of narrow band disturbances. In: Proceedings of the 27th IEEE Conference on Decision and Control 1988. IEEE, pp 1032–1037
22. Zhou K, Doyle J (1998) Essentials of robust control. Prentice-Hall International, Upper Saddle River
23. Alma M, Martinez J, Landau I, Buche G (2012) Design and tuning of reduced order H_∞ feedforward compensators for active vibration control. IEEE Trans Control Syst Technol 20(2):554–561. doi:10.1109/TCST.2011.2119485
24. Tharp H, Medanic J, Perkins W (1988) Parameterization of frequency weighting for a two-stage linear quadratic regulator based design. Automatica 24(5):415–418

Part IV
Feedback Attenuation of Narrow-Band Disturbances

Chapter 11
Robust Controller Design for Feedback Attenuation of Narrow-Band Disturbances

11.1 Introduction

To illustrate the design of a robust controller for active vibration control system, we will consider the case of multiple unknown and time-varying sinusoidal disturbances located within two distinct *relatively small frequency ranges*. To be specific, two situations will be considered

(1) The case of two time-varying tonal disturbances, located in two distinct frequency regions.
(2) The case of four simultaneous tonal disturbances, located two by two in two distinct frequency regions. In this context, a very important problem is to be able to counteract the very low frequency oscillation which is generated when the two frequencies are very close (vibrational interference). This phenomenon is considered for example in [1]. It also occurs often on ships with two thrusters which cannot be perfectly synchronized. A typical image of the phenomenon is shown in Fig. 11.1.

Since these disturbances are located within two relatively small frequency ranges, it is possible to consider a linear control design which will shape the output sensitivity function in such a way that a sufficient attenuation is introduced in these two frequency regions but avoiding significant amplification at other frequencies (both for performance and robustness reason). This problem in the context of active noise control has been considered in [2] and the shaping of the output sensitivity function has been achieved using the convex optimization procedure introduced in [3].[1] An H_∞ approach can also eventually be used but it will require a quite complicated procedure for defining the appropriate weighting functions.

In this chapter, it will be shown that the technique of shaping the sensitivity functions using band-stop (notch) filters (see [4] and also Chap. 7) can be efficiently

[1] See also Sect. 7.4.

I.D. Landau et al., *Adaptive and Robust Active Vibration Control*,
Advances in Industrial Control, DOI 10.1007/978-3-319-41450-8_11

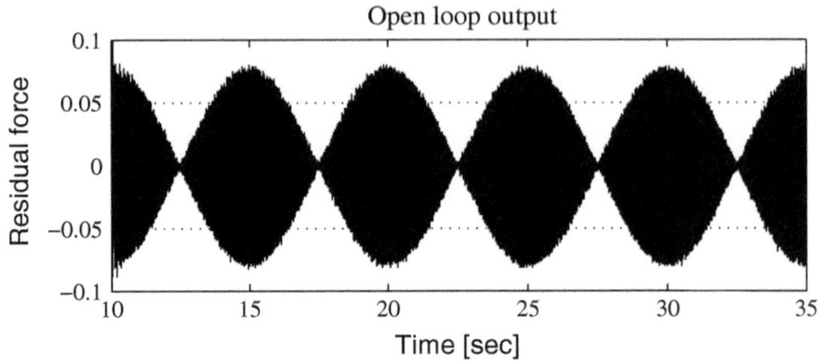

Fig. 11.1 Vibrational interference of two sinusoidal disturbances

used to design controllers for robustly attenuating single or multiple narrow-band disturbances varying within a relatively small frequency range.

Experimental validation of the design on the active vibration control system using an inertial actuator will conclude the chapter.

11.2 System Description

The linear time invariant (LTI) discrete time model of the secondary path, used for controller design is

$$G(z^{-1}) = \frac{z^{-d}B(z^{-1})}{A(z^{-1})} = \frac{z^{-d-1}B^*(z^{-1})}{A(z^{-1})}, \tag{11.1}$$

where

$$A(z^{-1}) = 1 + a_1 z^{-1} + \cdots + a_{n_A} z^{-n_A}, \tag{11.2}$$

$$B(z^{-1}) = b_1 z^{-1} + \cdots + b_{n_B} z^{-n_B} = z^{-1} B^*, \tag{11.3}$$

$$B^* = b_1 + \cdots + b_{n_B} z^{-n_B+1}, \tag{11.4}$$

and d is the plant pure time delay in number of sampling periods.[2] To illustrate the methodology, the active vibration control system using an inertial actuator described in Sect. 2.2, will be used. The identification of the secondary path model has been done in Sect. 6.2. The parameters of the identified model of the secondary path are given in Table 6.2 ($d = 0$).

[2]The complex variable z^{-1} will be used to characterize the system's behaviour in the frequency domain and the delay operator q^{-1} will be used for the time domain analysis.

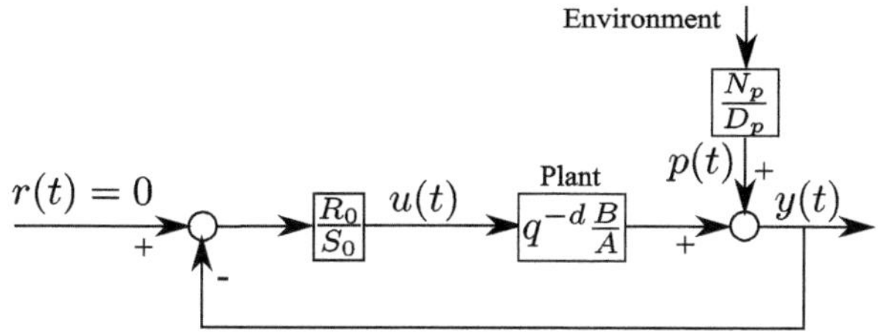

Fig. 11.2 Feedback regulation scheme for rejection of disturbances

The output of the plant $y(t)$ and the input $u(t)$ may be written as (see Fig. 11.2)

$$y(t) = \frac{q^{-d}B(q^{-1})}{A(q^{-1})} \cdot u(t) + p(t), \tag{11.5}$$

$$S_0(q^{-1}) \cdot u(t) = -R_0(q^{-1}) \cdot y(t). \tag{11.6}$$

In (11.5), $p(t)$ is the effect of the disturbances on the measured output[3] and $R_0(z^{-1})$, $S_0(z^{-1})$ are polynomials in z^{-1} having the following expressions[4]

$$S_0 = 1 + s_1^0 z^{-1} + \cdots + s_{n_S}^0 z^{-n_S} = S_0' \cdot H_{S_0}, \tag{11.7}$$

$$R_0 = r_0^0 + r_1^0 z^{-1} + \cdots + r_{n_R}^0 z^{-n_R} = R_0' \cdot H_{R_0}, \tag{11.8}$$

where $H_{S_0}(z^{-1})$ and $H_{R_0}(z^{-1})$ represent pre-specified parts of the controller (used for example to incorporate the internal model of a disturbance or to open the loop at certain frequencies). $S_0'(z^{-1})$ and $R_0'(z^{-1})$ are the results of the Bezout equation

$$P_0 = \left(A \cdot H_{S_0}\right) \cdot S_0' + \left(z^{-d}B \cdot H_{R_0}\right) \cdot R_0'. \tag{11.9}$$

In the last equation, $P_0(z^{-1})$ represents the characteristic polynomial, which specifies the desired closed-loop poles of the system.

The transfer functions between the disturbance $p(t)$ and the output of the system $y(t)$ and the control input $u(t)$, denoted respectively *output sensitivity function* and *input sensitivity function*, are given by

$$S_{yp}(z^{-1}) = \frac{A(z^{-1})S_0(z^{-1})}{P_0(z^{-1})} \tag{11.10}$$

and

$$S_{up}(z^{-1}) = -\frac{A(z^{-1})R_0(z^{-1})}{P_0(z^{-1})}, \tag{11.11}$$

[3]The disturbance passes through a so-called *primary path* and $p(t)$ is its output.

[4]The argument (z^{-1}) will be omitted in some of the following equations to make them more compact.

It is important to remark that one should only reject disturbances located in frequency regions where the plant model has enough gain. This can be seen by looking at Eq. (11.10) and noticing that perfect rejection at a certain frequency, ω_0, is obtained *iff* $S_{yp}(e^{-j\omega_0}) = 0$. On the other hand, from (11.9) and (11.11) one can see that, at ω_0

$$S_{up}(e^{-j\omega_0}) = -\frac{AR_0}{0 + e^{-dj\omega_0}BR_0} = -\frac{A}{e^{-dj\omega_0}B} = \frac{1}{G(e^{-j\omega_0})}. \tag{11.12}$$

Equation (11.12) corresponds to the inverse of the gain of the system to be controlled at the frequency ω_0. If the gain of the controlled system is too low, $|S_{up}|$ will be large at these frequencies. Therefore, the robustness versus additive plant model uncertainties will be reduced and the stress on the actuator will become important (see Sect. 7.2.5 and [4]). The implication of Eq. (11.12) is that cancellation (or in general an important attenuation) of disturbances on the output should be done only in frequency regions where the system gain is large enough. Equation (11.12) also implies that serious problems will occur if $B(z^{-1})$ has complex zeros close to the unit circle (stable or unstable zeros) at frequencies where an important attenuation of disturbances is required. It is mandatory to avoid attenuation of disturbances at these frequencies.

11.3 Robust Control Design

In this section, the design of a linear robust digital controller for disturbance attenuation is presented.

Before presenting the objectives for robustness and regulation, a few notions about feedback disturbance attenuation should be reminded. In the case of a feedback controlled system, the Bode integral constraint leads to a waterbed effect on the output sensitivity function (transfer function from disturbance $p(t)$ to output $y(t)$ in closed-loop, see Sects. 7.1 and 11.2). In other words, forcing the magnitude of the output sensitivity function at certain frequencies below 0 dB (in order to attenuate disturbances) has an inverse effect on neighbouring frequencies, where an amplification will be observed. Recalling from Sect. 7.2.4 that the minimal distance between the Nyquist plot of the open-loop transfer function and the critical point $-1 + 0i$ (also called *modulus margin*) corresponds to the inverse of the maximum of the output sensitivity function, it can be concluded that "too much" attenuation at some frequencies can have a bad effect on the robust stability of the closed-loop system.

To summarize, the attenuation surfaces should be equal to the amplification surfaces with the constraint that the maximum amplification be less or equal to 8 dB in order to assure a convenient modulus margin. This has to be verified on the operational frequency range. Outside this zone, the output sensitivity function is close to 0 dB since the input sensitivity function is forced to be very low (no gain in the controller) for robustness reasons and actuator solicitations.

Taking into consideration the secondary path frequency response in Fig. 6.18 and the fact that disturbances can only be attenuated where the system has enough gain (see Sect. 11.2), it has been concluded that only disturbances within the 50–95 Hz frequency band (operational frequency range) can be attenuated.

For the design of the linear robust digital controller, the following specifications are considered

- up to 4 sine disturbances are supposed to affect the output of the system (known structure of the disturbance model);
- their frequencies are not known exactly but they are varying within ± 2.5 Hz around 60 and 80 Hz;
- the controller should attenuate the disturbances by a minimum of 14 dB;
- the maximum allowed amplification of the output sensitivity function is 8 dB;
- the effect of disturbances on the control input should be attenuated above 100 Hz in order to improve robustness with respect to unmodeled dynamics and nonlinear phenomena ($S_{up}(e^{-j\omega}) < -20\,\text{dB},\ \forall\omega \in [100, 400\,\text{Hz}]$);
- the gain of the controller has to be zero at zero frequency (since the system has a double differentiation behaviour); and
- the gain of the controller should be zero at $0.5 f_s$ where the system has low gain and uncertainties exist.

It is shown in [4, Property 7, Sect. 3.6.1] and in Sect. 7.2.9 that very accurate shaping of the output or the input sensitivity function can be obtained by the use of band-stop filters (BSF). These are IIR notch filters obtained from the discretization of a continuous-time filter of the form

$$F(s) = \frac{s^2 + 2\zeta_{num}\omega_0 s + \omega_0^2}{s^2 + 2\zeta_{den}\omega_0 s + \omega_0^2} \tag{11.13}$$

using the bilinear transform $s = \frac{2}{Ts}\frac{1-z^{-1}}{1+z^{-1}}$. The use of BSFs introduces an attenuation

$$M = 20\log\left(\frac{\zeta_{num}}{\zeta_{den}}\right) \tag{11.14}$$

at the normalized discretized frequency

$$\omega_d = 2 \cdot \arctan\left(\frac{\omega_0 T_S}{2}\right). \tag{11.15}$$

Design details can be found in Sect. 7.2.9.

Depending on whether the filter is designed for shaping the output or the input sensitivity function, the numerator of the discretized filter is included in the fixed part of the controller denominator H_{S_0} or numerator H_{R_0}, respectively. *The filter*

Table 11.1 Band-stop filters for output and input sensitivity functions

	Frequency (Hz)	Amplification (dB)	Damping
S_{yp}	57.5	−17	0.1
	59.8	−25	0.5
	62	−15	0.1
	77.5	−13	0.05
	79.8	−20	0.2
	82	−12	0.05
S_{up}	155	−16	0.5

denominator is always included in the closed-loop characteristic polynomial. As such, the filter denominator influences the design of the controller indirectly since S_0' and R_0' are solutions of the Bezout equation (11.9). They will be used for a fine shaping of both the output and input sensitivity functions.

The steps for the design of the linear controller are[5]:

1. Include all (stable) secondary path poles in the closed-loop characteristic polynomial.
2. Open the loop at 0 Hz and at 400 Hz by setting the fixed part of the controller numerator
$$H_R = (1 + q^{-1}) \cdot (1 - q^{-1}). \tag{11.16}$$
3. 3 BSFs on S_{yp} have been used around each of the frequencies where attenuation is desired in order to assure the desired attenuation within ±2.5 Hz (see Table 11.1 for specifications).
4. 1 BSF has been used on S_{up} to reduce its magnitude above 100 Hz (see Table 11.1 for specifications).
5. To improve robustness, 2 complex conjugate poles have been added to the characteristic polynomial, one at 55 Hz and the second at 95 Hz, both of them with 0.1 damping factor.

The output and input sensitivity functions with this linear controller can be analyzed in Figs. 11.3 and 11.4, respectively. In Fig. 11.3, it can be observed that the attenuation of 14 dB and the maximum amplification of 8 dB on S_{yp} are achieved. This is a trade off between performance and robustness. The specification of −20 dB attenuation on S_{up} above 100 Hz is satisfied.

[5]The software iREG has been used for the design of this robust digital controller but the same results can be obtained using functions written in MATLAB/Scilab languages (see http://www.gipsa-lab.grenoble-inp.fr/~ioandore.landau/identificationandcontrol/).

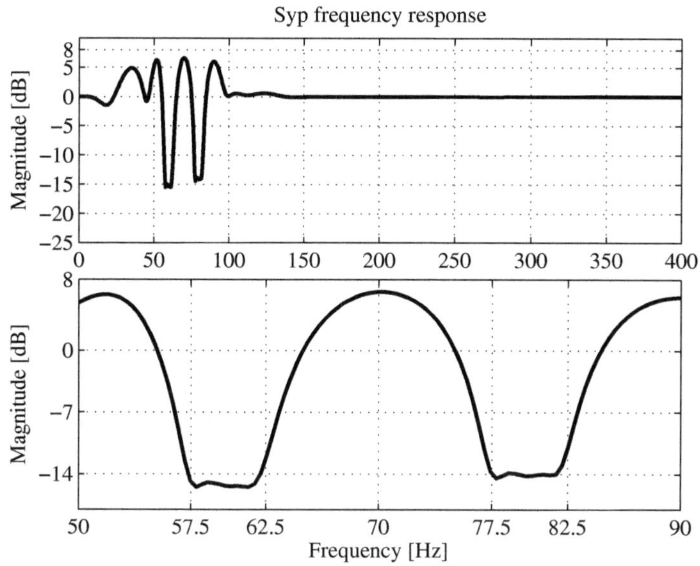

Fig. 11.3 Output sensitivity function with the linear controller (*upper figure*) and zoom in the 50–90 Hz frequency interval (*lower figure*)

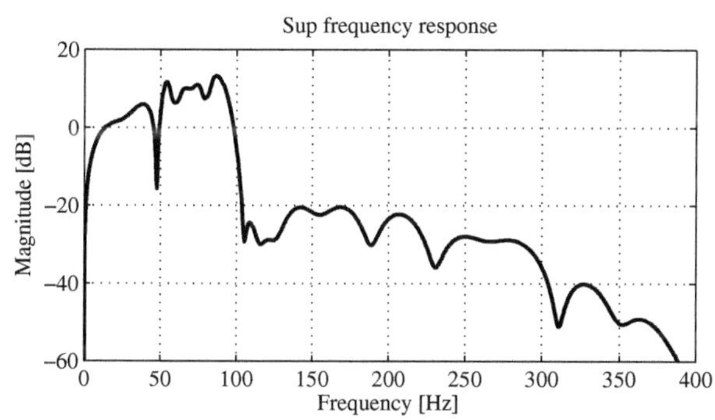

Fig. 11.4 Input sensitivity function with the linear controller

11.4 Experimental Results

The performance of the robust design will be illustrated on the test bench presented in Sect. 2.2. Comparison with the results obtained with an adaptive attenuation scheme will be given in Sect. 12.4.

11.4.1 Two Time-Varying Tonal Disturbances

The results in this subsection are obtained by considering 2 sinusoidal disturbances with time-varying frequencies on the system output. The time variations of the frequencies are obtained by using 2 independent pseudorandom binary sequences (PRBS). The 2 sinusoidal disturbances vary around 60 and 80 Hz, respectively, remaining inside the ± 2.5 Hz frequency intervals for which the robust linear controller introduces 14 dB of attenuation. Figure 11.5 shows the evolution of the frequencies and the corresponding PRBS generators.

Note that all subsequent experiments start at 10 s. This period has been introduced in order to give enough time to activate the electronic boards for real-time experimentation. Also, the system operates in open-loop for 5 s (from 10 to 15 s). Finally, 5 s before the end of the experiments, the system is switched back to open-loop and the system input and the disturbances are removed.

To avoid large transients when switching on the controllers, a bumpless transfer scheme from open to closed-loop has been used (see also [4, Chap. 8]).

In Fig. 11.6, time domain experimental results are shown for the open-loop and the closed-loop with the linear controller. The system operates in open-loop without disturbance during the last 5 s, from 35 to 40 s, so that the residual forces can be compared to the system noise.

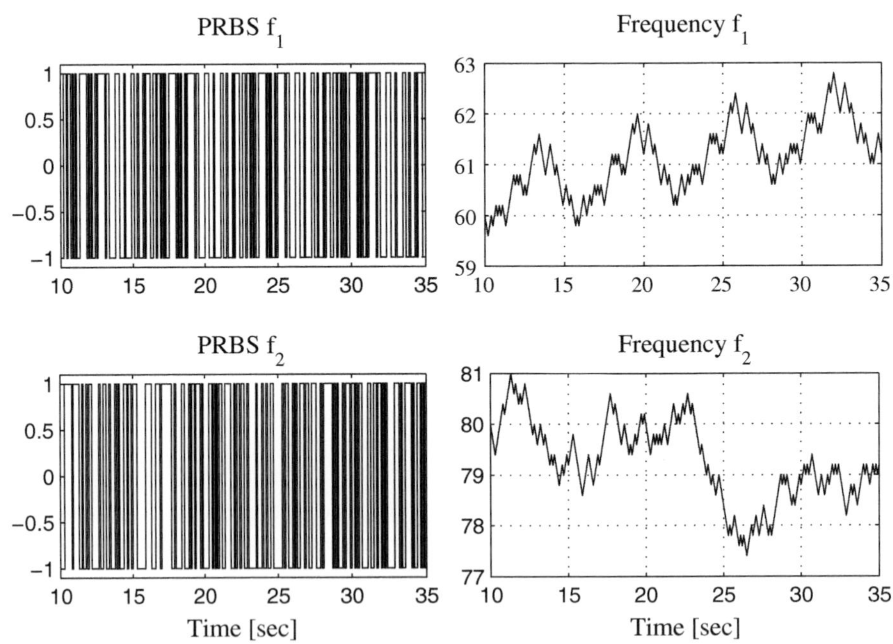

Fig. 11.5 Pseudorandom binary sequences (*left figures*) and evolution of the frequencies of the sinusoidal disturbances (*right figures*)

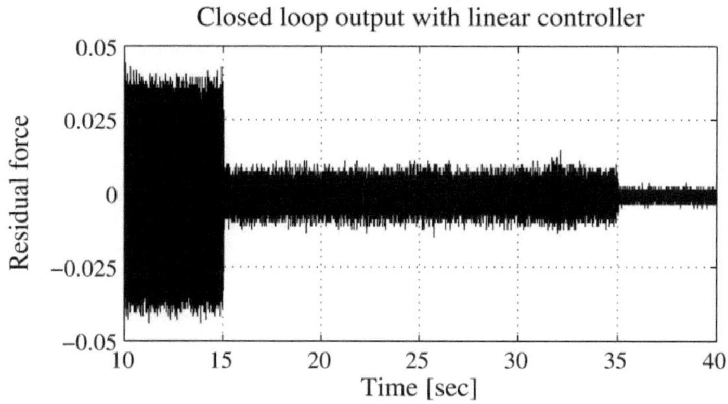

Fig. 11.6 Residual force in closed-loop with linear controller. The experiment is started in open-loop for 5 s. Range of frequency variations: ±2.5 Hz

The global attenuation is computed over the last 3 s of each closed-loop experimentation. For the robust linear controller the global attenuation is 25.70 dB.

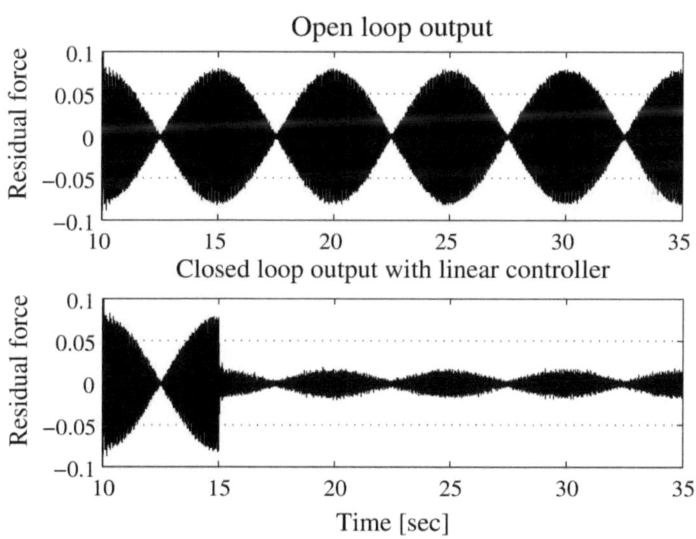

Fig. 11.7 Residual force in open-loop (*upper plot*) and in closed-loop with the linear controller (*lower plot*). The loop is closed at t = 15 s

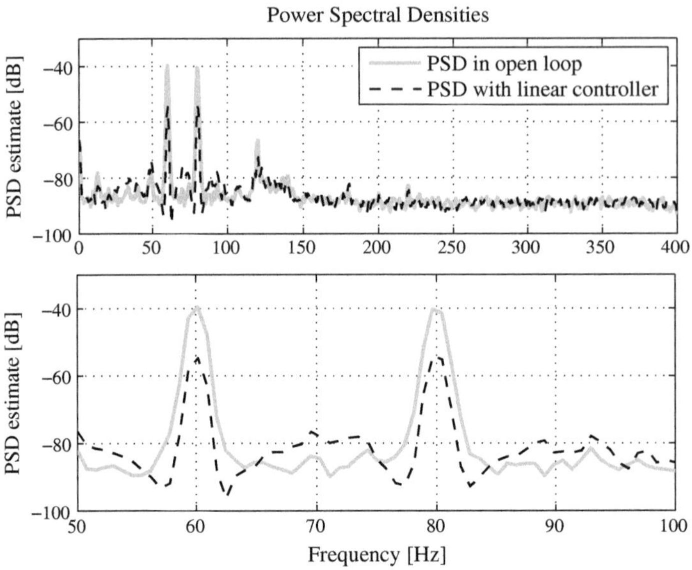

Fig. 11.8 Power spectral densities of the open-loop and the robust linear controller. Full frequency range in the *upper plot*, zoom between 50 and 100 Hz in the *lower plot*

11.4.2 Attenuation of Vibrational Interference

This subsection deals with the attenuation of vibrational interference (two-mode sinusoidal vibrations) on the active vibration control systems using an inertial actuator. It can be shown (see also [1]) that when two sinusoidal disturbances are close enough, a flapping phenomena appears due to the periodic cancellation of the two neighbouring sinusoidal disturbances (vibration interference). This phenomena is shown in Fig. 11.1 where 2 pairs of neighbouring sinusoidal disturbances are introduced, one pair around 60 Hz (at 59.9 and 60.1 Hz) and the second around 80 Hz (at 79.9 and 80.1 Hz). The same robust linear controller as described earlier can be used as its attenuation frequency band is large enough to accommodate the neighbouring disturbances.

Time domain results are shown in Fig. 11.7. The global attenuation for the robust linear controller is 27.50 dB.

The power spectral density (PSD) estimate for the robust linear controller is given in Fig. 11.8. The effective attenuation introduced by the controller action can be seen in Fig. 11.9. It can be observed that the attenuation introduced by the robust linear controller in the desired frequency zone is equal to 14 dB which is coherent with the design done in Sect. 11.3.

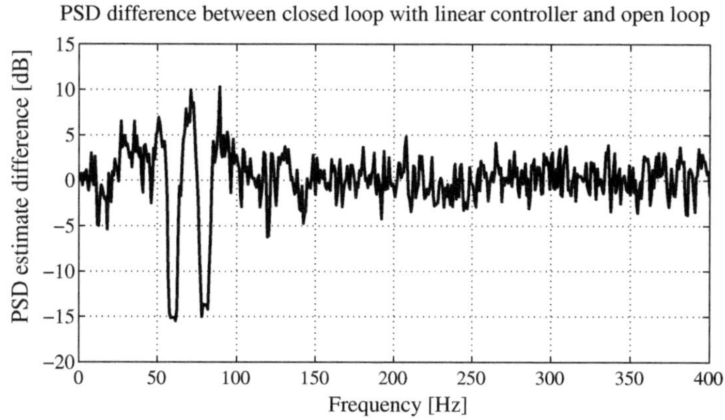

Fig. 11.9 Power spectral densities difference between the closed-loop with the linear controller and the open-loop

11.5 Concluding Remarks

- When the frequencies of single and multiple tonal disturbances vary within limited frequency regions, a robust linear controller design can be considered.
- The level of achievable attenuation depends upon the width of the uncertainty region in the frequency domain (as a consequence of the Bode integral).
- Shaping of the sensitivity functions is necessary in order to avoid unacceptable disturbance amplification in the neighbourhood of the attenuation zones and in order to assure acceptable values for the *modulus margin* and the *delay margin*.
- Pole placement combined with the use of band-stop filters (BSF) allow an efficient shaping of the sensitivity functions.

11.6 Notes and References

Other approaches to the design of robust linear controllers in the context of active vibration control are mentioned hereafter: H_∞ control in [5], H_∞ control with phase and gain policies [6], quantitative robust linear parameter varying (LPV) control [7]. Classical H_∞ and LQR controllers are compared in [8] taking into account the amount of power and energy consumed by the control device. LQR, improved H_∞ designs, and μ synthesis are evaluated in [9] for active vibration control of a flexible beam. Other approaches encountered in AVC systems design include also: repetitive control [10, 11] and synchrophasing in [12].

The technology for the implementation of the AVC in mechanical structures has evolved towards using inertial (electro-dynamic) actuators [13]. In many applications involving collocated actuators and sensors, piezoelectric materials are used (see [14] and various applications reported in [1, 15, 16]).

References

1. Li S, Li J, Mo Y (2014) Piezoelectric multimode vibration control for stiffened plate using ADRC-based acceleration compensation. IEEE Trans Ind Electron 61(12):6892–6902
2. Carmona JC, Alvarado VM (2000) Active noise control of a duct using robust control theory. IEEE Trans Control Syst Technol 8(6):930–938. doi:10.1109/87.880596
3. Langer J, Landau ID (1999) Combined pole placement/sensitivity function shaping method using convex optimization criteria. Automatica 35(6):1111–1120. doi:10.1016/S0005-1098(99)00013-8
4. Landau I, Zito G (2005) Digital control systems - design, identification and implementation. Springer, London
5. van Wingerden J, Hulskamp A, Barlas T, Houtzager I, Bersee H, van Kuik G, Verhaegen M (2011) Two-degree-of-freedom active vibration control of a prototyped “smart” rotor. IEEE Trans Control Syst Technol 19(2):284–296. doi:10.1109/TCST.2010.2051810
6. Zhang K, Scorletti G, Ichchou MN, Mieyeville F (2014) Robust active vibration control of piezoelectric flexible structures using deterministic and probabilistic analysis. J Intell Mater Syst Struct 25(6):665–679. doi:10.1177/1045389X13500574
7. Zhang K, Scorletti G, Ichchou M, Mieyeville F (2015) Quantitative robust linear parameter varying H_∞ vibration control of flexible structures for saving the control energy. J Intell Mater Syst Struct 26(8):1006–1027. doi:10.1177/1045389X14538529
8. Assadian F (2002) A comparative study of optimal linear controllers for vibration suppression. J Frankl Inst 339(3):347–360. The 2001 Franklin Institute Awards
9. Kumar R (2012) Enhanced active constrained layer damping (ACLD) treatment using stand-offlayer: robust controllers design, experimental implementation and comparison. J Vib Control. doi:10.1177/1077546311429148
10. Chen X, Tomizuka M (2014) New repetitive control with improved steady-state performance and accelerated transient. IEEE Trans Control Syst Technol 22(2):664–675. doi:10.1109/TCST.2013.2253102
11. Navalkar S, van Wingerden J, van Solingen E, Oomen T, Pasterkamp E, van Kuik G (2014) Subspace predictive repetitive control to mitigate periodic loads on large scale wind turbines. Mechatronics 24(8):916–925. doi:10.1016/j.mechatronics.2014.01.005
12. Dench M, Brennan M, Ferguson N (2013) On the control of vibrations using synchrophasing. J Sound Vib 332(20):4842–4855. doi:10.1016/j.jsv.2013.04.044. http://www.sciencedirect.com/science/article/pii/S0022460X1300401X
13. Marcos T (2000) The straight attraction. Motion Control 13:29–33
14. Moheimani SR, Fleming AJ (2006) Piezoelectric transducers for vibration control and damping. Springer, London
15. Trindade MA, Pagani CC Jr, Oliveira LP (2015) Semi-modal active vibration control of plates using discrete piezoelectric modal filters. J Sound Vib 351:17–28. doi:10.1016/j.jsv.2015.04.034. http://www.sciencedirect.com/science/article/pii/S0022460X15003867
16. Li S, Qiu J, Ji H, Zhu K, Li J (2011) Piezoelectric vibration control for all-clamped panel using DOB-based optimal control. Mechatronics 21(7):1213–1221. doi:10.1016/j.mechatronics.2011.07.005

Chapter 12
Direct Adaptive Feedback Attenuation of Narrow-Band Disturbances

12.1 Introduction

One of the basic problems in active vibration control and active noise control is the *strong* attenuation of single or multiple unknown and time-varying narrow-band disturbances[1] without measuring them. In this context, an adaptive feedback approach, termed as *adaptive regulation* is now generally used. In contrast with the feedforward compensation approach [1–4], the feedback approach, does not require an additional measurement highly correlated with the disturbance. This is a significant advantage. Feedback approaches avoid also the possible destabilizing positive feedback coupling between the compensator system and the measurement of the disturbance which occurs often in feedforward compensation schemes (see [5] and Sect. 1.5) and require less parameters to adapt.

A common assumption is that the disturbance is white noise or a Dirac impulse passed through a filter which characterizes the model of the disturbance.[2] To be more specific, the disturbances considered can be defined as "finite band disturbances." This includes single or multiple narrow-band disturbances or sinusoidal signals. For the purpose of this chapter, the disturbances are considered to be unknown and time-varying, in other words, their model has time-varying coefficients. This motivates the use of an adaptive regulation approach since the objective is to attenuate unknown disturbances without measuring them.

A popular methodology for this regulation problem in the case when the model of the disturbance is known, is to design a controller that incorporates the model of the disturbance (internal model principle—IMP). This technique has been introduced in Sect. 7.2.2 and has its roots described in the paper [6]. Additional references for the present context are [7–9]. The main problem, using the IMP principle, is that complete rejection of the disturbances is attempted (asymptotically) and this

[1]Called *tonal* disturbances.

[2]Throughout the chapter, it is assumed that the number of narrow-band disturbances is known (it can be estimated from data if necessary) but not their frequency characteristics.

I.D. Landau et al., *Adaptive and Robust Active Vibration Control*,
Advances in Industrial Control, DOI 10.1007/978-3-319-41450-8_12

may have a strong influence upon the sensitivity functions outside the frequency band in which attenuation is achieved. As long as rejection of a single narrow-band disturbance is considered [7, 9], the influence upon the output sensitivity functions does in general not pose problems. Nevertheless, if low damped complex zeros are located near the disturbance frequency, even in a single narrow-band disturbance context, the influence over $S_{yp}(z^{-1})$ represents a major challenge [10].

The IMP principle will be combined with a Youla–Kučera parametrization of the controller (see Sect. 7.2.3) which will allow to develop a direct adaptive regulation strategy. The parameters of the (Q) Youla–Kučera filter will be directly adapted in order to cancel the effect of the disturbance.

When multiple narrow-band disturbances are considered, the use of the (IMP) approach requires a very careful design of the central linear controller in order to avoid unacceptable water-bed effects (unwanted amplification on the output sensitivity function at certain frequencies). The problem of adaptive attenuation of multiple unknown narrow-band disturbance distributed over a large frequency range will be discussed in Chap. 13.

12.2 Direct Adaptive Feedback Attenuation of Unknown and Time-Varying Narrow-Band Disturbances

12.2.1 Introduction

The objective is to reject asymptotically or strongly attenuate single or multiple narrow-band disturbances which have unknown or time-varying spikes in the frequency domain. To asymptotically reject the disturbance, the internal model principle (IMP) has to be applied. As a consequence, the controller should include a model of the disturbance. Since the disturbances are unknown, two approaches can be considered

- Indirect adaptive regulation (one has to identify the model of the disturbance and recompute the controller which will include the estimated model of the disturbance).
- Direct adaptive regulation (the controller parameters will be directly adapted).

An important issue is the tuning of the controller as a function of the model of the disturbance but without affecting the stability of the closed-loop. It turns out that Youla–Kučera parametrization provides the good parametrization of the controller for decoupling the stability of the closed-loop from the attenuation problem. It also provides a disturbance observer. A rapprochement with the DOB control method [11, 12] can be established.

Indirect adaptive regulation solutions can be also build, however they are much more complex [7, 13] and their use should be justified (specific performance requirements). This approach will be discussed in Sect. 13.4.

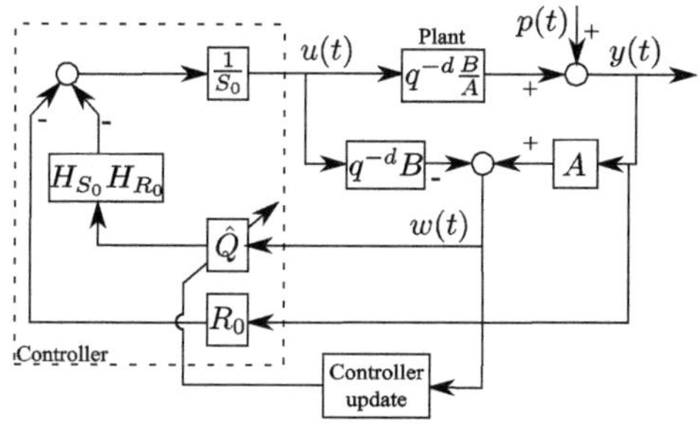

Fig. 12.1 Direct adaptive regulation scheme for rejection of unknown disturbances

Figure 12.1 gives the block diagram of the direct adaptive regulation scheme for attenuation of unknown narrow-band disturbances. $q^{-d}B/A$ defines the model of the secondary path (called also plant), $\hat{Q}$ represents the so-called *YK filter*. R_0 and S_0 represents, which is called, the *central controller*. The output of the plant $y(t)$ and the input $u(t)$ in the absence of the Youla–Kučera filters,[3] may be written as (consider Fig. 12.1 without the filter $\hat{Q}$)

$$y(t) = \frac{q^{-d}B(q^{-1})}{A(q^{-1})} \cdot u(t) + p(t), \tag{12.1}$$

$$S_0(q^{-1}) \cdot u(t) = -R_0(q^{-1}) \cdot y(t). \tag{12.2}$$

In (12.1), $p(t)$ is the effect of the disturbances on the measured output[4] and $R_0(z^{-1})$, $S_0(z^{-1})$ are polynomials in z^{-1} having the following expressions[5]

$$S_0 = 1 + s_1^0 z^{-1} + \cdots + s_{n_{S_0}}^0 z^{-n_{S_0}} = S_0'(z^{-1}) \cdot H_{S_0}(z^{-1}), \tag{12.3}$$

$$R_0 = r_0^0 + r_1^0 z^{-1} + \cdots + r_{n_{R_0}}^0 z^{-n_{R_0}} = R_0'(z^{-1}) \cdot H_{R_0}(z^{-1}), \tag{12.4}$$

where $H_{S_0}(q^{-1})$ and $H_{R_0}(q^{-1})$ represent prespecified parts of the controller (used for example to incorporate the internal model of a disturbance or to open the loop at certain frequencies) and $S_0'(q^{-1})$ and $R_0'(q^{-1})$ are computed using Pole Placement (see Chap. 7). The characteristic polynomial, which specifies the desired closed-loop poles of the system is given by (see also [14])[6]

$$P_0(z^{-1}) = A(z^{-1})S_0(z^{-1}) + z^{-d}B(z^{-1})R_0(z^{-1}), \tag{12.5}$$

[3]The Youla–Kučera parametrization has been presented in Chap. 7.

[4]The disturbance passes through a so called *primary path* which is not represented in this figure, and $p(t)$ is its output.

[5]The argument (z^{-1}) will be omitted in some of the following equations to make them more compact.

[6]It is assumed that a reliable model identification is achieved and therefore the estimated model is assumed to be equal to the true model.

Introducing the expressions of S_0 and R_0 given in Eqs. (12.3) and (12.4), R'_0 and S'_0 are solutions of

$$P_0(z^{-1}) = A(z^{-1})S'_0(z^{-1})H_{S_0}(q^{-1}) + z^{-d}B(z^{-1})R'_0(z^{-1})H_{R_0}(q^{-1}), \qquad (12.6)$$

In what follows the Youla–Kučera parametrization [15, 16] is used. Nevertheless, the Youla–Kučera parametrization is not unique. It depends on the right coprime factorization selected $G = ND^{-1}$. Four factorizations are mostly used [17]

$$N = G; \quad D = I. \qquad (12.7)$$

$$N = z^{-m}; \quad D = P_m \quad \text{with} \quad G \approx z^{-m}P_m^{-1}. \qquad (12.8)$$

$$N = q^{-d}B; \quad D = A \quad \text{with} \quad G = q^{-d}\frac{B}{A}. \qquad (12.9)$$

$$N = q^{-d}BF; D = AF \quad \text{with} \quad G = q^{-d}\frac{B}{A}; F = \frac{F_N}{F_D}, \qquad (12.10)$$

with F and F^{-1} asymptotically stable. More details can be found in [17]. Subsequently the parametrization (12.9) will be used.

Selecting a FIR structure for the Q filter associated to the Youla–Kučera parametrization, the controller's polynomials become

$$R = R_0 + AQH_{S_0}H_{R_0}, \qquad (12.11)$$

$$S = S_0 - z^{-d}BQH_{S_0}H_{R_0}, \qquad (12.12)$$

where R_0 and S_0 define the central controller which verifies the desired specifications in the absence of the disturbance. The characteristic polynomial of the closed-loop is still given by (12.6) (can be verified by simple calculations). The output sensitivity function (the transfer function between the disturbance $p(t)$ and the output of the system $y(t)$) is

$$S_{yp}(z^{-1}) = \frac{A(z^{-1})S(z^{-1})}{P_0(z^{-1})} \qquad (12.13)$$

and the input sensitivity function (the transfer function between the disturbance $p(t)$ and the control input $u(t)$) is

$$S_{up}(z^{-1}) = -\frac{A(z^{-1})R(z^{-1})}{P_0(z^{-1})}, \qquad (12.14)$$

A key aspect of this methodology is the use of the internal model principle (IMP) which has been discussed in Chap. 7, Sect. 7.2.2. It is supposed that $p(t)$ is a deterministic disturbance given by

$$p(t) = \frac{N_p(q^{-1})}{D_p(q^{-1})} \cdot \delta(t), \tag{12.15}$$

where $\delta(t)$ is a Dirac impulse and N_p, D_p are coprime polynomials of degrees n_{N_p} and n_{D_p}, respectively.[7] In the case of stationary narrow-band disturbances, the roots of $D_p(z^{-1})$ are on the unit circle.

Applying the internal model principle (IMP), the prespecified part of $S(z^{-1})$ should incorporate the denominator of the model of the disturbance D_p, i.e.,

$$H_S(z^{-1}) = D_p(z^{-1})H_{S_0}(z^{-1}).$$

The controller is computed solving

$$P = AD_pH_{S_0}S' + z^{-d}BH_{R_0}R', \tag{12.16}$$

where P, D_p, A, B, H_{R_0}, H_{S_0} and d are given.[8] In the context of the Youla–Kučera controller parametrization using a FIR Q filter,

$$Q(z^{-1}) = q_0 + q_1 z^{-1} + \cdots + q_{n_Q} z^{-n_Q}. \tag{12.17}$$

application of the internal model principle leads to the problem of finding Q such that

$$S = S_0' H_{S_0} - z^{-d} BQH_{S_0} H_{R_0} = D_p H_{S_0} S' \tag{12.18}$$

So in order to compute the corresponding Q-polynomial one has to solve the diophantine equation

$$S'D_p + z^{-d}BH_{R_0}Q = S_0', \tag{12.19}$$

where D_p, d, B, S_0', and H_{R_0} are known and S' and Q are unknown. Equation (12.19) has a unique solution for S' and Q with: $n_{S_0'} \leq n_{D_p} + n_B + d + n_{H_{R_0}} - 1$, $n_{S'} = n_B + d + n_{H_{R_0}} - 1$, $n_Q = n_{D_p} - 1$. One sees that the order n_Q of the polynomial Q depends upon the structure of the disturbance model and not upon the structure of the plant model.

The use of the Youla–Kučera parametrization, with Q given in (12.17), is interesting in this case because it allows to maintain the closed-loop poles as given by the central controller but at the same time it introduces the parameters of the internal model into the controller.

---

[7] Throughout the book, n_X denotes the degree of the polynomial X.

[8] Of course, it is assumed that D_p and B do not have common factors.

12.2.2 Direct Adaptive Regulation Using Youla–Kučera Parametrization

The objective is to find an estimation algorithm which will directly estimate the parameters of the internal model in the controller in the presence of an unknown disturbance (but of known structure) without modifying the closed-loop poles. Clearly, the Q-parametrization is a potential option since modifications of the Q-polynomial will not affect the closed-loop poles. In order to build an estimation algorithm it is necessary to define an error equation which will reflect the difference between the optimal Q-polynomial and its current value. Note also that in the time domain, the internal model principle can be interpreted as finding Q such that asymptotically y(t) becomes zero [16]. Using the Q-parametrization, the output of the system in the presence of a disturbance can be expressed as

$$y(t) = \frac{A[S_0 - q^{-d}BH_{S_0}H_{R_0}Q]}{P} \cdot \frac{N_p}{D_p} \cdot \delta(t) = \frac{S_0 - q^{-d}BH_{S_0}H_{R_0}Q}{P} \cdot w(t), \quad (12.20)$$

where $w(t)$ is given by (see also Fig. 12.1)

$$w(t) = \frac{AN_p}{D_p} \cdot \delta(t) = A \cdot y(t) - q^{-d}B \cdot u(t). \quad (12.21)$$

Taking into consideration that the adaptation of Q is done in order to obtain an output $y(t)$ which tends asymptotically to zero, one can define $\varepsilon^0(t+1)$ as the value of $y(t+1)$ obtained with $\hat{Q}(t, q^{-1})$ (the estimate of Q at time t, written also $\hat{Q}(t)$)

$$\varepsilon^\circ(t+1) = \frac{S_0}{P} \cdot w(t+1) - \hat{Q}(t)\frac{q^{-d}B^*H_{S_0}H_{R_0}}{P} \cdot w(t). \quad (12.22)$$

Similarly, the *a posteriori* error becomes (using $\hat{Q}(t+1)$) as[9]

$$\varepsilon(t+1) = \frac{S_0}{P} \cdot w(t+1) - \hat{Q}(t+1)\frac{q^{-d}B^*H_{S_0}H_{R_0}}{P} \cdot w(t). \quad (12.23)$$

Replacing S_0 from the last equation by its expression given in (12.3) and using (12.19) for S_0', one obtains

$$\varepsilon(t+1) = [Q - \hat{Q}(t+1)] \cdot \frac{q^{-d}B^*H_{S_0}H_{R_0}}{P} \cdot w(t) + \eta(t+1), \quad (12.24)$$

[9]In adaptive control and estimation the predicted output at $t+1$ can be computed either on the basis of the previous parameter estimates (*a priori*, time t) or on the basis of the current parameter estimates (*a posteriori*, time $t+1$).

where

$$\eta(t) = \frac{S' D_p H_{S_0}}{P} \cdot w(t) = \frac{S' H_{S_0} A N_p}{P} \cdot \delta(t) \tag{12.25}$$

is a signal which tends asymptotically towards zero since P is an asymptotically stable polynomial.

Define the estimated polynomial $\hat{Q}(t, q^{-1}) = \hat{q}_0(t) + \hat{q}_1(t) q^{-1} + \cdots + \hat{q}_{n_Q}(t) q^{-n_Q}$ and the associated estimated parameter vector $\hat{\theta}(t) = [\hat{q}_0(t)\ \hat{q}_1(t)\ \ldots\ \hat{q}_{n_Q}(t)]^T$. Define the fixed parameter vector corresponding to the optimal value of the polynomial Q as: $\theta = [q_0\ q_1\ \ldots\ q_{n_Q}]^T$.

Denote

$$w_2(t) = \frac{q^{-d} B^* H_{S_0} H_{R_0}}{P} \cdot w(t) \tag{12.26}$$

and define the following observation vector

$$\phi^T(t) = [w_2(t)\ w_2(t-1)\ \ldots\ w_2(t-n_Q)]. \tag{12.27}$$

Equation (12.24) becomes

$$\varepsilon(t+1) = [\theta^T - \hat{\theta}^T(t+1)] \cdot \phi(t) + v(t+1). \tag{12.28}$$

One can remark that $\varepsilon(t+1)$ corresponds to an *a posteriori* adaptation error [18] and therefore the basic adaptation algorithm given in Chap. 4 can be used. From (12.22), one obtains the *a priori* adaptation error

$$\varepsilon^\circ(t+1) = w_1(t+1) - \hat{\theta}^T(t)\phi(t), \tag{12.29}$$

with

$$w_1(t+1) = \frac{S_0(q^{-1})}{P(q^{-1})} \cdot w(t+1), \tag{12.30}$$

$$w(t+1) = A(q^{-1}) \cdot y(t+1) - q^{-d} B^*(q^{-1}) \cdot u(t), \tag{12.31}$$

where $B(q^{-1})u(t+1) = B^*(q^{-1})u(t)$.

The *a posteriori* adaptation error is obtained from (12.23)

$$\varepsilon(t+1) = w_1(t+1) - \hat{\theta}^T(t+1)\phi(t). \tag{12.32}$$

For the estimation of the parameters of $\hat{Q}(t, q^{-1})$ the following PAA (I-PAA) is used [18]

$$\hat{\theta}(t+1) = \hat{\theta}(t) + F(t)\phi(t)\varepsilon(t+1), \tag{12.33}$$

$$\varepsilon(t+1) = \frac{\varepsilon^\circ(t+1)}{1 + \phi^T(t) F(t) \phi(t)}, \tag{12.34}$$

$$\varepsilon^{\circ}(t+1) = w_1(t+1) - \hat{\theta}^T(t)\phi(t), \tag{12.35}$$

$$F(t+1) = \frac{1}{\lambda_1(t)}\left[F(t) - \frac{F(t)\phi(t)\phi^T(t)F(t)}{\frac{\lambda_1(t)}{\lambda_2(t)} + \phi^T(t)F(t)\phi(t)}\right], \tag{12.36}$$

$$1 \geq \lambda_1(t) > 0, \; 0 \leq \lambda_2(t) < 2, \tag{12.37}$$

where $\lambda_1(t)$, $\lambda_2(t)$ allow to obtain various profiles for the evolution of the adaptation gain $F(t)$ (for details see Sect. 4.3.1 and [14, 18]). Two modes of operation are considered

- *Adaptive operation* (the adaptation is performed continuously and the controller is updated at each sampling). In this case the adaptation gain never goes to zero; and
- *Self-tuning operation* (the adaptation procedure starts either on demand or when the performance is unsatisfactory). In this case the adaptation gain goes asymptotically towards 0.

Stability of the resulting scheme is a consequence of the results given in Chap. 4, Sect. 4.4.2. Considering (12.28) and neglecting the signal $v(t+1)$, which goes to 0 anyway, one concludes using Theorem 4.1 that $\varepsilon(t+1)$ goes to zero without any positive real condition to be satisfied. Furthermore, if the number of sinusoidal disturbances is n it can be shown that there is also parameter convergence if $n_Q = 2n - 1$. For a detailed stability proof under the hypothesis *model* = *plant* see [7, 18].

The following procedure is applied at each sampling time for adaptive operation:

1. Get the measured output $y(t+1)$ and the applied control $u(t)$ to compute $w(t+1)$ using (12.31).
2. Compute $w_1(t+1)$ and $w_2(t)$ using (12.30) and (12.26) with P given by (7.38).
3. Estimate the Q-polynomial using the parameter adaptation algorithm (12.33)–(12.36).
4. Compute and apply the control (see Fig. 12.1):

$$\begin{aligned} S_0(q^{-1})u(t+1) = &- R_0(q^{-1})y(t+1) \\ &- H_{S_0}(q^{-1})H_{R_0}(q^{-1})\hat{Q}(t+1, q^{-1})w(t+1) \end{aligned} \tag{12.38}$$

The explicit expression for the control $u(t)$ is given by

$$u(t) = -R_0(q^{-1})y(t) - S_0^*(q^{-1})u(t-1) - H_{S_0}(q^{-1})H_{R_0}(q^{-1})\hat{Q}(t, q^{-1})w(t), \tag{12.39}$$

12.2.3 Robustness Considerations

To avoid unacceptable high values of the modulus of the output sensitivity function when internal model principle is used, a robust design for the central controller should be considered assuming that the model of the disturbance and its domain of variation in the frequency domain are known. The objective is that for all situations, acceptable modulus margin ($|S_{yp}(e^{-j\omega})|^{-1}_{\max}$) and delay margin are obtained.

Furthermore, at the frequencies where perfect rejection of the disturbance is achieved one has $S_{yp}(e^{-j\omega}) = 0$ and

$$\left|S_{up}(e^{-j\omega})\right| = \left|\frac{A(e^{-j\omega})}{B(e^{-j\omega})}\right|. \tag{12.40}$$

Equation (12.40) corresponds to the inverse of the gain of the system to be controlled. The implication of Eq. (12.40) is that cancellation (or in general an important attenuation) of disturbances on the output should be done only in frequency regions where the system gain is large enough. If the gain of the controlled system is too low, $|S_{up}|$ will be large at these frequencies. Therefore, the robustness versus additive plant model uncertainties will be reduced and the stress on the actuator will become important [13].

Equation (12.40) also implies that serious problems will occur if $B(z^{-1})$ has complex zeros close to the unit circle (stable or unstable zeros) at frequencies where an important attenuation of disturbances is required. It is mandatory to avoid attenuation of disturbances at these frequencies and special attention should be given to the behaviour of the controller in the regions of attenuation close to these low damped complex zeros [9, 10].

12.3 Performance Evaluation Indicators for Narrow-Band Disturbance Attenuation

Before presenting the experimental results obtained it is important to clearly define the performance indices for narrow-band disturbance attenuation and the corresponding measurement methods.

Tuning Capabilities

Tuning capabilities concern evaluation of the performance in steady state operation after application of the disturbance once the adaptation transient settles. The corresponding indicators are evaluated in the presence of narrow-band disturbances with constant frequency. Three indicators are considered

1. **Global attenuation** (GA): measured in dB and defined by

$$GA = 20\log_{10}\frac{N^2Y_{ol}}{N^2Y_{cl}}, \tag{12.41}$$

where N^2Y_{ol} and N^2Y_{cl} correspond to the square of the truncated 2-norm of the measured residual force in open-loop and in closed-loop, respectively, evaluated

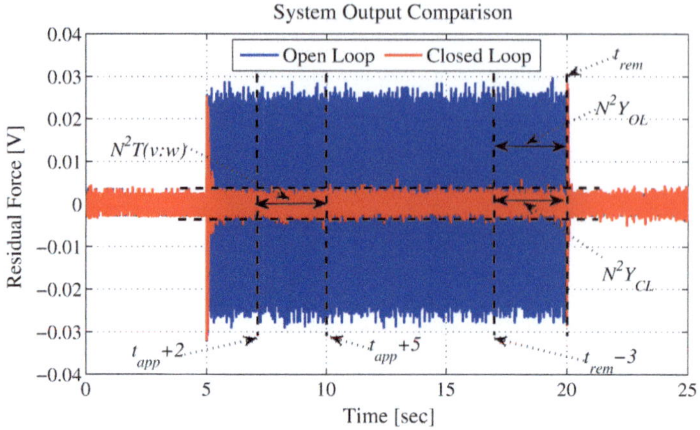

Fig. 12.2 Definitions of the time intervals for global attenuation (GA) and transient evaluation. The intervals of computation ($t_{app}+2$, $t_{app}+5$, $t_{rem}-3$, t_{rem}) are displayed. (t_{app}—time of application of the disturbance, t_{rem}—time of removal of the disturbance)

during the last part of the experiment before the disturbance is removed (between $t_{rem}-3s$ and t_{rem} in Fig. 12.2 which illustrates the procedure).
The truncated 2-norm has the following expression

$$N^2T = \sum_{i=1}^{m} y(i)^2, \tag{12.42}$$

where $y(i)$ is a sample of the discrete time signal (residual force or acceleration). This quantity indicates the energy contained in the measured signal.

2. **Disturbance attenuation** (DA): measured in dB. It is defined as the difference between the estimated Power Spectral Density (PSD) of the residual force in closed-loop and in open-loop at the frequency of the disturbance as shown in Fig. 12.3 for the case of two tonal disturbances. Its expression is

$$DA = \min\left(PSD_{cl} - PSD_{ol}\right), \tag{12.43}$$

3. **Maximum amplification** (MA): measured in dB, it is defined as the maximum value of the difference between the estimated PSD of the residual force in closed-loop and open-loop. It is defined by

$$MA = \max\left(PSD_{cl} - PSD_{ol}\right). \tag{12.44}$$

Note that the inverse of the maximum amplifications gives the *modulus margin*.

Through these three measurements, it is possible to assess the performance of the controller in terms of disturbance attenuation (global and disturbance attenuations) and to analyse the robustness (maximum amplification and modulus margin).

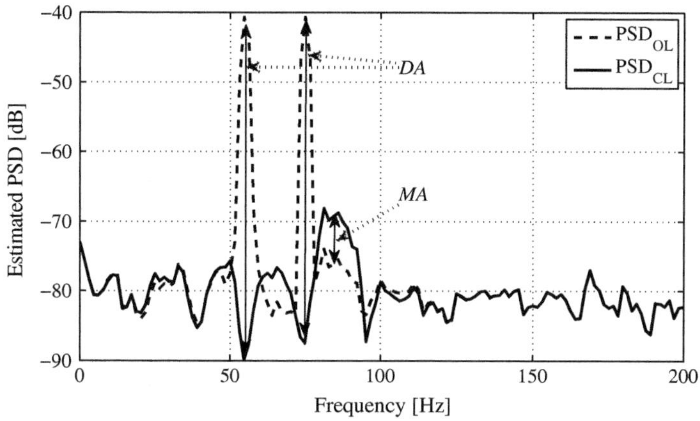

Fig. 12.3 Definition of disturbance attenuation (*DA*) and Maximum amplification (*MA*)

Transient Performance:

The transient performance is evaluated for a constant frequency step change.

- Transient time evaluation: It is required that the transient duration, when a disturbance is applied, be smaller than a certain value (in what follows one considers a desired transient duration of 2 s). A performance index is established for 100 % of fulfilment (transient duration equal or less than 2 s).[10] This means that 2 s after the application of a disturbance the square of the truncated 2-norm of the residual force (acceleration) has to be equal to or smaller than 1.21 of the steady-state value of the square of the truncated 2-norm of the residual force. The square of the truncated 2-norm is evaluated over an interval of 3 s both for transient and steady state. Taking into account the instant of application of the disturbance t_{app} and the instant when the disturbance is removed t_{rem}, the square of the truncated 2-norm is denoted as $N2T(v : w)$ where v and w define the interval of computation. If the square of the truncated 2-norm of the residual force (acceleration) is equal or higher than 2.42 of the square of the truncated 2-norm of the residual force then the value of the index is 0 %. One defines

$$\alpha = \frac{N^2T\left(t_{\text{app}} + 2 : t_{\text{app}} + 5\right)}{N^2T\left(t_{\text{rem}} - 3 : t_{\text{rem}}\right)} = \frac{N^2T\left(t_{\text{app}} + 2 : t_{\text{app}} + 5\right)}{N^2Y_{CL}} \tag{12.45}$$

[10] Of course the value of 2 s can be changed, but the principle of measurement remains the same.

and the transient duration index $\Delta Trans$ is given by

$$\Delta Trans = 100\,\% \text{ if } \alpha \leq 1.21 \tag{12.46a}$$

$$\Delta Trans = \frac{2.42 - \alpha}{1.21} 100\,\% \text{ if } \alpha > 1.21 \tag{12.46b}$$

$$\Delta Trans = 0\,\% \text{ if } \alpha \geq 2.42 \tag{12.46c}$$

12.4 Experimental Results: Adaptive Versus Robust

The experimental results presented in this section are related to the experimental results presented in Chap. 11 and obtained with a robust controller on the active vibration control system using an inertial actuator for the case of multiple narrow-band disturbances located in a limited region of the frequency domain.

12.4.1 Central Controller for Youla–Kučera Parametrization

The design of the central controller used in the Youla–Kučera parametrization is similar to the design of the robust linear controller described in Chap. 11 with the exception that the BSFs on S_{yp} have not been used and the resulting free auxiliary roots to be assigned have been moved from 0 to 0.2. Remark that the order of the characteristic polynomial is given by $n_P = n_A + n_B + n_{H_S} + n_{H_R} + d - 1$ which in this case gives $22 + 25 + 0 + 4 + 0 - 1 = 50$. Given the roots already specified (28 as can be concluded from the design of the robust controller excepting roots given by BSFs for S_{yp}), it follows that 22 roots can be selected. These 22 auxiliary poles at 0.2 have the effect of reducing the magnitude of S_{up} above 100 Hz. They were not used in the robust linear design.

12.4.2 Two Single-Mode Vibration Control

The results in this subsection are obtained by considering two sinusoidal disturbances with time-varying frequencies located in two distinct regions of the frequency domain. The time variations of the frequencies are obtained by using two independent pseudo random binary sequences (PRBS). The two sinusoidal disturbances vary around 60 and 80 Hz, respectively, remaining inside the ±2.5 Hz (like for the robust control design discussed in Chap. 11) or ±5 Hz frequency intervals. See Fig. 11.5 in Chap. 11.

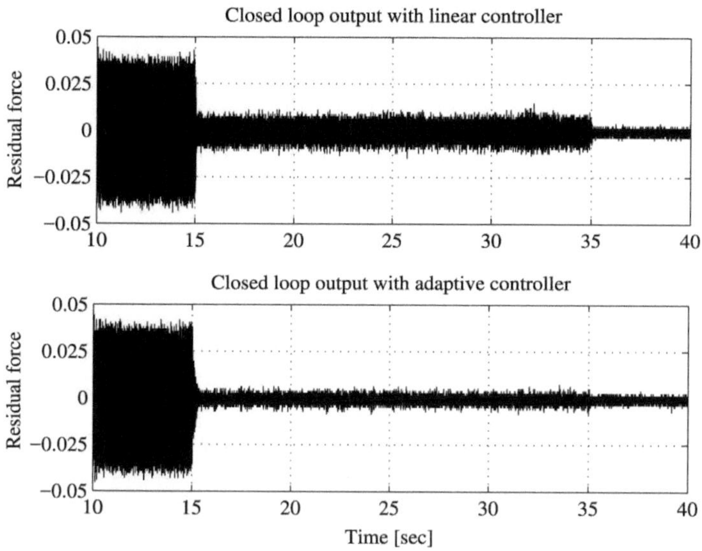

Fig. 12.4 Residual force in closed-loop with linear robust controller (*upper plot*) and with adaptive controller (*lower plot*). The experiments are started in open-loop for 5 s and the disturbances are removed at t = 35 s. Range of frequency variation: ±2.5 Hz

Note that all subsequent experiments (like for the robust controller case) start at 10 s. This period has been introduced in order to give enough time to activate the electronic boards for real-time experimentation. Also, the system operates in open-loop for 5 s (from 10 to 15 s). Finally, 5 s before the end of the experiments, the system is switched back to open-loop and the system input and the disturbances are removed (between 35 and 40 s). To avoid large transients when switching on the controllers, a bumpless transfer scheme from open to closed-loop has been used like in the experiments with the robust controller (see also [14, Chap. 8]).

In Fig. 12.4, time domain experimental results are shown for the open-loop, the closed-loop with the robust linear controller and for the closed-loop with the adaptive controller. As it can be observed, for the adaptive regulator, the residual force is almost the same level as the system noise.

For adaptive regulation, the I-PAA is used. The matrix $F(0)$ is chosen diagonal with 0.2 being the value of each diagonal element (trace = 0.8). A constant trace adaptation algorithm is used with constant trace of 0.8 (see Sect. 4.3.4 for further details on the choice of the adaptation gain). The order of the Q-polynomial has been chosen equal to 3 (4 adapted parameters). The evolution of the parameters of the Q-polynomial can be viewed in Fig. 12.5. As it can be observed, the vector of the estimated Q-parameters, $\hat{\theta}$ is initialized at zero. Once the loop is closed, the adaptive algorithm starts to adjust the parameters in order to reduce the residual force. It can be seen that the parameters of the Youla–Kučera filter evolve continuously during the experiments in order to adjust to the changing frequencies of the disturbances.

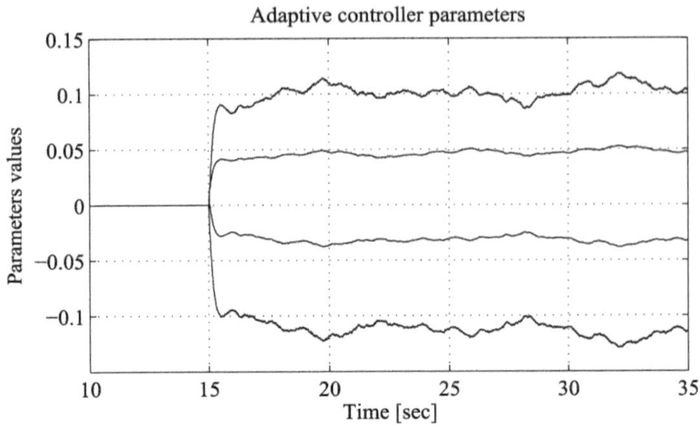

Fig. 12.5 Evolution of the Q-parameters

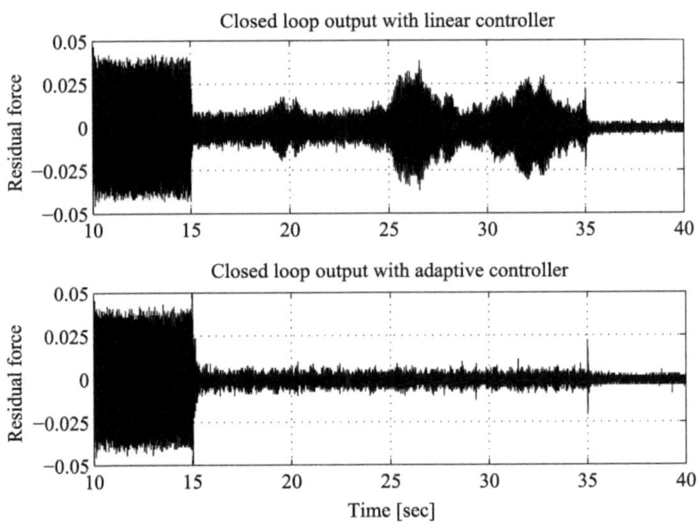

Fig. 12.6 Residual force in closed-loop with linear robust controller (*upper plot*) and with adaptive controller (*lower plot*). The experiments are started in open-loop for 5 s. Range of frequency variation: ±5 Hz

The global attenuation is computed over the last 3 s for each closed-loop experimentation. For the robust linear controller the global attenuation is 25.70 dB, while in the adaptive case it is 39.68 dB. A small additional improvement can be obtained by using the "Integral + Proportional" parameter adaptation algorithm (IP-PAA) described in Appendix E.

Finally, experimental results for frequencies variations of ±5 Hz intervals around 60 and 80 Hz are shown in Fig. 12.6. As expected the results provided by the robust linear controller are not good (we are outside of the domain considered for design). The last 5 s without disturbance are also plotted as reference.

Clearly on one hand the use of the adaptive regulator allows to improve the performance of the robust controller even if the domain of variations of the frequency of the disturbances is the one used for design and on the other hand it allows to expand the domain of variations of the frequencies of the disturbances for which the attenuation performances are assured.

12.4.3 Vibrational Interference

This subsection deals with the adaptive attenuation of two vibrational interferences located in two distinct frequency regions. This phenomenon is shown in Fig. 11.1, Chap. 11, where 2 pairs of neighbouring sinusoidal disturbances are introduced, one pair around 60 Hz (at 59.9 and 60.1 Hz) and the second around 80 Hz (at 79.9 and 80.1 Hz). The results obtained with the adaptive approach will be compared with those obtained with the robust linear controller designed in Chap. 11.

For adaptive regulation, the I-PAA has been used with an initial diagonal adaptation gain matrix $F(0) = \alpha \cdot I$, with $\alpha = 0.2$ and I the identity matrix (initial trace of 0.8), and a decreasing gain followed by constant trace adaptation. The constant trace is chosen equal to 0.02. The number of parameters for the Q-polynomial is also equal to 4 (order equal to 3). Augmenting the order of the polynomial Q to 7 (8 parameters—two for each sinusoidal disturbance) does not improve the performance (probably because the frequencies of the pair of sines are too close). Time domain

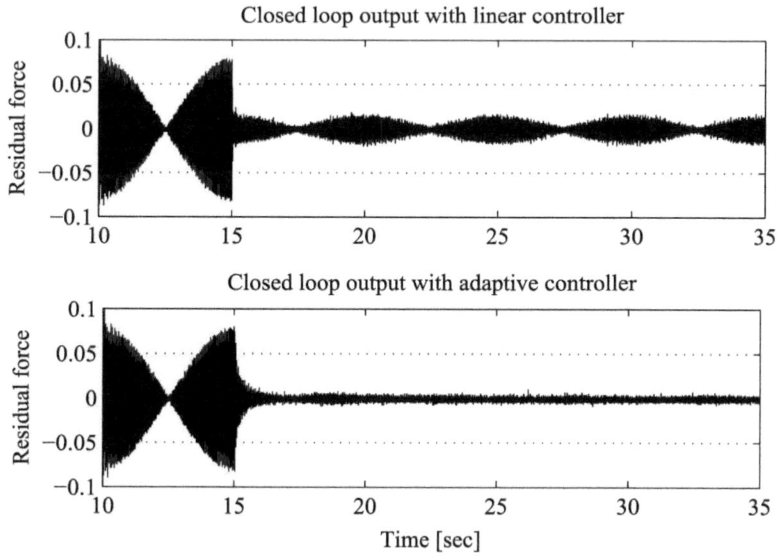

Fig. 12.7 Residual force in closed-loop with linear controller (*upper plot*) and with adaptive controller (*lower plot*). The loop is closed at t = 15 s

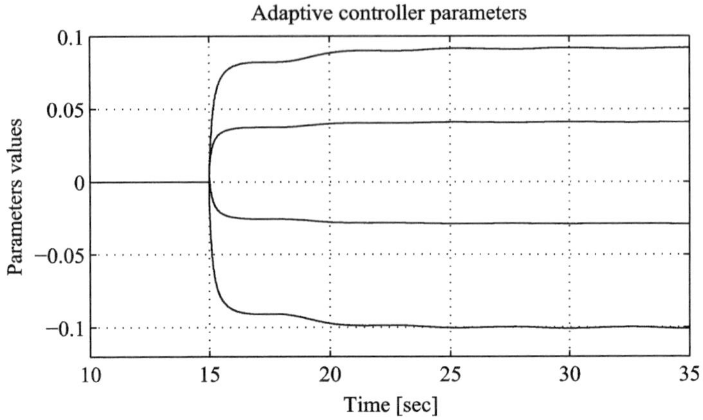

Fig. 12.8 Evolution of the Q-parameters

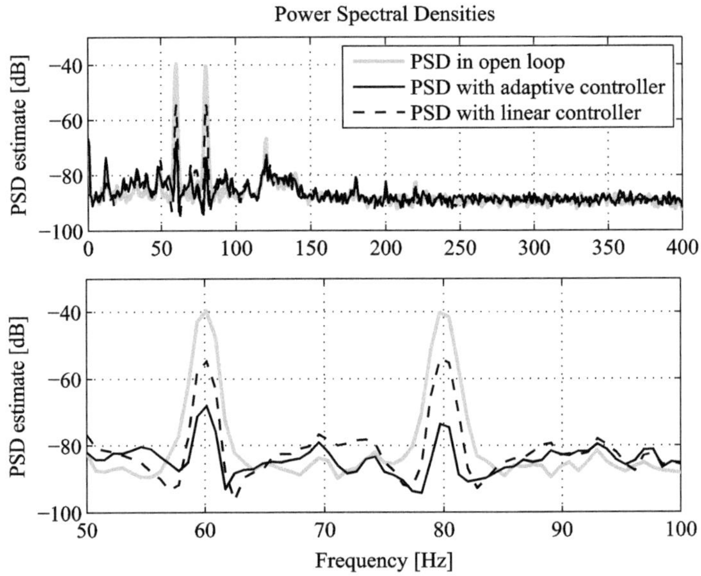

Fig. 12.9 Power spectral densities of the open-loop, robust linear controller, and adaptive regulator. Full frequency range in the *upper plot*, zoom between 50 and 100 Hz in the *lower plot*

results are shown in Figs. 12.7 and 12.8. The global attenuation for the robust linear controller is 27.50 dB and for the adaptive controller is 45.59 dB.

Power spectral densities (PSD) estimates of the two control schemes are given in Fig. 12.9. It can be observed that the attenuation introduced by the robust linear controller in the desired frequency zone is equal to 14 dB which is coherent with the design done in Sect. 11.3. The adaptive regulator assures a better attenuation of

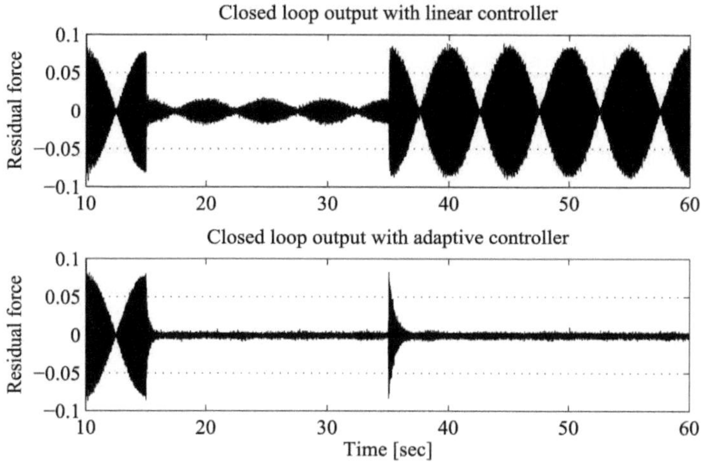

Fig. 12.10 Residual force with step frequency changes (+5 Hz) in closed-loop with linear controller (*upper plot*) and with adaptive controller (*lower plot*). In both cases, the system is in open-loop until t = 15 s

disturbances and also does not amplify at other frequencies more than the linear controller.

Adaptation capabilities are tested and the results are compared to the linear robust controller in Fig. 12.10. In this figure, all four sinusoidal disturbances are modified at 35 s by adding 5 Hz to their frequencies. As such the new disturbance frequencies are centred around 65 Hz (64.9 and 65.1 Hz) and 85 Hz (84.9 and 85.1 Hz). As expected the linear robust controller fails to provide an acceptable attenuation. The adaptation transient is about 1.5 s.

12.5 Adaptive Attenuation of an Unknown Narrow-Band Disturbance on the Active Hydraulic Suspension

The narrow-band disturbance rejection procedure using the direct adaptive control scheme proposed in Sect. 12.2.2 is illustrated in real time for the case of the control of an active hydraulic suspension (presented in Sect. 2.1). In this application the disturbance will be a single time-varying sinusoidal disturbance, so one should consider $n_{D_p} = 2$ and $n_Q = n_{D_p} - 1 = 1$.

The identification procedures for the active suspension have been discussed in Sect. 6.1 (identification in open-loop operation). The frequency characteristics of the identified primary path and secondary path models model (open-loop identification) are shown in Fig. 6.8. The first vibration mode of the primary path model is near 32 Hz. The model of the secondary path has the following complexity: $n_A = 14$,

$n_B = 16$, $d = 0$. The secondary path has several low damped vibration modes. The first one is at 31.8 Hz with a damping factor 0.07.

The central controller (without the internal model of the disturbance) has been designed using the pole placement method and the secondary path identified model. A pair of dominant poles has been fixed at the frequency of the first vibration mode (31.8 Hz), with a damping $\xi = 0.8$, and the other poles of the model have been considered as auxiliary desired closed-loop poles. In addition a prespecified part $H_R = 1 + q^{-1}$ ($R = H_R R'$) which assures the opening of the loop at $0.5 f_s$ has been introduced and 10 auxiliary poles at 0.7 have been added to the desired closed-loop poles. The resulting nominal controller has the following complexity: $n_R = 14$, $n_S = 16$ and it satisfies the imposed robustness constraints on the sensitivity functions.[11]

Only the results in adaptive operation will be presented. For results in self-tuning operations see [7]. In adaptive operation the PAA works permanently (once the loop is closed) and the controller is recomputed at each sampling. The adaptation gain in this case never goes to zero.

In order to evaluate the performances in real-time, time-varying frequency sinusoidal disturbances between 25 and 47 Hz have been used (the first vibration mode of the primary path is near 32 Hz). Two protocols have been considered:

- Step changes of the frequency of the disturbance
- Continuously time-varying frequency of the disturbance (chirp)

Step Changes of the Frequency of the Disturbance

Start up: the system is started in open-loop. After 5 s (4000 samples) a sinusoidal disturbance of 32 Hz is applied on the shaker and simultaneously the loop is closed. After the *start up* ends, every 15 s (8000 samples) sinusoidal disturbances of different frequency are applied (step change in frequency value). The sequence is as follows: 32, 25, 32, 47, 32 Hz.

The measured residual force obtained in direct adaptive operation is presented in Fig. 12.11. The I-PAA given in Eqs. (12.33) through (12.36) has been used. An adaptation gain with *variable forgetting factor combined with a constant trace* [18] has been used in order to be able to track automatically the changes of the disturbance characteristics. The low level threshold of the trace has been fixed to 3×10^{-9}.

The spectral densities of the measured residual forces obtained in adaptive operation in open and closed-loop for the three different frequencies considered (25, 32, 47 Hz) are presented in Fig. 12.12 for the direct adaptation method.

One observes the appearance of two harmonics of the first vibration mode of the primary path on the spectral density in open-loop when the frequency of the disturbance corresponds to the first resonance mode of the system (32 Hz). They appear in open-loop because of the nonlinearities of the system at large signals (there is an important amplification of the disturbance at the resonance frequency of the system in). The harmonics do not appear in closed-loop operation. The attenuations

[11] Any design method allowing to satisfy these constraints can be used.

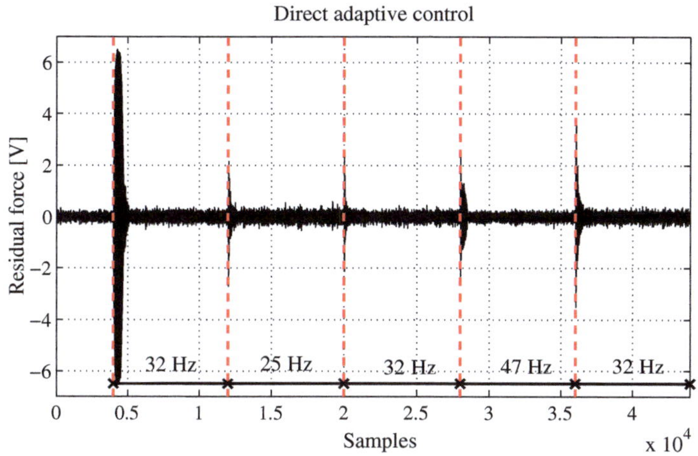

Fig. 12.11 Time domain results with the direct adaptation method (trace = 3×10^{-9})

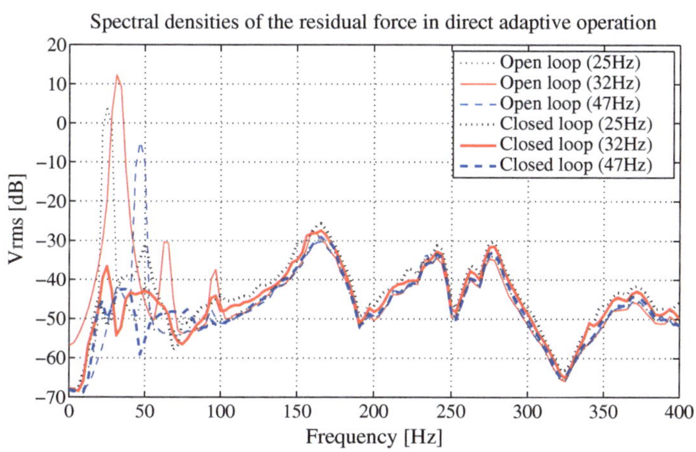

Fig. 12.12 Spectral densities of the residual force in open and closed-loop in direct adaptive operation)

Table 12.1 Real-time performance of the direct adaptive method

Method	Direct adaptation		
Disturbance frequency (Hz)	25	32	47
Disturbance attenuation (dB)	56.18	65.43	53.97

obtained are larger than 50 dB for the 3 different frequencies considered. The values of the attenuations are summarized in Table 12.1.[12]

The duration of the adaptation transients is less than 0.25 s [7].

Attenuation of Sinusoidal Disturbances with Continuously Time-Varying Frequency

[12]For results obtained with an indirect adaptive control scheme see [7].

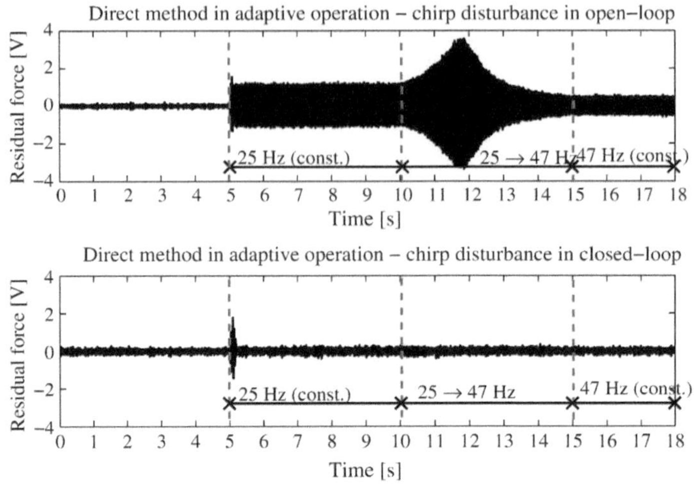

Fig. 12.13 Real-time results obtained with the direct adaptive method and a chirp disturbance: open-loop (*upper plot*), closed-loop (*lower plot*)

Consider now that the frequency of the sinusoidal disturbance varies continuously and let's use a chirp disturbance signal (linear swept-frequency signal) between 25 and 47 Hz.

The tests have been done as follows: Start up in closed-loop at $t = 0$ with the central controller. Once the loop is closed, the adaptation algorithm works permanently and the controller is updated (direct approach) at each sampling instant. After 5 s a sinusoidal disturbance of 25 Hz (constant frequency) is applied on the shaker. From 10 to 15 s a chirp between 25 and 47 Hz is applied. After 15 s a 47 Hz (constant frequency) sinusoidal disturbance is applied and the tests are stopped after 18 s. The time domain results obtained in open and in closed-loop (direct adaptive control) are presented in Fig. 12.13. One can remark that the performances obtained are very good.

12.6 Adaptive Attenuation of an Unknown Narrow-Band Disturbance on the Active Vibration Control System Using an Inertial Actuator

The narrow-band disturbance rejection procedure using the direct adaptive control scheme proposed in Sect. 12.2.2 is illustrated in real time on an active vibration control system using an inertial actuator. The case of one tonal disturbance will be considered.[13]

[13]The case of multiple unknown narrow-band disturbances will be discussed in Chap. 13.

The system has been presented in Sect. 2.2 and the identification procedure has been described in Sect. 6.2. The frequency characteristics of the identified model of the secondary path has been shown in Fig. 6.18. As discussed in Sect. 12.2.3 attenuation of the disturbances can be done only in the frequency regions where the system has enough gain. In particular this system will be able to attenuate disturbances located between 50 and 95 Hz. Note that the borders of the frequency region considered are very close to some complex low damped zeros (no gain at these frequencies) so this has to be taken in account when designing the central controller (the input sensitivity function should be low at these frequencies).

12.6.1 Design of the Central Controller

A key issue is the design of the central controller. It should assure that in the presence of disturbances with known frequency, using the internal model principle, the specifications are satisfied for all the possible frequencies of the disturbance. Specifically one should obtain a disturbance attenuation (DA) of 40 dB, a global attenuation (GA) of 30 dB and a maximum amplification (MA) with respect to open-loop operation of less than 6 dB (for more details see Table 13.2 and Sect. 13.2). The performance of the central controller in this context gives the best achievable performance. Adding the adaptation capabilities will only allow to approach this performance when the frequencies of the disturbances are unknown and time-varying.

The region of operation is between 50 and 95 Hz. As it can be seen from the zoom of the frequency characteristics of the secondary path shown are in Fig. 12.14,

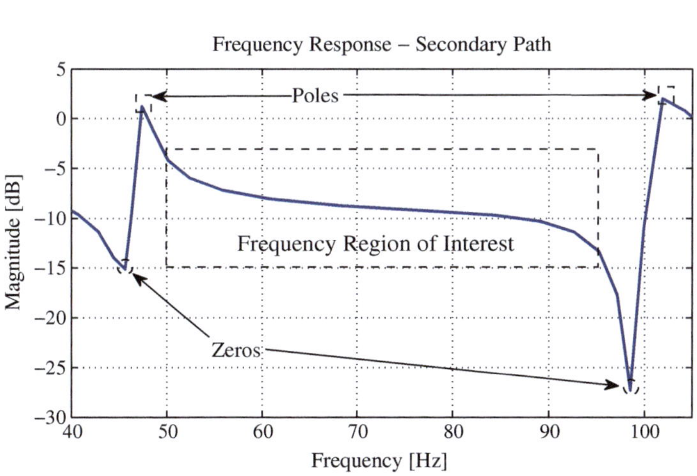

Fig. 12.14 Zoom of the magnitude of the frequency response of the secondary path between 40 and 105 Hz

the borders of the region of operation are quite close to low damped complex zeros located at 47.36 and 101.92 Hz.

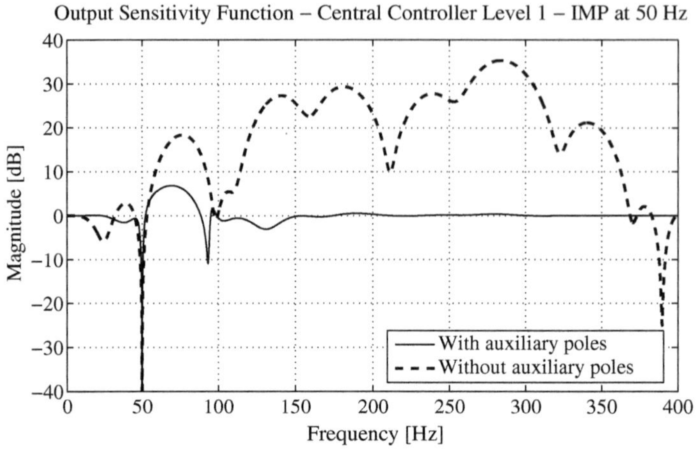

Fig. 12.15 Output sensitivity function with (*solid line*) and without (*dashed line*) auxiliary resonant poles at 50 and 95 Hz and with an internal model tuned for 50 Hz

All the poles of the system have been included as desired closed-loop poles (they are all stable), but to reduce the effect of the IMP on the borders two auxiliary low damped auxiliary poles have been introduced at 50 and 90 Hz, with damping 0.0629 and 0.0157 respectively. Figure 12.15 shows the effect of these auxiliary poles on the shape of the output sensitivity function.

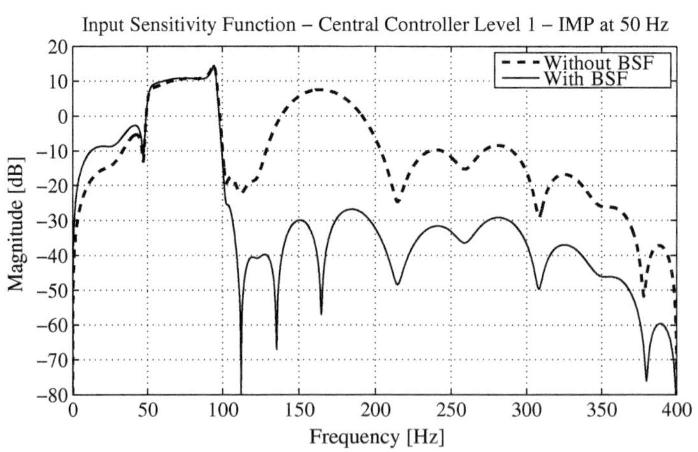

Fig. 12.16 Input sensitivity function with (*solid line*) and without (*dashed line*) BSF filters

Another objective was to reduce significantly the modulus of the input sensitivity function outside the region of operation (to improve robustness and reduce noise amplification). This has been achieved by shaping the input sensitivity function using band-stop filters (see Sect. 7.2.9 for details). Three band-stop filters located between 110 and 170 Hz have been used. Their effect upon the input sensitivity function is shown in Fig. 12.16.

12.6.2 Real-Time Results

I-PAA algorithm given in Eqs. (12.33)–(12.36) has been used with decreasing adaptation gain and constant trace. The initial trace of the matrix adaptation gain has been fixed at 2000 (2 parameters to adapt) and the desired constant trace at 2. Figure 12.17 shows the time results obtained for a simple step test (i.e., application of a 75 Hz disturbance) and for step changes in the frequency of the disturbance (the sequence

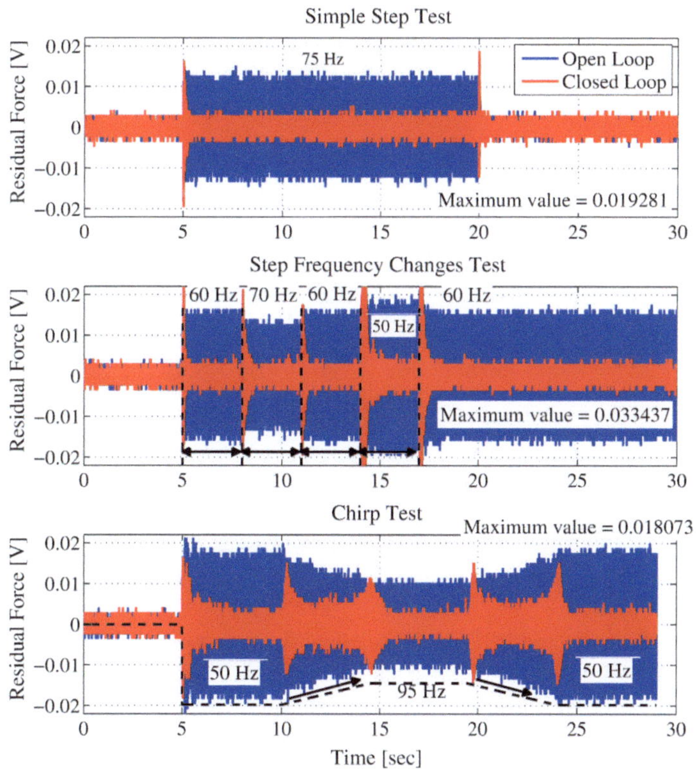

Fig. 12.17 Time response comparison between open-loop and adaptive closed-loop operation (*up* step disturbance application, *middle* step frequency changes, *bottom* chirp disturbance)

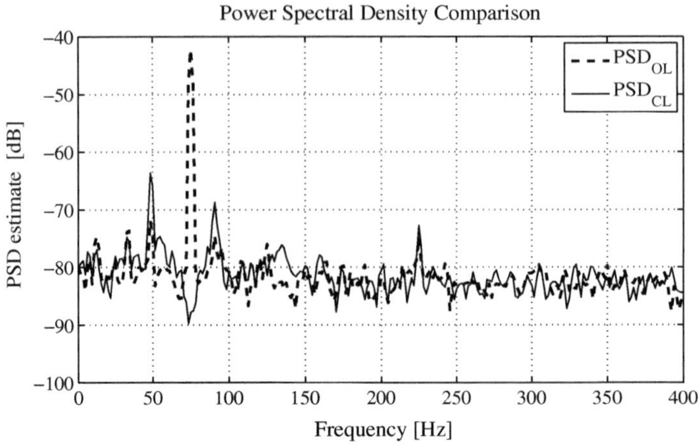

Fig. 12.18 Power spectral density comparison between open-loop (*dashed line*) and adaptive closed-loop (*solid line*) for a 75 Hz disturbance

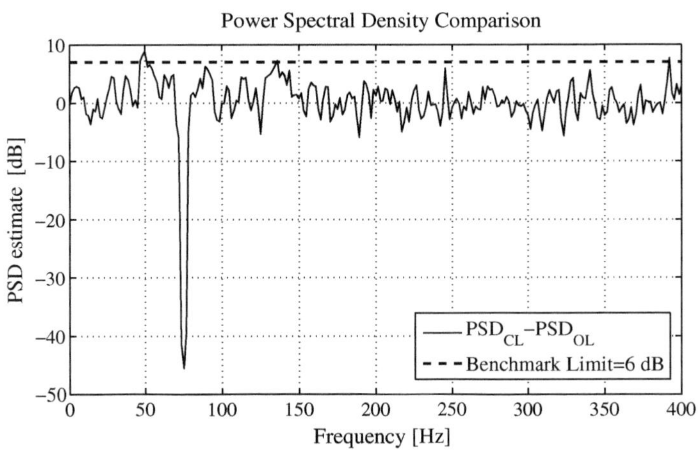

Fig. 12.19 Attenuation using adaptive closed-loop regulation for a 75 Hz disturbance

was: 60, 70, 60, 50, 60 Hz). Lower part of the figure shows the behaviour in the presence of a chirp disturbance varying between 50 and 95 Hz and from 95 to 50 Hz disturbance.

Figure 12.18 shows the PSD of the residual force in open-loop (dashed line) and in closed-loop (solid line) for a disturbance located at 75 Hz. Figure 12.19 shows the resulting attenuation/amplification using adaptive feedback regulation (the difference between the PSD in open-loop and the PSD with adaptive feedback regulation). Similar results are obtained for the other frequencies.

Table 12.2 gives a summary of the results obtained with the adaptive scheme for various frequencies of the disturbance. Column 1 gives the global attenuation (GA).

Table 12.2 Experimental results—simple step test (GA—global attenuation, DA—disturbance attenuation, MA—maximum amplification)

One single tonal disturbance			
Frequency (Hz)	GA (dB)	DA (dB)	MA (dB@Hz)
50	34.60	38.49	9.83@65.63
55	34.54	50.45	9.48@118.75
60	33.34	49.49	8.23@79.69
65	32.78	50.04	9.65@90.63
70	30.54	47.90	9.01@89.06
75	29.53	45.54	8.90@50.00
80	30.28	48.72	8.49@95.31
85	28.47	45.94	10.66@57.81
90	28.02	42.65	8.24@73.44
95	24.63	34.55	9.06@82.81

Column 2 gives the attenuation of the tonal disturbance (DA) and Column 3 gives the maximum amplification in closed-loop with respect to the open-loop operation and the frequency at which it occurs. Except the border of the domain of operation, the attenuation of the unknown tonal disturbance is more than 40 dB and the maximum amplification is below 10.66 dB for all frequencies.

12.7 Other Experimental Results

Experimental results obtained on the active flexible structure described in Sect. 2.3 for single and multiple unknown narrow-band disturbances can be found in [13, 19]. Results on a different active vibration control using an inertial actuator can be found in [13]. The use of this type of algorithm for adaptive suppression of main periodic disturbances in Blu-ray disc drives servomechanisms is presented in [13, 20]. The same type of algorithm has been used for a different type of disturbance (an exponential) in the context of fed-batch reactor [21, 22]. This approach has been used also for active noise control [23, 24].

12.8 Concluding Remarks

- The use of the internal model principle allows to provide solutions for suppressing the effect of tonal disturbances upon the output, provided that, either the model of the disturbance can be estimated or the internal model of the disturbance can be directly estimated in the controller.

- The use of the Youla–Kučera parametrization allows to build direct adaptive regulation schemes (one directly adapt the parameters of the Youla–Kučera filter included in the controller).
- The number of parameters to adapt depend upon the number of tonal disturbances to attenuate and not upon the complexity of the model of the system.
- The performance of the direct adaptive regulation using Youla–Kučera parametrization is better than those of the robust linear controller solutions in terms of performance and expansion of the domain of operation.

12.9 Notes and References

The problem discussed in this chapter belongs to the case of adaptive regulation, i.e., known plant and unknown disturbance model. The problem of known plant and unknown disturbance model has been addressed in a number of papers [7, 21, 23–29] among others. The following approaches considered for solving this problem may be mentioned

1. Use of the internal model principle [6–8, 16, 21–24, 28–30].
2. Use of an observer for the disturbance [26, 27, 31, 32].
3. Use of the "phase-locked" loop structure considered in communication systems [25, 33].

The use of the Youla–Kučera parametrization for the problem of disturbance rejection takes its roots from an idea of Tsypkin [34] who expressed the IMP in the time domain using the sensitivity functions and the Youla–Kučera parametrization.

An important issue which was addressed is to try to take in account the possible variation of the plant model by over parametrization of the Youla–Kučera filter. This has been considered in [21]. The Ref. [35] provides a stability proof in this context.

Over parametrization of the Youla–Kučera filter has been also considered in [36, 37] for improving the robustness of the central controller. A comparative evaluation of this approach and the design of the central controller using pole placement in the presence of low damped complex zeros can be found in [38] where also a combination of the two approaches is considered. Note that over parametrization leads to the increase of the computation load.

Applications of the approach presented in this chapter are reported in [12, 13, 22, 24, 39–44] using related structures and adaptation algorithms.

Extension to the multivariable case is considered in [45]. Solution for the continuous time formulation is provided in [46]. The rejection of sinusoidal disturbances in chaotic planar oscillators is discussed in [47].

The case of unknown plant model and known disturbance model is considered in [48, 49] among other references. The case of unknown plant and disturbance models is considered in [50].

References

1. Widrow B, Stearns S (1985) Adaptive signal processing. Prentice-Hall, Englewood Cliffs
2. Beranek L, Ver I (1992) Noise and vibration control engineering: principles and applications. Wiley, New York
3. Fuller C, Elliott S, Nelson P (1997) Active control of vibration. Academic Press, New York
4. Elliott S (2001) Signal processing for active control. Academic Press, San Diego
5. Landau I, Alma M, Airimitoaie T (2011) Adaptive feedforward compensation algorithms for active vibration control with mechanical coupling. Automatica 47(10):2185–2196. doi:10.1016/j.automatica.2011.08.015
6. Francis B, Wonham W (1976) The internal model principle of control theory. Automatica 12(5):457–465. doi:10.1016/0005-1098(76)90006-6
7. Landau I, Constantinescu A, Rey D (2005) Adaptive narrow band disturbance rejection applied to an active suspension - an internal model principle approach. Automatica 41(4):563–574
8. Bengtsson G (1977) Output regulation and internal models - a frequency domain approach. Automatica 13(4):333–345. doi:10.1016/0005-1098(77)90016-4
9. Landau I, Alma M, Martinez J, Buche G (2011) Adaptive suppression of multiple time-varying unknown vibrations using an inertial actuator. IEEE Trans Control Syst Technol 19(6):1327–1338. doi:10.1109/TCST.2010.2091641
10. Castellanos Silva A, Landau ID, Ioannou P (2014) Direct adaptive regulation in the vicinity of low damped complex zeros - application to active vibration control. In: 22nd Mediterranean conference on control and automation (MED), pp 255–260 (2014)
11. Li S, Qiu J, Ji H, Zhu K, Li J (2011) Piezoelectric vibration control for all-clamped panel using dob-based optimal control. Mechatronics 21(7):1213–1221. doi:10.1016/j.mechatronics.2011.07.005
12. Chen X, Tomizuka M (2015) Overview and new results in disturbance observer based adaptive vibration rejection with application to advanced manufacturing. Int J Adapt Control Signal Process 29(11):1459–1474. http://dx.doi.org/10.1002/acs.2546
13. Landau ID, Alma M, Constantinescu A, Martinez JJ, Noë M (2011) Adaptive regulation - rejection of unknown multiple narrow band disturbances (a review on algorithms and applications). Control Eng Pract 19(10):1168–1181. doi:10.1016/j.conengprac.2011.06.005
14. Landau I, Zito G (2005) Digital control systems - design, identification and implementation. Springer, London
15. Anderson B (1998) From Youla-Kučera to identification, adaptive and nonlinear control. Automatica 34(12):1485–1506. http://www.sciencedirect.com/science/article/pii/S0005109898800022
16. Tsypkin Y (1997) Stochastic discrete systems with internal models. J Autom Inf Sci 29(4&5):156–161
17. Landau ID, Silva AC, Airimitoaie TB, Buche G, Noé M (2013) Benchmark on adaptive regulation - rejection of unknown/time-varying multiple narrow band disturbances. Eur J Control 19(4):237–252. doi:10.1016/j.ejcon.2013.05.007
18. Landau ID, Lozano R, M'Saad M, Karimi A (2011) Adaptive control, 2nd edn. Springer, London
19. Landau ID, Airimitoaie TB, Castellanos Silva A (2015) Adaptive attenuation of unknown and time-varying narrow band and broadband disturbances. Int J Adapt Control Signal Process 29(11):1367–1390
20. Alma M, Martinez J, Landau I, Buche G (2012) Design and tuning of reduced order H_∞ feedforward compensators for active vibration control. IEEE Trans Control Syst Technol 20(2):554–561. doi:10.1109/TCST.2011.2119485
21. Valentinotti S (2001) Adaptive rejection of unstable disturbances: application to a fed-batch fermentation. Thèse de doctorat, École Polytechnique Fédérale de Lausanne
22. Valentinotti S, Srinivasan B, Holmberg U, Bonvin D, Cannizzaro C, Rhiel M, von Stockar U (2003) Optimal operation of fed-batch fermentations via adaptive control of overflow metabolite. Control Eng Pract 11(6):665–674. doi:10.1016/S0967-0661(02)00172-7

23. Ben Amara F, Kabamba P, Ulsoy A (1999) Adaptive sinusoidal disturbance rejection in linear discrete-time systems - Part I: theory. J Dyn Syst Meas Control 121:648–654
24. F Ben Amara, Kabamba P, Ulsoy A (1999) Adaptive sinusoidal disturbance rejection in linear discrete-time systems - Part II: experiments. J Dyn Syst Meas Control 121:655–659
25. Bodson M, Douglas S (1997) Adaptive algorithms for the rejection of sinusosidal disturbances with unknown frequency. Automatica 33:2213–2221
26. Marino R, Santosuosso G, Tomei P (2003) Robust adaptive compensation of biased sinusoidal disturbances with unknown frequency. Automatica 39:1755–1761
27. Ding Z (2003) Global stabilization and disturbance suppression of a class of nonlinear systems with uncertain internal model. Automatica 39(3):471–479. doi:10.1016/S0005-1098(02)00251-0
28. Gouraud T, Gugliemi M, Auger F (1997) Design of robust and frequency adaptive controllers for harmonic disturbance rejection in a single-phase power network. In: Proceedings of the European control conference, Bruxelles
29. Hillerstrom G, Sternby J (1994) Rejection of periodic disturbances with unknown period - a frequency domain approach. In: Proceedings of American control conference, Baltimore, pp 1626–1631
30. Johnson C (1976) Theory of disturbance-accomodating controllers. In: Leondes CT (ed) Control and dynamical systems, vol 12, pp 387–489
31. Serrani A (2006) Rejection of harmonic disturbances at the controller input via hybrid adaptive external models. Automatica 42(11):1977–1985. doi:10.1016/j.automatica.2006.06.014
32. Jia QW (2009) Disturbance rejection through disturbance observer with adaptive frequency estimation. IEEE Trans Magn 45(6):2675–2678. doi:10.1109/TMAG.2009.2018605
33. Bodson M (2005) Rejection of periodic distrubances of unknown and time-varying frequency. Int J Adapt Control Signal Process 19:67–88
34. Tsypkin Y (1991) Adaptive-invariant discrete control systems. In: Thoma M, Wyner A (eds) Foundations of adaptive control. Lecture notes in control and information science, vol 160. Springer, Heidelberg, pp 239–268
35. Mullhaupt Ph, Valentinotti S, Srinivasan B, Bonvin D (2012) Asymptotic rejection of nonvanishing disturbances despite plant-model mismatch. Int J Adapt Control Signal Process 26(12):1090–1110
36. Jafari S, Ioannou P, Fitzpatrick B, Wang Y (2013) Robust stability and performance of adaptive jitter supression in laser beam pointing. In: 52nd IEEE conference on decision and control. Florence, Italy
37. Jafari S, Ioannou P, Fitzpatrick B, Wang Y (2013) Robustness and performance of adaptive suppresion of unknown periodic disturbances. IEEE Trans Autom Control (Under review)
38. Castellanos-Silva A, Landau ID, Ioannou P (2015) Robust direct adaptive regulation of unknown disturbances in the vicinity of low-damped complex zeros-application to AVC. IEEE Trans Control Syst Technol 24(2):733–740. doi:10.1109/TCST.2015.2445859
39. Chen X, Tomizuka M (2012) A minimum parameter adaptive approach for rejecting multiple narrow-band disturbances with application to hard disk drives. IEEE Trans Control Syst Technol 20(2):408–415. doi:10.1109/TCST.2011.2178025
40. Martinez JJ, Alma M (2012) Improving playability of blu-ray disc drives by using adaptive suppression of repetitive disturbances. Automatica 48(4):638–644
41. Semba T, White M, Huang FY (2011) Adaptive cancellation of self-induced vibration. IEEE Trans Magn 47(7):1958–1963. doi:10.1109/TMAG.2011.2138685
42. Bohn C, Cortabarria A, Hrtel V, Kowalczyk K (2004) Active control of engine-induced vibrations in automotive vehicles using disturbance observer gain scheduling. Control Eng Pract 12(8):1029–1039. http://dx.doi.org/10.1016/j.conengprac.2003.09.008. http://www.sciencedirect.com/science/article/pii/S0967066103002144. Special Section on Emerging Technologies for Active Noise and Vibration Control Systems
43. Hara S, Yoshida K (1996) Simultaneous optimization of positioning and vibration control using a time-varying frequency-shaped criterion function. Control Eng Pract 4(4):553–561. doi:10.1016/0967-0661(96)00039-1. http://www.sciencedirect.com/science/article/pii/0967066196000391

44. Beltrán-Carbajal F, Silva-Navarro G (2014) Active vibration control in duffing mechanical systems using dynamic vibration absorbers. J Sound Vib 333(14):3019–3030. doi:10.1016/j.jsv.2014.03.002. http://www.sciencedirect.com/science/article/pii/S0022460X14001825
45. Ficocelli M, Ben Amara F (2009) Adaptive regulation of MIMO linear systems against unknown sinusoidal exogenous inputs. Int J Adapt Control Signal Process 23(6):581–603. http://dx.doi.org/10.1002/acs.1072
46. Jafari S, Ioannou PA (2016) Rejection of unknown periodic disturbances for continuous-time MIMO systems with dynamic uncertainties. Int J Adapt Control Signal Process. http://dx.doi.org/10.1002/acs.2683
47. Menini L, Possieri C, Tornambè A (2015) Sinusoidal disturbance rejection in chaotic planar oscillators. Int J Adapt Control Signal Process 29(12):1578–1590. http://dx.doi.org/10.1002/acs.2564
48. Sun Z, Tsao T (2000) Adaptive control with asymptotic tracking performance and its application to an electro-hydraulic servo system. J Dyn Syst Meas Control 122:188–195
49. Zhang Y, Mehta P, Bitmead R, Johnson C (1998) Direct adaptive control for tonal disturbance rejection. In: Proceedings of the American control conference, Philadelphia, pp 1480–1482
50. Feng G, Palaniswami M (1992) A stable adaptive implementation of the internal model principle. IEEE Trans Autom Control 37:1220–1225

Chapter 13
Adaptive Attenuation of Multiple Sparse Unknown and Time-Varying Narrow-Band Disturbances

13.1 Introduction

In this chapter, the focus is on the strong attenuation of multiple sparsely located unknown and time-varying disturbances. One assumes that the various tonal disturbances are distant to each other in the frequency domain by a distance in Hz at least equal to 10 % of the frequency of the disturbance and that the frequency of these disturbances vary over a wide frequency region.

The problem is to assure in this context a certain number of performance indices such as global attenuation, disturbance attenuation at the frequency of the disturbances, a tolerated maximum amplification (water bed effect), a good adaptation transient (see Sect. 12.3). The most difficult problem is to be sure that in all the configurations the maximum amplification is below a specified value. There is first a fundamental problem to solve: one has to be sure that in the known frequency case, for any combination of disturbances the attenuation and the maximum amplification specifications are achieved. The adaptive approach will only try to approach the performances of a linear controller for the case of known disturbances. So before discussing the appropriate adaptation schemes one has to consider the design methods to be used in order to achieve these constraints for the known frequencies case. This will be discussed in Sect. 13.2.

13.2 The Linear Control Challenge

In this section, the linear control challenge will be presented for the case of rejection of multiple narrow-band disturbances taking also into account the possible presence of low damped complex zeros in the vicinity of the border of the operational zone. Considering that in a linear context all the information is available, the objective is to set up the best achievable performance for the adaptive case.

I.D. Landau et al., *Adaptive and Robust Active Vibration Control*,
Advances in Industrial Control, DOI 10.1007/978-3-319-41450-8_13

Assuming that only one tonal vibration has to be cancelled in a frequency region far from the presence of low damped complex zeros and that the models of the plant and of the disturbance are known, the design of a linear regulator is relatively straightforward, using the internal model principle (see Chaps. 7 and 12).

The problem becomes much more difficult if several tonal vibrations (sinusoidal disturbances) have to be attenuated simultaneously since the water bed effect may become significant without a careful shaping of the sensitivity function when using the internal model principle. Furthermore, if the frequencies of the disturbance may be close to those of some of very low damped complex zeros of the plant, the use of the internal model principle should be used with care even in the case of a single disturbance (see Sect. 12.5).

This section will examine the various aspects of the design of a linear controller in the context of multiple tonal vibrations and the presence of low damped complex zeros. It will review various linear controller strategies.

To be specific these design aspects will be illustrated in the context of the active vibration control system using an inertial actuator, described in Sect. 2.2 and which has been already used for the case of a single tonal disturbance.

In this system, the tonal vibrations are located in the range of frequencies between 50 and 95 Hz. The frequency characteristics of the secondary path are given in Sect. 6.2.

Assume that a tonal vibration (or a narrow-band disturbance) $p(t)$ is introduced into the system affecting the output $y(t)$. The effect of this disturbance is centred at a specific frequency. As mentioned in Sect. 12.2.3, the IMP can be used to asymptotically reject the effects of a narrow-band disturbance at the system's output if the system has enough gain in this region.

It is important also to take into account the fact that the secondary path (the actuator path) has no gain at very low frequencies and very low gain in high frequencies near $0.5\ f_s$. Therefore, the control system has to be designed such that the gain of the controller be very low (or zero) in these regions (preferably 0 at 0 Hz and $0.5\ f_s$). Not taking into account these constraints can lead to an undesirable stress on the actuator.

In order to assess how *good* the controller is, it is necessary to define some control objectives that have to be fulfilled. For the remaining of this section, the narrow-band disturbance is supposed to be known and composed of 3 sinusoidal signals with 55, 70 and 85 Hz frequencies. The control objective is to attenuate each component of the disturbance by a minimum of 40 dB, while limiting the maximum amplification at 9 dB within the frequency region of operation. Furthermore it will be required that low values of the modulus of the input sensitivity function be achieved outside the operation region.

The use of the IMP principle completed with the use of auxiliary real (aperiodic) poles which have been used in Chap. 11 as a basic design for adaptive attenuation of one unknown disturbance may not work satisfactory for the case of multiple unknown disturbances even if it may provide good performance in some situations [1]. Even in the case of a single tonal disturbance, if low damped complex zeros near the border of the operation region are present, this simple design is not satisfactory. Auxiliary low damped complex poles have to be added. See Chap. 12, Sect. 12.6.

One can say in general, that the IMP is doing too much in terms of attenuation of tonal disturbances which of course can generate in certain case unacceptable water bed effects. In fact in practice one does not need a full rejection of the disturbance, but just a certain level of attenuation.

Three linear control strategy for attenuation of multiple narrow-band disturbances will be considered

1. Band-stop filters (BSF) centred at the frequencies of the disturbances
2. IMP combined with tuned notch filters
3. IMP with additional fixed resonant poles

The controller design will be done in the context of pole placement. The initial desired closed-loop poles for the design of the central controller defined by the characteristic polynomial P_0 include all the stable poles of the secondary path model and the free auxiliary poles are all set at 0.3. The fixed part of the central controller numerator is chosen as $H_R(z^{-1}) = (1 - z^{-1}) \cdot (1 + z^{-1})$ in order to open the loop at 0 Hz and $0.5\, f_s$.

13.2.1 Attenuation of Multiple Narrow-Band Disturbances Using Band-Stop Filters

The purpose of this method is to allow the possibility of choosing the desired attenuation and bandwidth of attenuation for each of the narrow-band component of the disturbance. Choosing the level of attenuation and the bandwidth allows to preserve acceptable characteristics of the sensitivity functions outside the attenuation bands and this is very useful in the case of multiple narrow-band disturbances. This is the main advantage with respect to classical internal model principle which in the case of several narrow-band disturbances, as a consequence of complete cancellation of the disturbances, may lead to unacceptable values of the modulus of the output sensitivity function outside the attenuation regions. The controller design technique uses the shaping of the output sensitivity function in order to impose the desired attenuation of narrow-band disturbances. This shaping techniques has been presented in Sect. 7.2.

The process output can be written as[1]

$$y(t) = G(q^{-1}) \cdot u(t) + p(t), \tag{13.1}$$

where

$$G(q^{-1}) = q^{-d} \frac{B(q^{-1})}{A(q^{-1})} \tag{13.2}$$

[1]The complex variable z^{-1} is used to characterize the system's behaviour in the frequency domain and the delay operator q^{-1} will be used for the time domain analysis.

is called the secondary path of the system.

As specified in the introduction, the hypothesis of constant dynamic characteristics of the AVC system is considered (similar to [2, 3]). The denominator of the secondary path model is given by

$$A(q^{-1}) = 1 + a_1 q^{-1} + \cdots + a_{n_A} q^{-n_A}, \tag{13.3}$$

the numerator is given by

$$B(q^{-1}) = b_1 q^{-1} + \cdots + b_{n_B} q^{-n_B} = 1 + q^{-1} B^*(q^{-1}) \tag{13.4}$$

and d is the integer delay (number of sampling periods).[2]

The control signal is given by

$$u(t) = -R(q^{-1}) \cdot y(t) - S^*(q^{-1}) \cdot u(t-1), \tag{13.5}$$

with

$$\begin{aligned} S(q^{-1}) &= 1 + q^{-1} S^*(q^{-1}) = 1 + s_1 q^{-1} + \cdots + s_{n_S} q^{-n_S} \\ &= S'(q^{-1}) \cdot H_S(q^{-1}), \end{aligned} \tag{13.6}$$

$$R(q^{-1}) = r_0 + r_1 q^{-1} + \cdots + r_{n_R} q^{-n_R} = R'(q^{-1}) \cdot H_R(q^{-1}), \tag{13.7}$$

where $H_S(q^{-1})$ and $H_R(q^{-1})$ represent fixed (imposed) parts in the controller and $S'(q^{-1})$ and $R'(q^{-1})$ are computed.

The basic tool is a digital filter $S_{BSF_i}(z^{-1})/P_{BSF_i}(z^{-1})$ with the numerator included in the controller polynomial S and the denominator as a factor of the desired closed-loop characteristic polynomial, which will assure the desired attenuation of a narrow-band disturbance (index $i \in \{1, \cdots, n\}$).

The BSFs have the following structure

$$\frac{S_{BSF_i}(z^{-1})}{P_{BSF_i}(z^{-1})} = \frac{1 + \beta_1^i z^{-1} + \beta_2^i z^{-2}}{1 + \alpha_1^i z^{-1} + \alpha_2^i z^{-2}} \tag{13.8}$$

resulting from the discretization of a continuous filter (see also [4, 5])

$$F_i(s) = \frac{s^2 + 2\zeta_{n_i} \omega_i s + \omega_i^2}{s^2 + 2\zeta_{d_i} \omega_i s + \omega_i^2} \tag{13.9}$$

using the bilinear transformation. This filter introduces an attenuation of

[2] As indicated earlier, it is assumed that a reliable model identification is achieved and therefore the estimated model is assumed to be equal to the true model.

$$M_i = -20 \cdot \log_{10} \left(\frac{\zeta_{n_i}}{\zeta_{d_i}} \right) \tag{13.10}$$

at the frequency ω_i. Positive values of M_i denote attenuations ($\zeta_{n_i} < \zeta_{d_i}$) and negative values denote amplifications ($\zeta_{n_i} > \zeta_{d_i}$). Details on the computation of the corresponding digital BSF have been given in Chap. 7.[3]

Remark The design parameters for each BSF are the desired attenuation (M_i), the central frequency of the filter (ω_i) and the damping of the denominator (ζ_{d_i}). The denominator damping is used to adjust the frequency bandwidth of the BSF. For very small values of the frequency bandwidth the influence of the filters on frequencies other than those defined by ω_i is negligible. Therefore, the number of BSFs and subsequently that of the narrow-band disturbances that can be compensated can be large.

For n narrow-band disturbances, n BSFs will be used

$$H_{BSF}(z^{-1}) = \frac{S_{BSF}(z^{-1})}{P_{BSF}(z^{-1})} = \frac{\prod_{i=1}^{n} S_{BSF_i}(z^{-1})}{\prod_{i=1}^{n} P_{BSF_i}(z^{-1})} \tag{13.11}$$

As stated before, the objective is that of shaping the output sensitivity function. $S(z^{-1})$ and $R(z^{-1})$ are obtained as solutions of the Bezout equation

$$P(z^{-1}) = A(z^{-1})S(z^{-1}) + z^{-d}B(z^{-1})R(z^{-1}), \tag{13.12}$$

where

$$S(z^{-1}) = H_S(z^{-1})S'(z^{-1}), \quad R(z^{-1}) = H_{R_1}(z^{-1})R'(z^{-1}), \tag{13.13}$$

and $P(z^{-1})$ is given by

$$P(z^{-1}) = P_0(z^{-1})P_{BSF}(z^{-1}). \tag{13.14}$$

In the last equation, P_{BSF} is the product of the denominators of all the BSFs, (13.11), and P_0 defines the initial imposed poles of the closed-loop system in the absence of the disturbances (allowing also to satisfy robustness constraints). The fixed part of the controller denominator H_S is in turn factorized into

$$H_S(z^{-1}) = S_{BSF}(z^{-1})H_{S_1}(z^{-1}), \tag{13.15}$$

where S_{BSF} is the combined numerator of the BSFs, (13.11), and H_{S_1} can be used if necessary to satisfy other control specifications. H_{R_1} is similar to H_{S_1} allowing to introduce fixed parts in the controller's numerator if needed (like opening the loop at certain frequencies). It is easy to see that the output sensitivity function becomes

[3]For frequencies bellow $0.17\ f_S$ (f_S is the sampling frequency) the design can be done with a very good precision directly in discrete time [5].

$$S_{yp}(z^{-1}) = \frac{A(z^{-1})S'(z^{-1})H_{S_1}(z^{-1})}{P_0(z^{-1})} \frac{S_{BSF}(z^{-1})}{P_{BSF}(z^{-1})} \tag{13.16}$$

and the shaping effect of the BSFs upon the sensitivity functions is obvious. The unknowns S' and R' are solutions of

$$P(z^{-1}) = P_0(z^{-1})P_{BSF}(z^{-1}) = A(z^{-1})S_{BSF}(z^{-1})H_{S_1}(z^{-1})S'(z^{-1}) + \\ + z^{-d}B(z^{-1})H_{R_1}(z^{-1})R'(z^{-1}). \tag{13.17}$$

and can be computed by putting (13.17) into matrix form (see also [5]). The size of the matrix equation that needs to be solved is given by

$$n_{Bez} = n_A + n_B + d + n_{H_{S_1}} + n_{H_{R_1}} + 2 \cdot n - 1, \tag{13.18}$$

where n_A, n_B and d are respectively the order of the plant's model denominator, numerator and delay (given in (13.3) and (13.4)), $n_{H_{S_1}}$ and $n_{H_{R_1}}$ are the orders of $H_{S_1}(z^{-1})$ and $H_{R_1}(z^{-1})$ respectively and n is the number of narrow-band disturbances. Equation (13.17) has an unique minimal degree solution for S' and R', if $n_P \leq n_{Bez}$, where n_P is the order of the pre-specified characteristic polynomial $P(q^{-1})$. Also, it can be seen from (13.17) and (13.15) that the minimal orders of S' and R' will be:

$$n_{S'} = n_B + d + n_{H_{R_1}} - 1, \quad n_{R'} = n_A + n_{H_{S_1}} + 2 \cdot n - 1.$$

In Fig. 13.1, one can see the improvement obtained using BSF with respect to the case when IMP with real auxiliary poles is used. The dominant poles are the same in both cases. The input sensitivity function is tuned before introducing the BSFs.

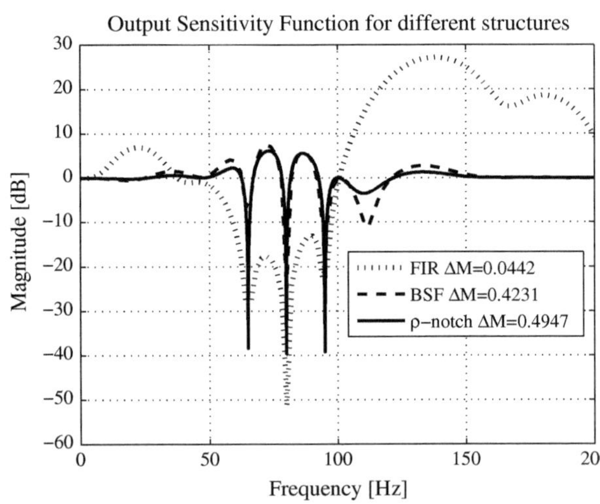

Fig. 13.1 Output sensitivity function for various controller designs: using IMP with auxiliary real poles (*dotted line*), using band-stop filters (*dashed line*), and using tuned ρ notch filters (*continuous line*)

13.2.2 IMP with Tuned Notch Filters

This approach is based on the idea of considering an optimal attenuation of the disturbance taking into account both the zeros and poles of the disturbance model. It is assumed that the model of the disturbance is a notch filter and the disturbance is represented by

$$p(t) = \frac{D_p(\rho q^{-1})}{D_p(q^{-1})} e(t) \tag{13.19}$$

where $e(t)$ is a zero mean white Gaussian noise sequence and

$$D_p(z^{-1}) = 1 + \alpha z^{-1} + z^{-2}, \tag{13.20}$$

is a polynomial with roots on the unit circle.[4]

In (13.20), $\alpha = -2\cos(2\pi\omega_1 T_s)$, ω_1 is the frequency of the disturbance in Hz, and T_s is the sampling time. $D_p(\rho z^{-1})$ is given by:

$$D_p(\rho z^{-1}) = 1 + \rho\alpha z^{-1} + \rho^2 z^{-2}, \tag{13.21}$$

with $0 < \rho < 1$. The roots of $D_p(\rho z^{-1})$ are in the same radial line as those of $D_p(z^{-1})$ but inside of the unitary circle, and therefore stable [6].

This model is pertinent for representing narrow-band disturbances as shown in Fig. 13.2, where the frequency characteristics of this model for various values of ρ are shown.

Using the output sensitivity function, the output of the plant in the presence of the disturbance can be expressed as

$$y(t) = \frac{AS'}{P_0}\frac{H_S}{P_{aux}}\frac{D_p(\rho q^{-1})}{D_p(q^{-1})} e(t) \tag{13.22}$$

or alternatively as

$$y(t) = \frac{AS'}{P_0}\beta(t) \tag{13.23}$$

where

$$\beta(t) = \frac{H_S}{P_{aux}}\frac{D_p(\rho q^{-1})}{D_p(q^{-1})} e(t) \tag{13.24}$$

In order to minimize the effect of the disturbance upon $y(t)$, one should minimize the variance of $\beta(t)$. One has two tuning devices H_S and P_{aux}. Minimization of the variance of $\beta(t)$ is equivalent of searching H_S and P_{aux} such that $\beta(t)$ becomes a

[4]Its structure in a mirror symmetric form guarantees that the roots are always on the unit circle.

white noise [5, 7]. The obvious choices are $H_S = D_p$ (which corresponds to the IMP) and $P_{aux} = D_p(\rho z^{-1})$. Of course this development can be generalized for the case of multiple narrow-band disturbances. Figure 13.1 illustrates the effect of this choice upon the output sensitivity function. As it can be seen, the results are similar to those obtained with BSF.

13.2.3 IMP Design Using Auxiliary Low Damped Complex Poles

The idea is to add a number of fixed auxiliary resonant poles which will act effectively as ρ-filters for few frequencies and as an approximation of the ρ-filters at the other frequencies. This means that a number of the real auxiliary poles used in the basic IMP design will be replaced by a number of resonant complex poles. The basic ad-hoc rule is that the number of these resonant poles is equal to the number of the low damped complex zeros located near the border of the operation region plus $n - 1$ (n is the number of tonal disturbances).

For the case of 3 tonal disturbances located in the operation region 50 to 95 Hz taking also into account the presence of the low damped complex zeros, the locations and the damping of these auxiliary resonant poles are summarized in Table 13.1. The poles at 50 and 90 Hz are related to the presence in the neighbourhood of low damped complex zeros. The poles at 60 and 80 Hz are related to the 3 tonal disturbances to

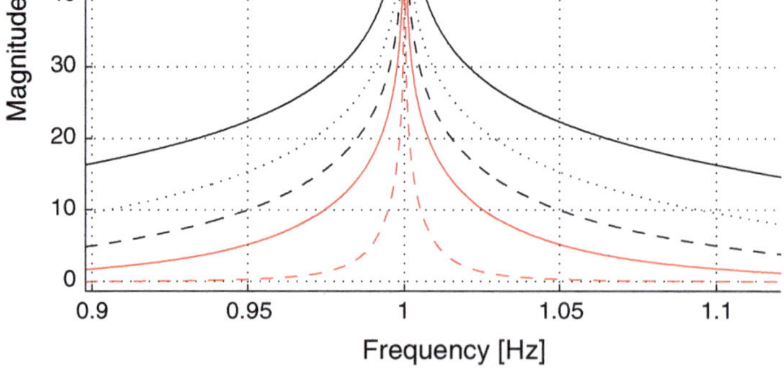

Fig. 13.2 Magnitude plot frequency responses of a notch filter for various values of the parameter ρ

Table 13.1 Auxiliary low damped complex poles added to the closed-loop characteristic polynomial

Closed-loop poles	$p_{1,2}$	$p_{3,4}$	$p_{5,6}$	$p_{7,8}$
Frequency [Hz]	50	60	80	90
Damping	0.1	0.3	0.135	0.1

be attenuated. The effect of this design with respect to the basic design using real auxiliary poles is illustrated in Fig. 13.3.

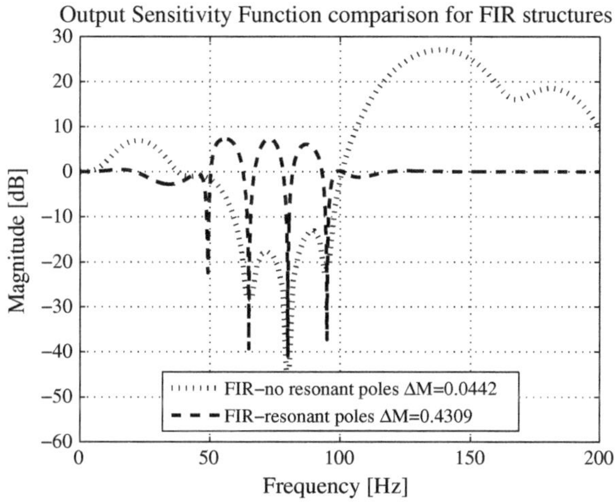

Fig. 13.3 Output sensitivity function for IMP design with real auxiliary poles and with resonant auxiliary poles

13.3 Interlaced Adaptive Regulation Using Youla–Kučera IIR Parametrization

The adaptive algorithm developed in Chap. 12 uses an FIR structure for the Q-filter. In this section, a new algorithm is developed, using an IIR structure for the Q filter in order to implement the linear control strategies using tuned notch filters (tuned auxiliary resonant poles). The use of this strategy is mainly dedicated to the case of multiple unknown tonal disturbances.

As indicated previously, since $D_p(\rho z^{-1})$ will define part of the desired closed-loop poles, it is reasonable to consider an IIR Youla–Kučera filter of the form $B_Q(z^{-1})/A_Q(z^{-1})$ with $A_Q(z^{-1}) = D_p(\rho q^{-1})$ (which will automatically introduce $D_p(\rho q^{-1})$ as part of the closed-loop poles). B_Q will introduce the internal model of the disturbance. In this context, the controller polynomials R and S are defined by

$$R(z^{-1}) = A_Q(z^{-1})R_0(z^{-1}) + H_{R_0}(z^{-1})H_{S_0}(z^{-1})A(z^{-1})B_Q(z^{-1}), \quad (13.25)$$
$$S(z^{-1}) = A_Q(z^{-1})S_0(z^{-1}) - H_{R_0}(z^{-1})H_{S_0}(z^{-1})z^{-d}B(z^{-1})B_Q(z^{-1}), \quad (13.26)$$

and the poles of the closed-loop are given by:

$$P(z^{-1}) = A_Q(z^{-1})P_0(z^{-1}). \quad (13.27)$$

$R_0(z^{-1})$, $S_0(z^{-1})$ are the numerator and denominator of the central controller

$$R_0(z^{-1}) = H_{R_0}(z^{-1})R_0'(z^{-1}), \quad (13.28)$$
$$S_0(z^{-1}) = H_{S_0}(z^{-1})S_0'(z^{-1}), \quad (13.29)$$

and the closed-loop poles defined by the central controller are the roots of

$$P_0(z^{-1}) = A(z^{-1})S_0(z^{-1})H_{S_0}(z^{-1}) + q^{-d}B(z^{-1})R_0(z^{-1})H_{R_0}(z^{-1}). \quad (13.30)$$

It can be seen from (13.25) and (13.26) that the new controller polynomials conserve the fixed parts of the central controller.

Using the expression of the output sensitivity function (AS/P) the output of the system can be written as follows:

$$y(t) = \frac{A\left[A_Q S_0 - H_{R_0}H_{S_0}q^{-d}BB_Q\right]}{P}p(t), \quad (13.31)$$
$$y(t) = \frac{\left[A_Q S_0 - H_{R_0}H_{S_0}q^{-d}BB_Q\right]}{P}w(t), \quad (13.32)$$

where the closed-loop poles are defined by (13.27) and where $w(t)$ is defined as:

$$w(t) = A(q^{-1})y(t) - q^{-d}B(q^{-1})u(t) \quad (13.33)$$
$$= A(q^{-1})p(t) \quad (13.34)$$

Comparing (13.32) with (12.20) from Chap. 12, one can see that they are similar except that S_0 is replaced by $A_Q S_0$ and P_0 by $A_Q P_0$. Therefore if A_Q is known, the algorithm given in Chap. 12 for the estimation of the Q FIR filter can be used for the estimation of B_Q. In fact this will be done using an estimation of A_Q. A block diagram of the interlaced adaptive regulation using the Youla–Kučera parametrization is shown in Fig. 13.4. The estimation of A_Q is discussed next.

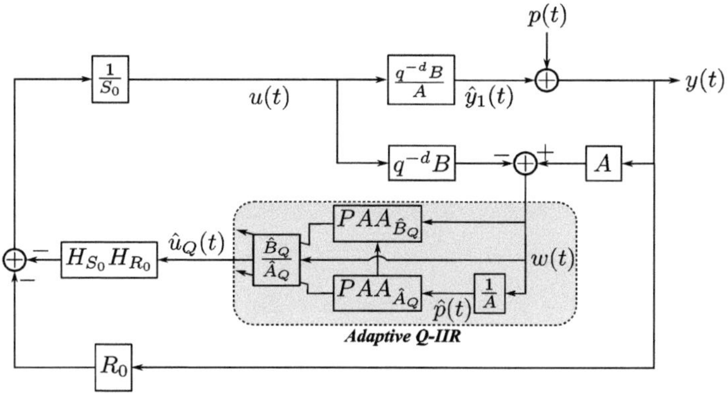

Fig. 13.4 Interlaced adaptive regulation using an IIR YK controller parametrization

13.3.1 Estimation of A_Q

Assuming that *plant model* = *true plant* in the frequency range, where the narrow-band disturbances are introduced, it is possible to get an estimation of $p(t)$, named $\hat{p}(t)$, using the following expression

$$\hat{p}(t) = \frac{1}{A(q^{-1})} w(t) \tag{13.35}$$

where $w(t)$ was defined in (13.33). The main idea behind this algorithm is to consider the signal $\hat{p}(t)$ as

$$\hat{p}(t) = \sum_{i=1}^{n} c_i \sin(\omega_i t + \beta_i) + \eta(t), \tag{13.36}$$

where $\{c_i, \omega_i, \beta_i\} \neq 0$, n is the number of narrow-band disturbances and η is a noise affecting the measurement. It can be verified that, after two steps of transient $\left(1 - 2\cos(2\pi\omega_i T_s)q^{-1} + q^{-2}\right) \cdot c_i \sin(\omega_i t + \beta_i) = 0$ [8]. Then the objective is to find the parameter $\{\alpha\}_{i=1}^{n}$ that makes $D_p(q^{-1})\hat{p}(t) = 0$.

The previous product can be equivalently written as $D_p(q^{-1})\hat{p}(t+1) = 0$ and its expression is

$$\begin{aligned} x(t+1) &= D_p(q^{-1})\hat{p}(t+1), \\ &= \hat{p}(t+1) + \sum_{i=n}^{n-1} \alpha_i \left[\hat{p}(t+1-i) + \hat{p}(t+1-2n+i)\right] + \cdots \\ &\cdots + \alpha_n \hat{p}(t+1-n) + \hat{p}(t+1-2n). \end{aligned} \tag{13.37}$$

where n is the number of narrow-band disturbances.

Defining the parameter vector as

$$\theta_{D_p} = [\alpha_1, \alpha_2, \ldots, \alpha_n]^T , \tag{13.38}$$

and the observation vector at time t as:

$$\phi_{D_p}(t) = \left[\phi_1^{D_p}(t), \phi_2^{D_p}(t), \ldots, \phi_n^{D_p}(t)\right]^T , \tag{13.39}$$

where

$$\phi_j^{D_p}(t) = \hat{p}(t+1-j) + \hat{p}(t+1-2n+j), \ j = 1, \ldots, n-1 \tag{13.40}$$

$$\phi_n^{D_p}(t) = \hat{p}(t+1-n). \tag{13.41}$$

Equation (13.37) can then be simply represented by

$$x(t+1) = \theta_{D_p}^T \phi_{D_p}(t) + \left(\hat{p}(t+1) + \hat{p}(t+1-2n)\right) . \tag{13.42}$$

Assuming that an estimation of $\hat{D}_p(q^{-1})$ is available at the instant t, the estimated product is written as follows:

$$\begin{aligned}\hat{x}(t+1) &= \hat{D}_p(q^{-1})\hat{p}(t+1), \\ &= \hat{p}(t+1) + \sum_{i=n}^{n-1} \hat{\alpha}_i \left[\hat{p}(t+1-i) + \hat{p}(t+1-2n+i)\right] + \cdots \\ &\cdots + \hat{\alpha}_n \hat{p}(t+1-n) + \hat{p}(t+1-2n) \qquad (13.43) \\ &= \hat{\theta}_{D_p}^T(t)\phi_{D_p}(t) + \left(\hat{p}(t+1) + \hat{p}(t+1-2n)\right) \qquad (13.44)\end{aligned}$$

where $\hat{\theta}_{D_p}(t)$ is the estimated parameter vector at time t. Then the *a priori* prediction error is given by

$$\varepsilon_{D_p}^{\circ}(t+1) = x(t+1) - \hat{x}(t+1) = \left[\theta_{D_p}^T - \hat{\theta}_{D_p}^T(t)\right] \cdot \phi_{D_p}(t), \tag{13.45}$$

and the *a posteriori* adaptation error using the estimation at $t+1$

$$\varepsilon_{D_p}(t+1) = \left[\theta_{D_p}^T - \hat{\theta}_{D_p}^T(t+1)\right] \cdot \phi_{D_p}(t), \tag{13.46}$$

Equation (13.46) has the standard form of an *a posteriori* adaptation error [9] which allows to associate the standard parameter adaptation algorithm (PAA) introduced in Chap. 4 (Eqs. (4.121)–(4.123)):

$$\hat{\theta}_{D_p}(t+1) = \hat{\theta}_{D_p}(t) + \frac{F_2(t)\phi_{D_p}(t)\varepsilon^{\circ}_{D_p}(t+1)}{1+\phi_{D_p}(t)^T F_2(t)\phi_{D_p}(t)} \tag{13.47}$$

$$\varepsilon^{\circ}_{D_p}(t+1) = x(t+1) - \hat{x}(t+1) \tag{13.48}$$

$$\hat{x}(t+1) = \hat{\theta}^T_{D_p}(t)\phi_{D_p}(t) + \big(\hat{p}(t+1) + \hat{p}(t+1-2n)\big) \tag{13.49}$$

$$F_2(t+1)^{-1} = \lambda_1(t)F_2(t)^{-1} - \lambda_2(t)\phi_{D_p}(t)\phi_{D_p}(t)^T \tag{13.50}$$

$$0 < \lambda_1(t) \leq 1; \quad 0 \leq \lambda_2(t) < 2; \quad F_2(0) > 0$$

The PAA defined in (4.121)–(4.123) is used with $\phi(t) = \phi_{D_p}(t), \hat{\theta}(t) = \hat{\theta}_{D_p}(t)$ and $\varepsilon^{\circ}(t+1) = \varepsilon^{\circ}_{D_p}(t+1)$. For implementation, since the objective is to make $x(t+1) \to 0$, the implementable *a priori* adaptation error is defined as follows:

$$\begin{aligned}\varepsilon^{\circ}_{D_p}(t+1) &= 0 - \hat{D}_p(q^{-1},t)\hat{p}(t+1)\\ &= -\hat{\theta}^T_{D_p}(t)\phi_{D_p}(t) - \big(\hat{p}(t+1) + \hat{p}(t-2n+1)\big)\,.\end{aligned} \tag{13.51}$$

Additional filtering can be applied on $\hat{p}(t)$ to improve the signal-noise ratio. Since a frequency range of interest was defined, a bandpass filter can be used on $\hat{p}(t)$. Once an estimation of D_p is available, $A_Q = D_p(\rho q^{-1})$ is immediately generated. Since the estimated $\hat{A}_Q$ will be used for the estimation of the parameters of B_Q one needs to show that: $\lim_{t\to\infty} \hat{A}_Q(z^{-1}) = A_Q(z^{-1})$. This is shown in Appendix C.

13.3.2 Estimation of $B_Q(q^{-1})$

Taking into account (13.12), (13.15), (13.16), and (13.17), it remains to compute $B_Q(z^{-1})$ such that

$$S(z^{-1}) = D_p(z^{-1})H_{S_0}(z^{-1})S'(z^{-1}). \tag{13.52}$$

Turning back to (13.26) one obtains

$$S_0 A_Q = D_p H_{S_0} S' + z^{-d} B H_{R_0} H_{S_0} B_Q. \tag{13.53}$$

and taking into consideration also (13.29) it results

$$S_0' A_Q = D_p S' + z^{-d} B H_{R_0} B_Q. \tag{13.54}$$

Once an estimation algorithm is developed for polynomial $\hat{A}_Q(q^{-1})$, the next step is to develop the estimation algorithm for $\hat{B}_Q(q^{-1})$. Assuming that the estimation $\hat{A}_Q(t)$ of $A_Q(z^{-1})$ is available, one can incorporate this polynomial to the adaptation algorithm defined in Sect. 12.2.2. Using (13.32) and (13.27) and assuming that an estimation of $\hat{B}_Q(q^{-1})$ is available at the instant t, the *a priori* error is defined as the output of the closed-loop system written as follows[5]

$$\varepsilon^\circ(t+1) = \frac{S_0\hat{A}_Q(t) - q^{-d}BH_{S_0}H_{R_0}\hat{B}_Q(t)}{P_0\hat{A}_Q(t)}w(t+1)$$

$$= \frac{S_0}{P_0}w(t+1) - \frac{q^{-d}B^*H_{S_0}H_{R_0}}{P_0}\frac{\hat{B}_Q(t)}{\hat{A}_Q(t)}w(t) \tag{13.55}$$

$$= w_1(t+1) - \frac{\hat{B}_Q(t)}{\hat{A}_Q(t)}w^f(t) \tag{13.56}$$

where the notations[6]

$$w(t+1) = A\frac{D_p(\rho)}{D_p}\delta(t+1) \tag{13.57}$$

$$w_1(t+1) = \frac{S_0}{P_0}w(t+1) \tag{13.58}$$

$$w^f(t) = \frac{q^{-d}B^*H_{S_0}H_{R_0}}{P_0}w(t) \tag{13.59}$$

have been introduced.

Substituting (13.53) in (13.55) one gets:

$$\varepsilon^\circ(t+1) = \frac{H_{S_0}D_pS'}{P_0A_Q}w(t+1) + \frac{q^{-d}B^*H_{S_0}H_{R_0}}{P_0}\frac{B_Q}{A_Q}w(t) - \frac{q^{-d}B^*H_{S_0}H_{R_0}}{P_0}\frac{\hat{B}_Q(t)}{\hat{A}_Q(t)}w(t) \tag{13.60}$$

$$= \upsilon(t+1) + \frac{q^{-d}B^*H_{S_0}H_{R_0}}{P_0}\left[\frac{B_Q}{A_Q} - \frac{\hat{B}_Q(t)}{\hat{A}_Q(t)}\right]w(t) \tag{13.61}$$

where

$$\upsilon(t+1) = \frac{H_{S_0}D_pS'}{P_0A_Q}\frac{AD_p(\rho)}{D_p}\delta(t+1) = \frac{H_{S_0}S'A}{P_0}\delta(t+1) \tag{13.62}$$

tends asymptotically to zero since it is the output of an asymptotically stable filter whose input is a Dirac pulse.

The equation for the *a posteriori* error takes the form[7]

[5]The argument (q^{-1}) will be dropped in some of the following equations.

[6]For the development of the equation for the adaptation error one assumes that the estimated parameters have constant values which allows to use the commutativity property of the various operators.

[7]The details of the developments leading to this equation are given in the Appendix C.

$$\varepsilon(t+1) = \frac{1}{A_Q}\left[\theta_1^T - \hat{\theta}_1^T(t+1)\right]\phi_1(t) + \upsilon^f(t+1) + \upsilon_1(t+1), \tag{13.63}$$

where

$$\upsilon^f(t+1) = \frac{1}{A_Q}\upsilon(t+1) \to 0, \text{ since } A_Q \text{ is } a.s. \tag{13.64}$$

$$\upsilon_1(t+1) = \frac{1}{A_Q}\left(A_Q^* - \hat{A}_Q^*(t+1)\right)\left(-\hat{u}_Q^f(t)\right) \to 0, \tag{13.65}$$

$$\theta_1 = \left[b_0^Q, \ldots, b_{2n-1}^Q\right]^T \tag{13.66}$$

$$\hat{\theta}_1(t+1) = \left[\hat{b}_0^Q(t+1), \ldots, \hat{b}_{2n-1}^Q(t+1)\right]^T \tag{13.67}$$

$$\phi_1(t) = \left[w^f(t), \ldots, w^f(t+1-2n)\right]^T \tag{13.68}$$

$$w^f(t) = \frac{q^{-d}B^* H_{S_1} H_{R_1}}{P_0} w(t) \tag{13.69}$$

and n is the number of narrow-band disturbances. The convergence towards zero for the signal $\upsilon_1(t+1)$ is assured by the fact that $\lim_{t\to\infty}\hat{A}_Q(t,z^{-1}) = A_Q(z^{-1})$ (see Appendix C). Since $\upsilon^f(t+1)$ and $\upsilon_1(t+1)$ tend towards zero, (13.63) has the standard form of an adaptation error equation (see Chap. 4 and [9]), and the following PAA is proposed:

$$\hat{\theta}_1(t+1) = \hat{\theta}_1(t) + F_1(t)\Phi_1(t)\nu(t+1) \tag{13.70}$$

$$\nu(t+1) = \frac{\nu^\circ(t+1)}{1 + \Phi_1^T(t)F_1(t)\Phi_1(t)} \tag{13.71}$$

$$F_1(t+1)^{-1} = \lambda_1(t)F_1(t)^{-1} - \lambda_2(t)\Phi_1(t)\Phi_1^T(t) \tag{13.72}$$

$$0 < \lambda_1(t) \le 1;\ 0 \le \lambda_2(t) < 2;\ F_1(0) > 0 \tag{13.73}$$

There are several possible choices for the regressor vector $\Phi_1(t)$ and the adaptation error $\nu(t+1)$, because there is a strictly positive real condition for stability related to the presence of the term $\frac{1}{A_Q}$ in (13.63). For the case where $\nu(t+1) = \varepsilon(t+1)$, one has $\nu^\circ(t+1) = \varepsilon^\circ(t+1)$, where

$$\varepsilon^\circ(t+1) = w_1(t+1) - \hat{\theta}_1^T(t)\Phi_1(t). \tag{13.74}$$

For the case where $\nu(t+1) = \hat{A}_Q\varepsilon(t+1)$:

$$\nu^{\circ}(t+1) = \varepsilon^{\circ}(t+1) + \sum_{i=1}^{n_{A_Q}} \hat{a}_i^Q \varepsilon(t+1-i). \tag{13.75}$$

These various choices result from the stability analysis given in Appendix C. They are detailed below and summarized in Table 13.2.

- $\Phi_1(t) = \phi_1(t)$. In this case, the prediction error $\varepsilon(t+1)$ is chosen as adaptation error $\nu(t+1)$ and the regressor vector $\Phi_1(t) = \phi_1(t)$. Therefore, the stability condition is: $H' = \frac{1}{A_Q} - \frac{\lambda_2}{2}$ ($\max_t \lambda_2(t) \leq \lambda_2 < 2$) should be strictly positive real (SPR).
- $\nu(t+1) = \hat{A}_Q \varepsilon(t+1)$. The adaptation error is considered as the filtered prediction error $\varepsilon(t+1)$ through a filter $\hat{A}_Q$. The regressor vector is $\Phi_1(t) = \phi_1(t)$ and the stability condition is modified to: $H' = \frac{\hat{A}_Q}{A_Q} - \frac{\lambda_2}{2}$ ($\max_t \lambda_2(t) \leq \lambda_2 < 2$) should be SPR where $\hat{A}_Q$ is a fixed estimation of A_Q.
- $\Phi_1(t) = \phi_1^f(t)$. Instead of filtering the adaptation error, the observations can be filtered to relax the stability condition.[8] By filtering the observation vector $\phi_1(t)$ through $\frac{1}{\hat{A}_Q}$ and using $\nu(t+1) = \varepsilon(t+1)$, the stability condition is: $H' = \frac{\hat{A}_Q}{A_Q} - \frac{\lambda_2}{2}$ ($\max_t \lambda_2(t) \leq \lambda_2 < 2$) should be SPR, where $\phi_1^f(t) = \frac{1}{\hat{A}_Q}\phi_1(t)$ ($\hat{A}_Q$ is a fixed estimation of A_Q).
- $\Phi_1(t) = \phi_1^f(t) = \frac{1}{\hat{A}_Q(t)}$ where $\hat{A}_Q = \hat{A}_Q(t)$ is the current estimation of A_Q. When filtering through a current estimation $\hat{A}_Q(t)$ the condition is similar to the previous case except that it is only valid locally [9].

It is this last option which is used in [10] and in Sect. 13.5.

The following procedure is applied at each sampling time for adaptive operation:

1. Get the measured output $y(t+1)$ and the applied control $u(t)$ to compute $w(t+1)$ using (13.33).
2. Obtain the filtered signal $\hat{p}(t+1)$ from (13.35).

Table 13.2 Comparison of algorithms for the adaptation of the numerator parameters $B_Q(z^{-1})$

Adaptation error $\nu(t+1)$	Prediction error $\varepsilon(t+1)$	Regressor vector $\Phi_1(t)$	Positive real cond. $H'(z^{-1})$	Stability
$\varepsilon(t+1)$	Eq. (13.63)	$\phi_1(t)$	$\frac{1}{A_Q} - \frac{\lambda_2}{2}$	Global
$\hat{A}_Q\varepsilon(t+1)$	Eq. (13.63)	$\phi_1(t)$	$\frac{\hat{A}_Q}{A_Q} - \frac{\lambda_2}{2}$	Global
$\varepsilon(t+1)$	Eq. (13.63)	$\phi_1^f(t)$	$\frac{\hat{A}_Q}{A_Q} - \frac{\lambda_2}{2}$	Global
$\varepsilon(t+1)$	Eq. (13.63)	$\phi_1^f(t)$	$\frac{\hat{A}_Q(t)}{A_Q} - \frac{\lambda_2}{2}$	Local

[8]Neglecting the non-commutativity of the time-varying operators.

3. Compute the implementable *a priori* adaptation error with (13.48).
4. Estimate $\hat{D}_p(q^{-1})$ using the PAA and compute at each step $\hat{A}_Q(q^{-1})$.
5. Compute $w^f(t)$ with (13.69).
6. Compute $w_1(t+1)$ with (13.58).
7. Put the filtered signal $w_2^f(t)$ in the observation vector, as in (13.68).
8. Compute the *a priori* adaptation error defined in (13.74).
9. Estimate the B_Q polynomial using the parametric adaptation algorithm (13.70)–(13.72).
10. Compute and apply the control (see Fig. 13.4):

$$S_0 u(t) = -R_0 y(t+1) - H_{S_0} H_{R_0} \left(\hat{B}_Q(t) w(t+1) - \hat{A}_Q^* \hat{u}_Q(t) \right). \quad (13.76)$$

13.4 Indirect Adaptive Regulation Using Band-Stop Filters

In this section, an indirect adaptive regulation scheme will be developed for implementing the attenuation of multiple unknown narrow-band disturbances using band-stop filters centred at the frequencies corresponding to spikes in the spectrum of the disturbance. The principle of the linear design problem has been discussed in Sect. 13.2.1.

The design of the BSF for narrow-band disturbance attenuation is further simplified by considering a Youla–Kučera parametrization of the controller [2, 11–13]. By doing this, the dimension of the matrix equation that has to be solved is reduced significantly and therefore the computation load will be much lower in the adaptive case.

In order to implement this approach in the presence of unknown narrow-band disturbances, one needs to estimate in real time the frequencies of the spikes contained in the disturbance. System identification techniques can be used to estimate the ARMA model of the disturbance [3, 14]. Unfortunately, to find the frequencies of the spikes from the estimated model of the disturbance requires computation in real time of the roots of an equation of order $2 \cdot n$, where n is the number of spikes. Therefore, this approach is applicable in the case of one eventually two narrow-band disturbances [1, 2]. What is needed is an algorithm which can directly estimate the frequencies of the various spikes of the disturbance. Several methods have been proposed [15]. The adaptive notch filter (ANF) is particularly interesting and has been reviewed in a number of articles [6, 16–21]. In this book, the estimation approach presented in [22, 23] will be used. Combining the frequency estimation procedure and the control design procedure, an indirect adaptive regulation system for attenuation of multiple unknown and/or time-varying narrow-band disturbances is obtained.

In the present context, the hypothesis of constant dynamic characteristics of the AVC system is made (like in [3]). Furthermore, the corresponding control model is supposed to be accurately identified from input/output data.

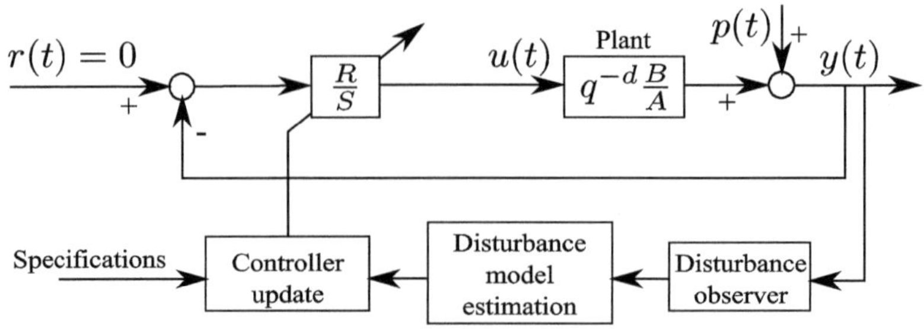

Fig. 13.5 Basic scheme for indirect adaptive regulation

13.4.1 Basic Scheme for Indirect Adaptive Regulation

The equation describing the system has been given in Sect. 13.2. The basic scheme for indirect adaptive regulation is presented in Fig. 13.5. In the context of unknown and time-varying disturbances, a disturbance observer followed by a disturbance model estimation block have to be used in order to obtain information on the disturbance characteristics needed to update the controller parameters.

With respect to Eq. (13.1), it is supposed that

$$p(t) = \frac{D(\rho q^{-1})}{D(q^{-1})}\delta(t), \quad \rho \in (0, 1) \text{ is a fixed constant,} \tag{13.77}$$

represents the effect of the disturbance on the measured output.[9]

Under the hypothesis that the plant model parameters are constant and that an accurate identification experiment can be run, a reliable estimate $\hat{p}(t)$ of the disturbance signal can be obtained using the following disturbance observer

$$\begin{aligned}\hat{p}(t+1) &= y(t+1) - \frac{q^{-d}B^*(q^{-1})}{A(q^{-1})}u(t) \\ &= \frac{1}{A(q^{-1})}\left(A(q^{-1})y(t+1) - q^{-d}B^*(q^{-1})u(t)\right)\end{aligned} \tag{13.78}$$

A disturbance model estimation block can then be used to identify the frequencies of the sines in the disturbance. With this information, the control parameters can directly be updated using the procedure described in Sect. 13.2.1. To deal with time-varying disturbances, the Bezout equation (13.17) has to be solved at each sampling instant in order to adjust the output sensitivity function. Nevertheless, given the size of this equation (see (13.18)), a significant part of the controller computation time would be consumed to solve this equation. To reduce the complexity of this equation,

[9]The disturbance passes through a so called "primary path" which is not represented in Fig. 13.5.

a solution based on the Youla–Kučera parametrization is described in the following section.

13.4.2 Reducing the Computational Load of the Design Using the Youla–Kučera Parametrization

The attenuation of narrow-band disturbances using band-stop filters (BSF) has been presented in Sect. 13.2.1 in the context of linear controllers.

In an indirect adaptive regulation scheme, the Diophantine equation (13.17) has to be solved either at each sampling time (adaptive operation) or each time when a change in the narrow-band disturbances' frequencies occurs (self-tuning operation). The computational complexity of (13.17) is significant (in the perspective of its use in adaptive regulation). In this section, we show how the computation load of the design procedure can be reduced by the use of the Youla–Kučera parametrization.

As before, a multiple band-stop filter, (13.11), should be computed based on the frequencies of the multiple narrow-band disturbance (the problem of frequencies estimation will be discussed in Sect. 13.4.3).

Suppose that a nominal controller is available, as in (13.28) and (13.29), that assures nominal performances for the closed-loop system in the absence of narrow-band disturbances. This controller satisfies the Bezout equation

$$P_0(z^{-1}) = A(z^{-1})S_0(z^{-1}) + q^{-z}B(z^{-1})R_0(z^{-1}). \tag{13.79}$$

Since $P_{BSF}(z^{-1})$ will define part of the desired closed-loop poles, it is reasonable to consider an IIR Youla–Kučera filter of the form $\frac{B_Q(z^{-1})}{P_{BSF}(z^{-1})}$ (which will automatically introduce $P_{BSF}(z^{-1})$ as part of the closed-loop poles). For this purpose, the controller polynomials are factorized as

$$R(z^{-1}) = R_0(z^{-1})P_{BSF}(z^{-1}) + A(z^{-1})H_{R_0}(z^{-1})H_{S_0}(z^{-1})B_Q(z^{-1}), \tag{13.80}$$
$$S(z^{-1}) = S_0(z^{-1})P_{BSF}(z^{-1}) - z^{-d}B(z^{-1})H_{R_0}(z^{-1})H_{S_0}(z^{-1})B_Q(z^{-1}), \tag{13.81}$$

where $B_Q(z^{-1})$ is an FIR filter that should be computed in order to satisfy

$$P(z^{-1}) = A(z^{-1})S(z^{-1}) + z^{-d}B(z^{-1})R(z^{-1}), \tag{13.82}$$

for $P(z^{-1}) = P_0(z^{-1})P_{BSF}(z^{-1})$, and $R_0(z^{-1})$, $S_0(z^{-1})$ given by (13.28) and (13.29), respectively. It can be seen from (13.80) and (13.81), using (13.28) and (13.29), that the new controller polynomials conserve the fixed parts of the nominal controller.

Equation (13.18) gives the size of the matrix equation to be solved if the Youla–Kučera parametrization is not used. Using the previously introduced YK parametrization, it will be shown here that a smaller size matrix equation can be found that allows to compute the $B_Q(z^{-1})$ filter so that the same shaping be introduced on the output sensitivity function (13.16). This occurs if the controller denominator $S(z^{-1})$

in (13.81) is the same as the one given in (13.13), i.e.,

$$S(z^{-1}) = S_{BSF}(z^{-1})H_{S_0}(z^{-1})S'(z^{-1}), \tag{13.83}$$

where $H_S(z^{-1})$ has been replaced by (13.15).

Replacing $S(z^{-1})$ in the left term with its formula given in (13.81) and rearranging the terms, one obtains

$$S_0 P_{BSF} = S_{BSF} H_{S_0} S' + z^{-d} B H_{R_0} H_{S_0} B_Q. \tag{13.84}$$

and taking into consideration also (13.29) it results

$$S'_0 P_{BSF} = S_{BSF} S' + q^{-d} B H_{R_0} B_Q, \tag{13.85}$$

which is similar to (13.54) except that band-stop filters are used instead of notch filters.

In the last equation, the left side of the equal sign is known and on its right side only $S'(z^{-1})$ and $B_Q(z^{-1})$ are unknown. This is also a Bezout equation which can be solved by finding the solution to a matrix equation of dimension

$$n_{Bez_{YK}} = n_B + d + n_{H_{R_0}} + 2 \cdot n - 1. \tag{13.86}$$

As it can be observed, the size of the new Bezout equation is reduced in comparison to (13.18) by $n_A + n_{H_{S_0}}$. For systems with large dimensions, this has a significant influence on the computation time. Taking into account that the nominal controller is an unique and minimal degree solution the Bezout equation (13.79), we find that the left hand side of (13.85) is a polynomial of degree

$$n_{S'_0} + 2 \cdot n = 2 \cdot n + n_B + d + n_{H_{R_0}} - 1, \tag{13.87}$$

which is equal to the quantity given in (13.86). Therefore, the solution of the simplified Bezout equation (13.85) is unique and of minimal degree. Furthermore, the order of the B_Q FIR filter is equal to $2 \cdot n$.

Figure 13.6 summarizes the implementation of the Youla–Kučera parametrized indirect adaptive controller.

13.4.3 Frequency Estimation Using Adaptive Notch Filters

In order to use the presented control strategy in the presence on unknown and/or time-varying narrow-band disturbances, one needs an estimation in real time of the spikes' frequencies in the spectrum of the disturbance. Based on this estimation in real time of the frequencies of the spikes, the band-stop filters will be designed in real time.

In the framework of narrow-band disturbance rejection, it is usually supposed that the disturbances are in fact sinusoidal signals with variable frequencies. It is assumed that the number of narrow-band disturbances is known (similar to [2, 3, 8]). A technique based on ANFs (adaptive notch filters) will be used to estimate the frequencies of the sinusoidal signals in the disturbance (more details can be found in [6, 23]).

The general form of an ANF is

$$H_f(z^{-1}) = \frac{A_f(z^{-1})}{A_f(\rho z^{-1})}, \tag{13.88}$$

where the polynomial $A_f(z^{-1})$ is such that the zeros of the transfer function $H_f(z^{-1})$ lie on the unit circle. A necessary condition for a monic polynomial to satisfy this property is that its coefficients have a mirror symmetric form

$$A_f(z^{-1}) = 1 + a_1^f z^{-1} + \cdots + a_n^f z^{-n} + \cdots + a_1^f z^{-2n+1} + z^{-2n}. \tag{13.89}$$

Another requirement is that the poles of the ANF should be on the same radial lines as the zeros but slightly closer to the origin of the unit circle. Using filter denominators of the general form $A_f(\rho z^{-1})$ with ρ a positive real number smaller but close to 1, the poles have the desired property and are in fact located on a circle of radius ρ [6].

The estimation algorithm will be detailed next. It is assumed that the disturbance signal (or a good estimation) is available.

A cascade construction of second-order ANF filters is considered. Their number is given by the number of narrow-band signals, whose frequencies have to be estimated. The main idea behind this algorithm is to consider the signal $\hat{p}(t)$ as having the form

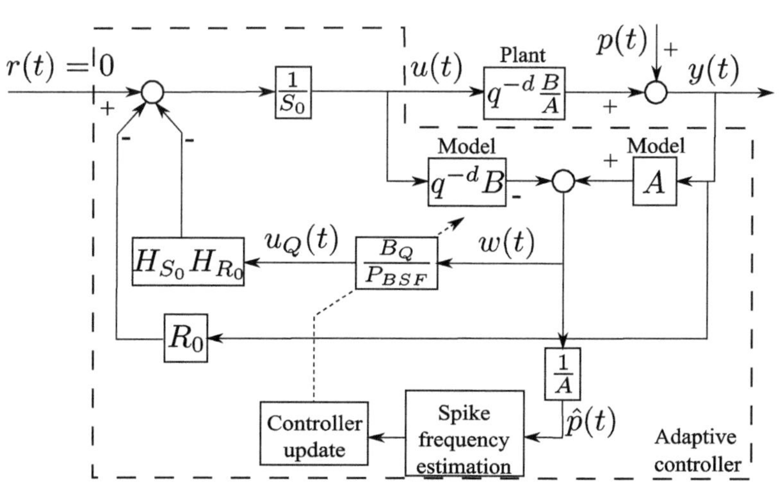

Fig. 13.6 Youla–Kučera schema for indirect adaptive control

$$\hat{p}(t) = \sum_{i=1}^{n} c_i \sin(\omega_i \cdot t + \beta_i) + \eta(t), \tag{13.90}$$

where $\eta(t)$ is a noise affecting the measurement and n is the number of narrow-band signals with different frequencies.

The ANF cascade form will be given by (this is an equivalent representation of Eqs. (13.88) and (13.89))

$$H_f(z^{-1}) = \prod_{i=1}^{n} H_f^i(z^{-1}) = \prod_{i=1}^{n} \frac{1 + a^{f_i} z^{-1} + z^{-2}}{1 + \rho a^{f_i} z^{-1} + \rho^2 z^{-2}}. \tag{13.91}$$

Next, the estimation of one spike's frequency is considered, assuming convergence of the other $n - 1$, which can thus be filtered out of the estimated disturbance signal, $\hat{p}(t)$, by applying

$$\hat{p}^j(t) = \prod_{\substack{i=1 \\ i \neq j}}^{n} \frac{1 + a^{f_i} z^{-1} + z^{-2}}{1 + \rho a^{f_i} z^{-1} + \rho^2 z^{-2}} \hat{p}(t). \tag{13.92}$$

The prediction error is obtained from

$$\varepsilon(t) = H_f(z^{-1}) \hat{p}(t) \tag{13.93}$$

and can be computed based on one of the $\hat{p}^j(t)$ to reduce the computation complexity. Each cell can be adapted independently after prefiltering the signal by the others. Following the Recursive Prediction Error (RPE) technique, the gradient is obtained as

$$\psi^j(t) = -\frac{\partial \varepsilon(t)}{\partial a^{f_j}} = \frac{(1-\rho)(1-\rho z^{-2})}{1 + \rho a^{f_j} z^{-1} + \rho^2 z^{-2}} \hat{p}^j(t). \tag{13.94}$$

The parameter adaptation algorithm can be summarized as

$$\hat{a}^{f_j}(t) = \hat{a}^{f_j}(t-1) + \alpha(t-1) \cdot \psi^j(t) \cdot \varepsilon(t) \tag{13.95}$$

$$\alpha(t) = \frac{\alpha(t-1)}{\lambda + \alpha(t-1)\psi^j(t)^2}. \tag{13.96}$$

where $\hat{a}^{f_j}$ are estimations of the true a^{f_j}, which are connected to the narrow-band signals' frequencies by $\omega_{f_j} = f_s \cdot \arccos(-\frac{a^{f_j}}{2})$, where f_s is the sampling frequency.

13.4.3.1 Implementation of the Algorithm

The design parameters that need to be provided to the algorithm are the number of narrow-band spikes in the disturbance (n), the desired attenuations and damping of the BSFs, either as unique values ($M_i = M,\ \zeta_{d_i} = \zeta_d,\ \forall i \in \{1, \ldots, n\}$) or as individual values for each of the spikes (M_i and ζ_{d_i}), and the central controller (R_0, S_0) together with its fixed parts (H_{R_0}, H_{S_0}) and of course the estimation of the spikes' frequencies. The control signal is computed by applying the following procedure at each sampling time:

1. Get the measured output $y(t+1)$ and the applied control $u(t)$ to compute the estimated disturbance signal $\hat{p}(t+1)$ as in (13.78).
2. Estimate the disturbances' frequencies using adaptive notch filters, Eqs. (13.92)–(13.96).
3. Calculate $S_{BSF}(z^{-1})$ and $P_{BSF}(z^{-1})$ as in (13.8)–(13.11).
4. Find $Q(z^{-1})$ by solving the reduced order Bezout equation (13.85).
5. Compute and apply the control using (13.5) with R and S given respectively by (13.80) and (13.81) (see also Fig. 13.6):

$$S_0 u(t) = -R_0 y(t+1) - H_{S_0} H_{R_0} \left(B_Q(t) w(t+1) - P^*_{BSF} u_Q(t) \right). \quad (13.97)$$

13.4.4 Stability Analysis of the Indirect Adaptive Scheme

The stability analysis of this scheme can be found in [24].

13.5 Experimental Results: Attenuation of Three Tonal Disturbances with Variable Frequencies

Samples of the experimental results obtained with the direct adaptive regulation scheme (see Sect. 13.2.3 and [25]), with the interlaced adaptive regulation scheme (see Sect. 13.3) and with the indirect adaptive regulation scheme (see Sect. 13.4) on the test bench described in Chap. 2, Sect. 2.2 are given in this section. A step change in the frequencies of three tonal disturbances is considered (with return to the initial values of the frequencies). Figures 13.7, 13.8 and 13.9 show the time responses of the residual force. Figures 13.10, 13.11 and 13.12 show the difference between the PSD in open-loop and in closed-loop as well as the estimated output sensitivity function. Figure 13.13 shows the evolution of the parameters of the FIR adaptive Youla–Kučera filter used in the direct adaptive regulation scheme. Figures 13.14 and 13.15 show the evolution of the estimated parameters of D_p (used to compute A_Q—the denominator of the IIR Youla–Kučera filter) and of the numerator B_Q of the IIR Youla–Kučera filter used in the interlaced adaptive regulation scheme. Figure 13.16 shows the evolution of the estimated frequencies of the three tonal disturbances used to compute the band-stop filters in the indirect adaptive regulation scheme.

For this particular experiment, the interlaced adaptive regulation scheme offers the best compromise disturbance attenuation/maximum amplification. Nevertheless, a global evaluation requires to compare the experimental results on a number of situations and this is done in the next section.

13.6 Experimental Results: Comparative Evaluation of Adaptive Regulation Schemes for Attenuation of Multiple Narrow-Band Disturbances

13.6.1 Introduction

Three schemes for adaptive attenuation of single and multiples sparsely located unknown and time-varying narrow-band disturbances have been presented in Chap. 12, Sect. 12.2.2 and in Sects. 13.3 and 13.4 of this chapter. They can be summarized as follows:

(1) Direct adaptive regulation using FIR Youla–Kučera parametrization
(2) Interlaced adaptive regulation using IIR Youla–Kučera parametrization
(3) Indirect adaptive regulation using band-stop filters

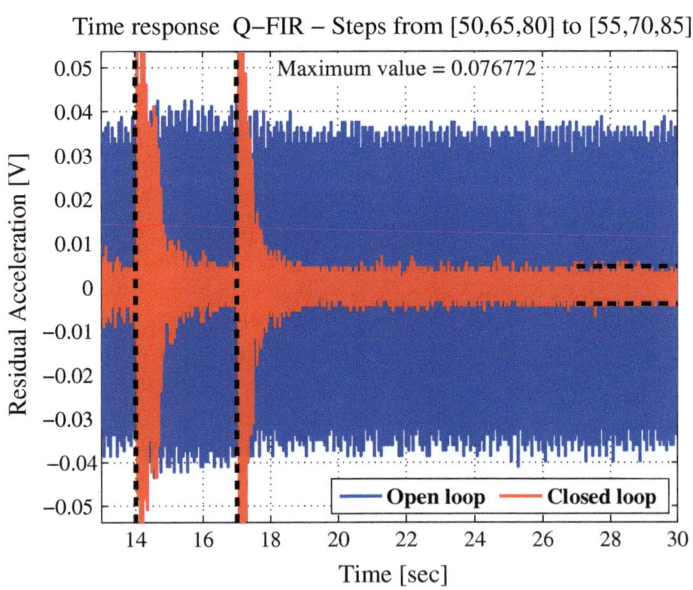

Fig. 13.7 Time response of the direct adaptive regulation scheme using a FIR Youla–Kučera filter for a step change in frequencies (three tonal disturbances)

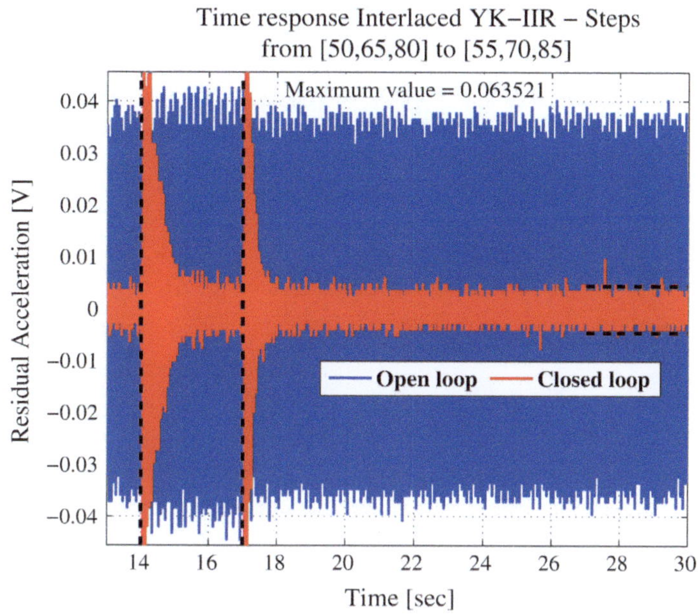

Fig. 13.8 Time response of the interlaced adaptive regulation scheme using an IIR Youla–Kučera filter for a step change in frequencies (three tonal disturbances)

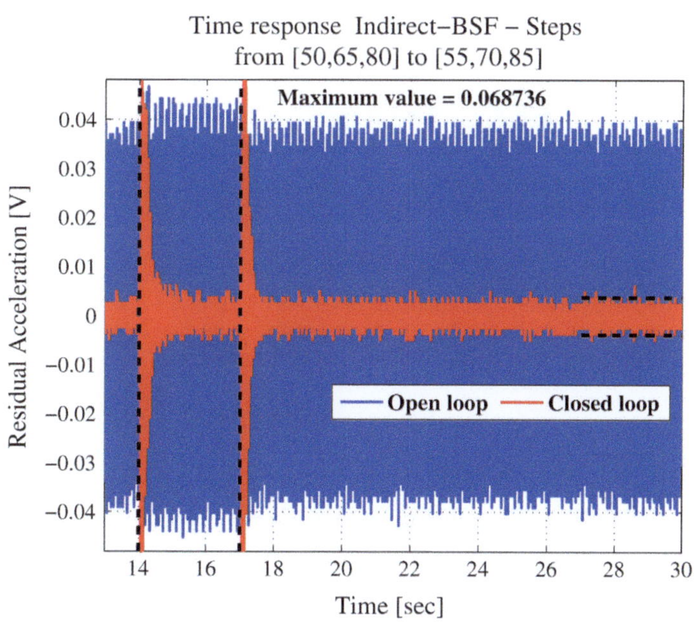

Fig. 13.9 Time response of the indirect adaptive regulation scheme using BSF filters for a step change in frequencies (three tonal disturbances)

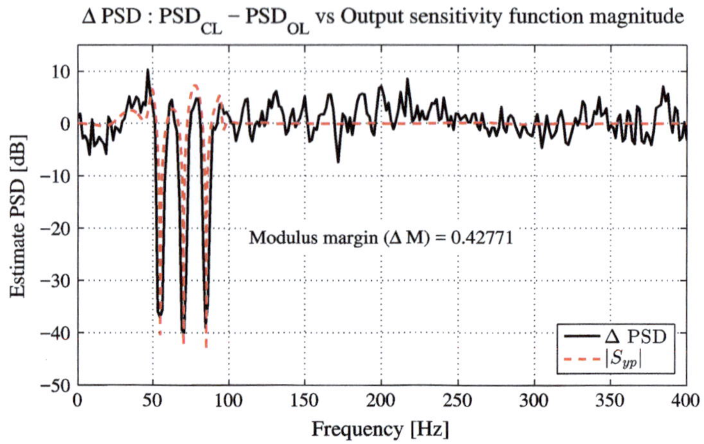

Fig. 13.10 Difference between open-loop and closed-loop PSD of the residual force and the estimated output sensitivity function for the direct adaptive regulation scheme

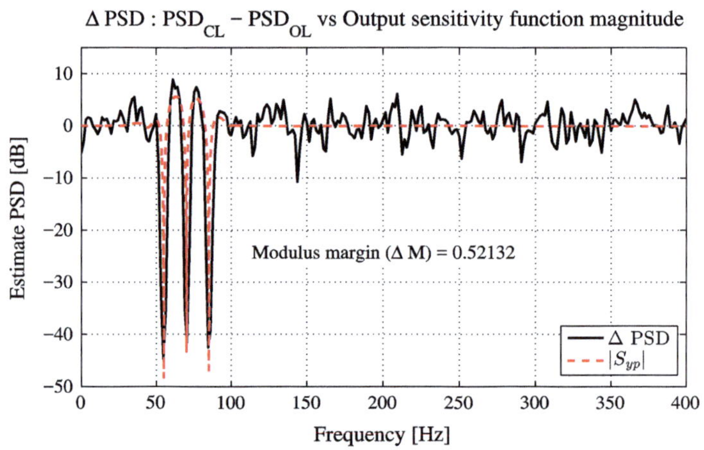

Fig. 13.11 Difference between open-loop and closed-loop PSD of the residual force and the estimated output sensitivity function for the interlaced adaptive regulation scheme

The objective is to comparatively evaluate these three approaches in a relevant experimental environment.

An *international benchmark* on adaptive regulation of sparse distributed unknown and time-varying narrow-band disturbances has been organized in 2012–2013. The summary of the results can be found in [26]. The various contributions can be found in [25, 27–32]. Approaches 1 and 3 have been evaluated in this context. The approach 2, which is *posterior* to the publication of the benchmark results has been also evaluated in the same context. Detailed results can be found in [33]. Approaches 1 and 3 provided some of the best results for the fulfilment of the benchmark specifications

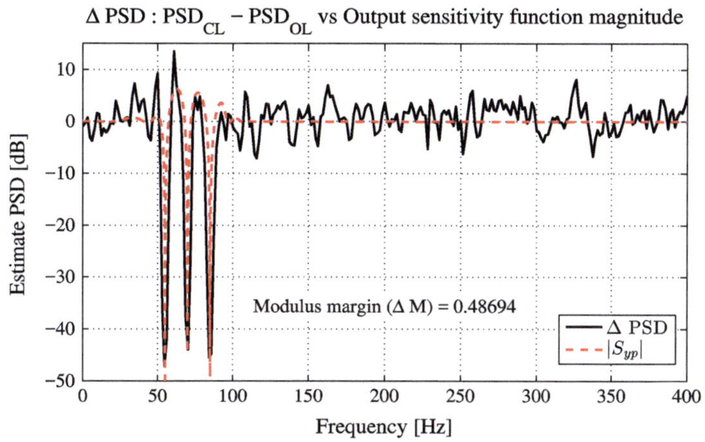

Fig. 13.12 Difference between open-loop and closed-loop PSD of the residual force and the estimated output sensitivity function for the indirect adaptive regulation scheme

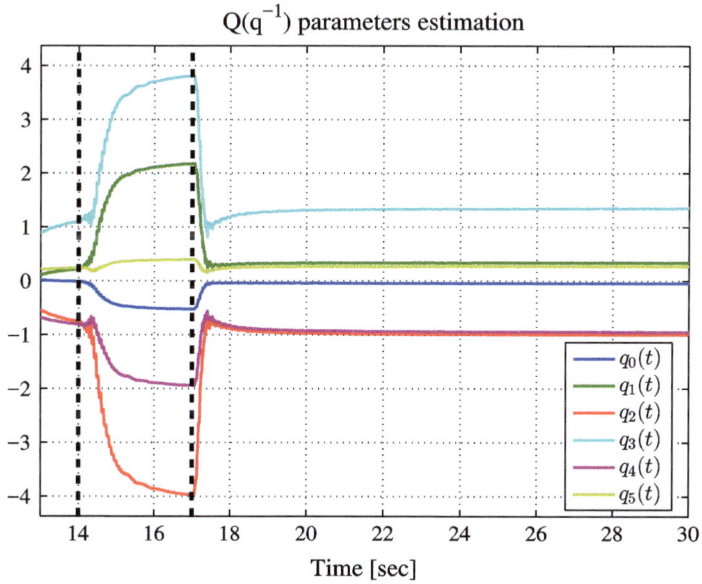

Fig. 13.13 Evolution of the parameters of the FIR Youla–Kučera filter for a step change in frequencies (direct adaptive regulation scheme)

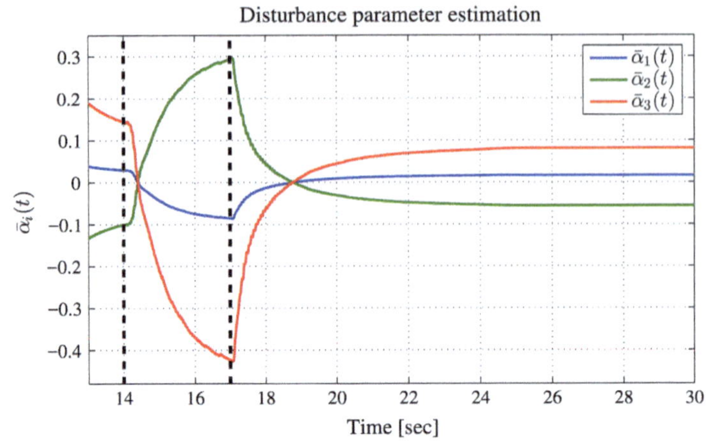

Fig. 13.14 Evolution of the estimated parameters of the D_P polynomial (disturbance model) during a step change of the disturbance frequencies (interlaced adaptive regulation scheme)

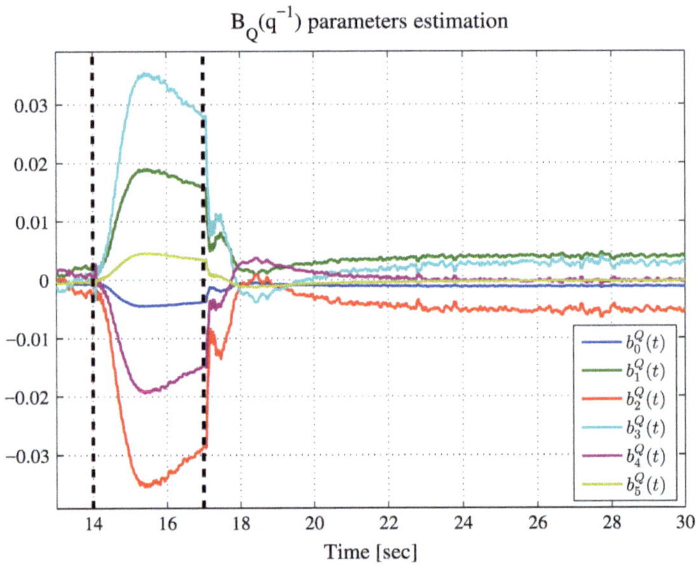

Fig. 13.15 Evolution of the parameters of the numerator of the IIR Youla–Kučera filter during a step change of the disturbance frequencies (interlaced adaptive regulation scheme)

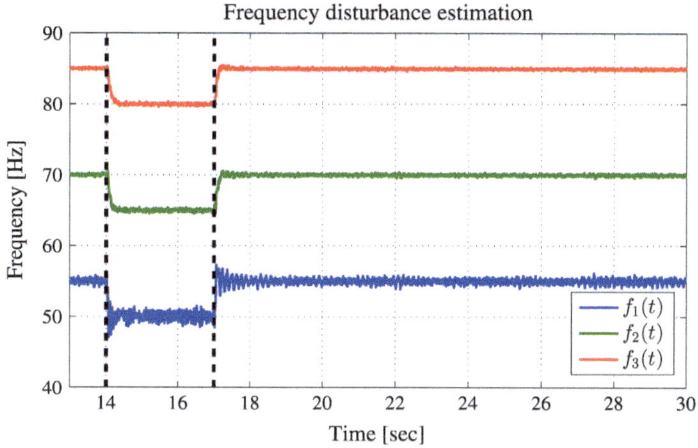

Fig. 13.16 Evolution of the estimated frequencies of the disturbance during a step change of disturbance frequencies (indirect adaptive regulation scheme)

(see [26]). Therefore a comparison of the second approach with the first and third approach is relevant for assessing its potential.

In what follows a comparison of the three approaches will be made in the context of the mentioned benchmark. The objective will be to assess their potential using some of the global indicators used in benchmark evaluation.

In Chap. 12, Sect. 12.3, some of the basic performance indicators have been presented. In the benchmark evaluation process, several protocols allowing to test the performance for various environmental conditions have been defined. Based on the results obtained for the various protocols, global performance indicator have been defined and they will be presented in the next section. This will allow later to present in a compact form the comparison of the real-time performance of the three approaches considered in Chap. 12 and this chapter. Further details can be found in [25, 28, 33].

The basic benchmark specification are summarized in Table 13.3 for the three levels of difficulty (range of frequencies variations: 50 to 95 Hz):

- **Level 1**: Rejection of a single time-varying sinusoidal disturbance.
- **Level 2**: Rejection of two time-varying sinusoidal disturbances.
- **Level 3**: Rejection of three time-varying sinusoidal disturbances.

13.6.2 Global Evaluation Criteria

Evaluation of the performances will be done for both simulation and real-time results. The simulation results will give us information upon the potential of the design methods under the assumption: *design model* = *true plant model*. The real-time

Table 13.3 Control specifications in the frequency domain

Control specifications	Level 1	Level 2	Level 3
Transient duration	≤2 s	≤2 s	≤2 s
Global attenuation	≥30 dB[a]	≥30 dB	≥30 dB
Minimum disturbance attenuation	≥40 dB	≥40 dB	≥40 dB
Maximum amplification	≤6 dB	≤7 dB	≤9 dB

[a]For level 1, the specification of 30 dB is for the range between 50 and 85 Hz, for 90 Hz is 28 dB and for 95 Hz is 24 dB

results will tell us in addition what is the robustness of the design with respect to plant model uncertainties and real noise.

Steady-State Performance (Tuning capabilities)

As mentioned earlier, these are the most important performances. Only if a good tuning for the attenuation of the disturbance can be achieved, it makes sense to examine the transient performance of a given scheme. For the steady-state performance, which is evaluated only for the simple step change in frequencies, the variable k, with $k = 1, \ldots, 3$, will indicate the *level* of the benchmark. In several criteria a mean of certain variables will be considered. The number of distinct experiments, M, is used to compute the mean. This number depends upon the level of the benchmark ($M = 10$ if $k = 1$, $M = 6$ if $k = 2$, and $M = 4$ if $k = 3$).

The performances can be evaluated with respect to the benchmark specifications. The benchmark specifications will be in the form: XXB, where XX will denote the evaluated variable and B will indicate the benchmark specification. ΔXX will represent the error with respect to the benchmark specification.

Global Attenuation—GA

The benchmark specification corresponds to $GAB_k = 30$ dB, for all the levels and frequencies, except for 90 and 95 Hz at $k = 1$, for which GAB_1 is 28 and 24 dB respectively.

Error:

$$\Delta GA_i = GAB_k - GA_i \quad \text{if} \quad GA_i < GAB_k$$
$$\Delta GA_i = 0 \quad \text{if} \quad GA_i \geq GAB_k$$

with $i = 1, \ldots, M$.

Global Attenuation Criterion:

$$J_{\Delta GA_k} = \frac{1}{M} \sum_{j=1}^{M} \Delta GA_i. \tag{13.98}$$

Disturbance Attenuation—DA

The benchmark specification corresponds to $DAB = 40$ dB, for all the levels and frequencies.

Error:

$$\Delta DA_{ij} = DAB - DA_{ij} \quad \text{if} \quad DA_{ij} < DAB$$
$$\Delta DA_{ij} = 0 \quad \text{if} \quad DA_{ij} \geq DAB$$

with $i = 1, \ldots, M$ and $j = 1, \ldots, j_{max}$, where $j_{max} = k$.

Disturbance Attenuation Criterion

$$J_{\Delta DA_k} = \frac{1}{kM} \sum_{i=1}^{M} \sum_{j=1}^{k} \Delta DA_{ij} \tag{13.99}$$

Maximum Amplification—MA

The benchmark specifications depend on the level, and are defined as

$$MAB_k = 6\,\text{dB}, \quad \text{if} \quad k = 1$$
$$MAB_k = 7\,\text{dB}, \quad \text{if} \quad k = 2$$
$$MAB_k = 9\,\text{dB}, \quad \text{if} \quad k = 3$$

Error:

$$\Delta MA_i = MA_i - MAB_k, \quad \text{if} \quad MA_i > MAB_k$$
$$\Delta MA_i = 0, \quad \text{if} \quad MA_i \leq MAB_k$$

with $i = 1, \ldots, M$.

Maximum Amplification Criterion

$$J_{\Delta MA_k} = \frac{1}{M} \sum_{i=1}^{M} \Delta MA_i. \tag{13.100}$$

Global Criterion of Steady-State Performance for One Level

$$J_{SS_k} = \frac{1}{3}[J_{\Delta GA_k} + J_{\Delta DA_k} + J_{\Delta MA_k}]. \tag{13.101}$$

Benchmark Satisfaction Index for Steady-State Performance

The *Benchmark Satisfaction Index* is a performance index computed from the *average* criteria $J_{\Delta GA_k}$, $J_{\Delta DA_k}$ and $J_{\Delta MA_k}$. The *Benchmark Satisfaction Index* is 100 %, if these quantities are "0" (full satisfaction of the benchmark specifications) and it is 0 % if the corresponding quantities are half of the specifications for GA, and DA or twice the specifications for MA. The corresponding reference error quantities are summarized below:

$$\begin{aligned} \Delta GA_{index} &= 15, \\ \Delta DA_{index} &= 20, \\ \Delta MA_{index,1} &= 6, \quad \text{if} \quad k = 1, \\ \Delta MA_{index,2} &= 7, \quad \text{if} \quad k = 2, \\ \Delta MA_{index,3} &= 9, \quad \text{if} \quad k = 3. \end{aligned}$$

The computation formulas are

$$\begin{aligned} GA_{index,k} &= \left(\frac{\Delta GA_{index} - J_{\Delta GA_k}}{\Delta GA_{index}} \right) 100\,\% \\ DA_{index,k} &= \left(\frac{\Delta DA_{index} - J_{\Delta DA_k}}{\Delta DA_{index}} \right) 100\,\% \\ MA_{index,k} &= \left(\frac{\Delta MA_{index,k} - J_{\Delta MA_k}}{\Delta MA_{index,k}} \right) 100\,\%. \end{aligned}$$

Then the *Benchmark Satisfaction Index* (BSI), is defined as

$$BSI_k = \frac{GA_{index,k} + DA_{index,k} + MA_{index,k}}{3}. \tag{13.102}$$

The results for BSI_k obtained both in simulation and real-time for each approach and all the levels are summarized in Tables 13.4 and 13.5 respectively and represented graphically in Fig. 13.17. The YK IIR scheme provides the best results in simulation for all the levels but the indirect approach provides very close results. In real time it is the YK IIR scheme which gives the best results for level 1 and the YK FIR which gives the best results for levels 2 and 3. Nevertheless, one has to mention that the results of the YK FIR scheme are highly dependent on the design of the central controller.

The results obtained in simulation allows the characterization of the performance of the proposed design under the assumption that *design model = true plant model*. Therefore in terms of capabilities of a design method to meet the benchmark specification the simulation results are fully relevant. It is also important to recall that Level 3 of the benchmark is the most important. The difference between the simulation

Table 13.4 Benchmark satisfaction index for steady-state performance (simulation results)

Method	Level 1	Level 2	Level 3
	BSI_1 (%)	BSI_2 (%)	BSI_3 (%)
Indirect	98.69	98.38	99.44
FIR	93.30	97.29	99.13
IIR	**99.07**	**99.84**	**100**

Table 13.5 Benchmark satisfaction index for steady-state performance (real-time results)

Method	Level 1	Level 2	Level 3
	BSI_1 (%)	BSI_2 (%)	BSI_3 (%)
Indirect	81.11	88.51	90.64
FIR	80.87	**89.56**	**97.56**
IIR	**89.37**	87.38	96.39

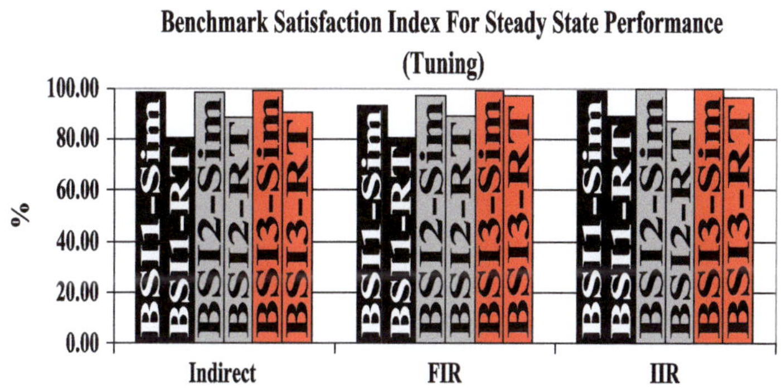

Fig. 13.17 Benchmark Satisfaction Index (BSI) for all levels for simulation and real-time results

results and real-time results, allows one to characterize the robustness in performance with respect to uncertainties on the plant and noise models used for design.

To assess the performance loss passing from simulation to real-time results the Normalized Performance Loss and its global associated index is used. For each level one defines the *Normalized Performance Loss* as

$$NPL_k = \left(\frac{BSI_{ksim} - BSI_{kRT}}{BSI_{ksim}} \right) 100\,\% \tag{13.103}$$

and the global NPL is given by

$$NPL = \frac{1}{M} \sum_{k=1}^{M} NPL_k \tag{13.104}$$

Table 13.6 Normalized performance loss

Method	NPL_1 (%)	NPL_2 (%)	NLP_3 (%)	NPL (%)
Indirect	17.81	10.03	8.85	12.23
FIR	13.32	**7.95**	**1.58**	**7.62**
IIR	**9.79**	12.48	3.61	8.63

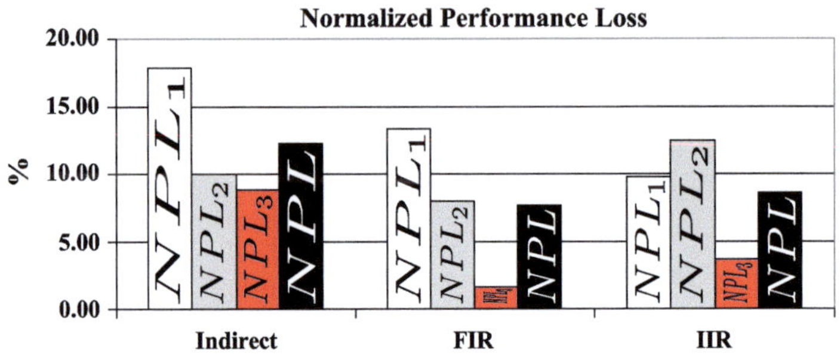

Fig. 13.18 Normalized Performance Loss (NPL) for all levels (smaller = better)

where $N = 3$.

Table 13.6 gives the normalized performance loss for the three schemes. Figure 13.18 summarizes in a bar graph these results. The YK IIR scheme assures a minimum loss for level 1, while the YK FIR scheme assures the minimum loss for level 2 and 3.

Global Evaluation of Transient Performance

For evaluation of the transient performance an indicator has been defined by Eq. 12.46. From this indicator, a global criterion can be defined as follows

$$J_{\Delta Trans_k} = \frac{1}{M}\sum_{j=1}^{M} \Delta Trans_i, \tag{13.105}$$

where $M = 10$ if $k = 1$, $M = 6$ if $k = 2$, and $M = 4$ if $k = 3$.

Transient performance are summarized in Table 13.7. All the schemes assures in most of the cases the 100 % of the satisfaction index for transient performance, which means that the adaptation transient duration is less or equal to 2 s in most of the cases (except the indirect scheme for level 2 in simulation).

Evaluation of the Complexity

For complexity evaluation, the measure of the *Task Execution Time* (TET) in the xPC Target environment will be used. This is the time required to perform all the calculations on the host target PC for each method. Such process has to be done on each sample time. The more complex is the approach, the bigger is the TET. One can

Table 13.7 Benchmark satisfaction index for transient performance (for simple step test)

Index	BSI_{Trans1}		BSI_{Trans2}		BSI_{Trans3}	
Method	Sim (%)	RT (%)	Sim (%)	RT (%)	Sim (%)	RT (%)
Indirect	**100**	99.17	83.33	**100**	**100**	100
FIR	**100**	96.45	**100**	95.74	**100**	**100**
IIR	**100**	**99.20**	**100**	**100**	92.74	95.23

argue that the TET depends also on the programming of the algorithm. Nevertheless, this may change the TET by a factor of 2 to 4 but not by an order of magnitude. The xPC Target MATLAB environment delivers an *average* of the TET ($ATET$). It is however interesting to asses the TET specifically associated to the controller by subtracting from the measured TET in closed-loop operation, the average TET in open-loop operation.

The following criteria to compare the complexity between all the approaches are defined.

$$\Delta TET_{Simple,k} = ATET_{Simple,k} - ATET_{OL_{Simple,k}} \quad (13.106)$$
$$\Delta TET_{Step,k} = ATET_{Step,k} - ATET_{OL_{Step,k}} \quad (13.107)$$
$$\Delta TET_{Chirp,k} = ATET_{Chirp,k} - ATET_{OL_{Chirp,k}} \quad (13.108)$$

where $k = 1, \ldots, 3$. The symbols *Simple, Step* and *Chirp*[10] are associated respectively to Simple Step Test (application of the disturbance), Step Changes in Frequency and Chirp Changes in Frequency. The global ΔTET_k for one level is defined as the average of the above computed quantities:

$$\Delta TET_k = \frac{1}{3}\left(\Delta TET_{Simple,k} + \Delta TET_{Step,k} + \Delta TET_{Chirp,k}\right) \quad (13.109)$$

where $k = 1, \ldots, 3$. Table 13.8 and Fig. 13.19 summarize the results obtained for the three schemes. All the values are in microseconds. Higher values indicate higher complexity. The lowest values (lower complexity) are highlighted.

As expected, the YK FIR algorithm has the smallest complexity. YK IIR has a higher complexity than the YK FIR (This is due to the incorporation of the estimation of $A_Q(z^{-1})$) but still significantly less complex than the indirect approach using BSF.

Tests with a different experimental protocol have been done. The results obtained are coherent with the tests presented above. Details can be found in [10, 34].

[10]The chirp will be considered only for complexity evaluation, for other results concerning chirp disturbance see [33, 34].

Table 13.8 Task execution time

Method	ΔTET		
	L1	L2	L3
Indirect	254.24	203.83	241.22
FIR	**3.26**	**3.90**	**5.60**
IIR	19.42	31.63	44.95

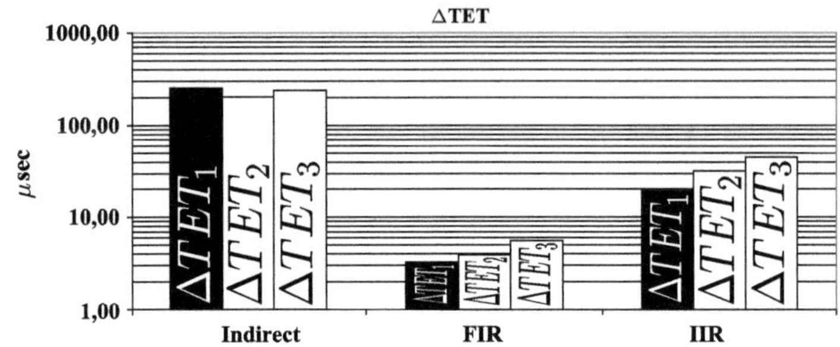

Fig. 13.19 The controller average task execution time (ΔTET)

13.7 Concluding Remarks

It is difficult to decide what is the best scheme for adaptive attenuation of multiple narrow-band disturbances. There are several criteria to be taken into account:

- If an individual attenuation level should be fixed for each spike, the indirect adaptive scheme using BSF is the most appropriate since it allows to achieve specific attenuation for each spike.
- If the objective is to have a very simple design of the central controller, YK IIR scheme and the indirect adaptive scheme have to be considered.
- If the objective is to have the simplest scheme requiring the minimum computation time, clearly the YK FIR has to be chosen.
- If the objective is to make a compromise between the various requirements mentioned above, it is the YK IIR adaptive scheme which has to be chosen.

13.8 Notes and References

The reference [34] gives a thorough view of the various solutions for adaptive attenuation of multiple narrow-band disturbances. The specific references are [25, 27–32] to which the reference [10] has to be added.

References

1. Landau ID, Alma M, Constantinescu A, Martinez JJ, Noë M (2011) Adaptive regulation - rejection of unknown multiple narrow band disturbances (a review on algorithms and applications). Control Eng Pract 19(10):1168–1181. doi:10.1016/j.conengprac.2011.06.005
2. Landau I, Constantinescu A, Rey D (2005) Adaptive narrow band disturbance rejection applied to an active suspension - an internal model principle approach. Automatica 41(4):563–574
3. Landau I, Alma M, Martinez J, Buche G (2011) Adaptive suppression of multiple time-varying unknown vibrations using an inertial actuator. IEEE Trans Control Syst Technol 19(6):1327–1338. doi:10.1109/TCST.2010.2091641
4. Procházka H, Landau ID (2003) Pole placement with sensitivity function shaping using 2nd order digital notch filters. Automatica 39(6):1103–1107. doi:10.1016/S0005-1098(03)00067-0
5. Landau I, Zito G (2005) Digital control systems - design, identification and implementation. Springer, London
6. Nehorai A (1985) A minimal parameter adaptive notch filter with constrained poles and zeros. IEEE Trans Acoust Speech Signal Process ASSP-33:983–996
7. Astrom KJ, Wittenmark B (1984) Computer controlled systems. Theory and design. Prentice-Hall, Englewood Cliffs
8. Chen X, Tomizuka M (2012) A minimum parameter adaptive approach for rejecting multiple narrow-band disturbances with application to hard disk drives. IEEE Trans Control Syst Technol 20(2):408–415. doi:10.1109/TCST.2011.2178025
9. Landau ID, Lozano R, M'Saad M, Karimi A (2011) Adaptive control, 2nd edn. Springer, London
10. Castellanos-Silva A, Landau ID, Dugard L, Chen X (2016) Modified direct adaptive regulation scheme applied to a benchmark problem. Eur J Control 28:69–78. doi:10.1016/j.ejcon.2015.12.006
11. Tsypkin Y (1997) Stochastic discrete systems with internal models. J Autom Inf Sci 29(4&5):156–161
12. de Callafon RA, Kinney CE (2010) Robust estimation and adaptive controller tuning for variance minimization in servo systems. J Adva Mech Design Syst Manuf 4(1):130–142
13. Tay TT, Mareels IMY, Moore JB (1997) High performance control. Birkhauser, Boston
14. Airimitoaie TB, Landau I, Dugard L, Popescu D (2011) Identification of mechanical structures in the presence of narrow band disturbances - application to an active suspension. In: 2011 19th mediterranean conference on control automation (MED), pp 904–909. doi:10.1109/MED.2011.5983076
15. Tichavský P, Nehorai A (1997) Comparative study of four adaptive frequency trackers. IEEE Trans Autom Control 45(6):1473–1484
16. Rao D, Kung SY (1984) Adaptive notch filtering for the retrieval of sinusoids in noise. IEEE Trans Acoust Speech Signal Process 32(4):791–802. doi:10.1109/TASSP.1984.1164398
17. Regalia PA (1991) An improved lattice-based adaptive IIR notch filter. IEEE Trans Signal Process 9(9):2124–2128. doi:10.1109/78.134453
18. Chen BS, Yang TY, Lin BH (1992) Adaptive notch filter by direct frequency estimation. Signal Process 27(2):161–176. doi:10.1016/0165-1684(92)90005-H
19. Li G (1997) A stable and efficient adaptive notch filter for direct frequency estimation. IEEE Trans Signal Process 45(8):2001–2009. doi:10.1109/78.611196
20. Hsu L, Ortega R, Damm G (1999) A globally convergent frequency estimator. IEEE Trans Autom Control 4(4):698–713
21. Obregon-Pulido G, Castillo-Toledo B, Loukianov A (2002) A globally convergent estimator for n-frequencies. IEEE Trans Autom Control 47(5):857–863
22. Stoica P, Nehorai A (1988) Performance analysis of an adaptive notch filter with constrained poles and zeros. IEEE Trans Acoust Speech Signal Process 36(6):911–919

23. M'Sirdi N, Tjokronegoro H, Landau I (1988) An RML algorithm for retrieval of sinusoids with cascaded notch filters. In: 1988 International conference on acoustics, speech, and signal processing, 1988. ICASSP-88, vol 4, pp 2484–2487. doi:10.1109/ICASSP.1988.197147
24. Airimitoaie TB, Landau ID (2014) Indirect adaptive attenuation of multiple narrow-band disturbances applied to active vibration control. IEEE Trans Control Syst Technol 22(2):761–769. doi:10.1109/TCST.2013.2257782
25. Castellanos-Silva A, Landau ID, Airimitoaie TB (2013) Direct adaptive rejection of unknown time-varying narrow band disturbances applied to a benchmark problem. Eur J Control 19(4):326– 336. doi:10.1016/j.ejcon.2013.05.012 (Benchmark on Adaptive Regulation: Rejection of unknown/time-varying multiple narrow band disturbances)
26. Landau ID, Silva AC, Airimitoaie TB, Buche G, Noé M (2013) An active vibration control system as a benchmark on adaptive regulation. In: Control conference (ECC), 2013 European, pp 2873–2878 (2013)
27. Aranovskiy S, Freidovich LB (2013) Adaptive compensation of disturbances formed as sums of sinusoidal signals with application to an active vibration control benchmark. Eur J Control 19(4):253–265. doi:10.1016/j.ejcon.2013.05.008 (Benchmark on Adaptive Regulation: Rejection of unknown/time-varying multiple narrow band disturbances)
28. Airimitoaie TB, Silva AC, Landau ID (2013) Indirect adaptive regulation strategy for the attenuation of time varying narrow-band disturbances applied to a benchmark problem. Eur J Control 19(4):313–325. doi:10.1016/j.ejcon.2013.05.011 (Benchmark on Adaptive Regulation: Rejection of unknown/time-varying multiple narrow band disturbances)
29. de Callafon RA, Fang H (2013) Adaptive regulation via weighted robust estimation and automatic controller tuning. Eur J Control 19(4):266–278. doi:10.1016/j.ejcon.2013.05.009 (Benchmark on Adaptive Regulation: Rejection of unknown/time-varying multiple narrow band disturbances)
30. Karimi A, Emedi Z (2013) H_∞ gain-scheduled controller design for rejection of time-varying narrow-band disturbances applied to a benchmark problem. Eur J Control 19(4):279–288. doi:10.1016/j.ejcon.2013.05.010 (Benchmark on Adaptive Regulation: Rejection of unknown/time-varying multiple narrow band disturbances)
31. Chen X, Tomizuka M (2013) Selective model inversion and adaptive disturbance observer for time-varying vibration rejection on an active-suspension benchmark. Eur J Control 19(4):300–312 (Benchmark on Adaptive Regulation: Rejection of unknown/time-varying multiple narrow band disturbances)
32. Wu Z, Amara FB (2013) Youla parameterized adaptive regulation against sinusoidal exogenous inputs applied to a benchmark problem. Eur J Control 19(4):289–299 (Benchmark on Adaptive Regulation: Rejection of unknown/time-varying multiple narrow band disturbances)
33. Castellanos-Silva A (2014) Compensation adaptative par feedback pour le contrôle actif de vibrations en présence d'incertitudes sur les paramétres du procédé. Ph.D. thesis, Université de Grenoble
34. Landau ID, Silva AC, Airimitoaie TB, Buche G, Noé M (2013) Benchmark on adaptive regulation - rejection of unknown/time-varying multiple narrow band disturbances. Eur J Control 19(4):237–252. doi:10.1016/j.ejcon.2013.05.007

Part V
Feedforward-Feedback Attenuation of Broad-Band Disturbances

Chapter 14
Design of Linear Feedforward Compensation of Broad-band Disturbances from Data

14.1 Introduction

Feedforward compensation of disturbances comes in view when the "waterbed" effect in feedback control cannot allow to assure the desired performance. This occurs systematically if the disturbance has a broad-band character and the attenuation imposed is too important. The use of the feedforward compensation requires the use of an additional transducer which is expected to provide a reliable information upon the disturbance ($w(t)$ in Fig. 14.1).

If such transducer is available in a specific application, feedforward compensation of broad-band disturbances can be implemented. It is important to remind (as indicated in Chap. 1) that feedforward compensation induces an internal coupling (positive) between the control of the secondary path (the compensator system) and the measurement of the disturbance. See the test bench described in Sect. 2.3.

The design of a linear compensator can be viewed (see Fig. 14.1) as finding a linear compensator such that the lower part of Fig. 14.1 has the same transfer function as the primary path but with the reverse sign. With the notation of the figure, we are looking to find N (the compensator) such that:

$$D = -\frac{N}{(1 - NM)}G. \tag{14.1}$$

This is a pure algebraic problem and requires the assumption:

(*Perfect matching condition*) There exists a filter $N(z^{-1})$ of finite dimension such that[1]:

$$D = -\frac{N}{(1 - NM)}G \tag{14.2}$$

[1] In many cases, the argument q^{-1} or z^{-1} will be dropped out.

I.D. Landau et al., *Adaptive and Robust Active Vibration Control*, Advances in Industrial Control, DOI 10.1007/978-3-319-41450-8_14

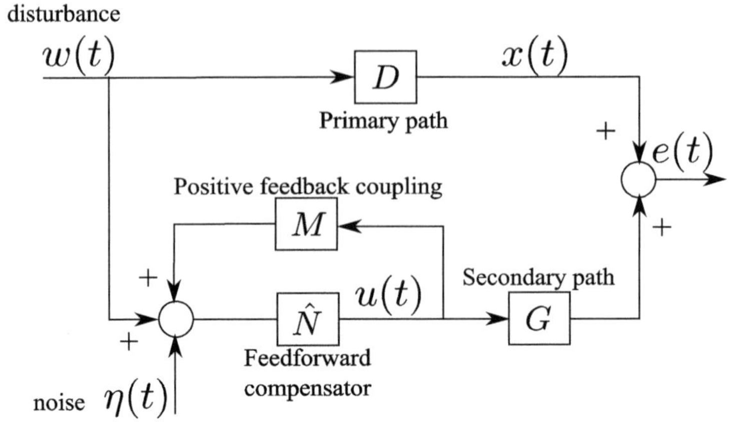

Fig. 14.1 A linear feedforward compensation scheme

and the characteristic polynomial of the "internal" feedback loop:

$$P(z^{-1}) = A_M(z^{-1})S(z^{-1}) - B_M(z^{-1})R(z^{-1}) \tag{14.3}$$

is a Hurwitz polynomial.

This hypothesis means also that D can be equivalently represented by (14.2) where N is unknown. In practice, one can consider however a less strong requirement, i.e., a good fit of the two transfer functions in the frequency region where the disturbance is significant. This problem can be formulated as an H_2 or an H_∞ problem.

Assuming that the control design problem is formulated as a perfect matching objective or as the minimization of a H_2 or H_∞ criterion, in order to compute N one need the models of the primary path, secondary path and reverse path. So one has to solve first the identification problem. The techniques for identification of such systems have been described in Chap. 5 and illustrated in Chap. 6 by the identification of the test bench described in Sect. 2.3.

Assuming that these models are available and knowing the power spectral distribution of the disturbance $w(t)$ (by analyzing the data captured by the additional transducer), the computation of the compensator N can be transformed in a frequency weighted error minimization since one would like to have a good fit between the two transfer functions in the frequency zone where the disturbance is located.

Therefore, if one has reliable identified models for the primary path, secondary path reverse path and the power spectral density of the disturbance, one can formulate this problem as an H_2 or an H_∞ problem.

Basically for an H_∞ approach one considers

- disturbance-output sensitivity function:

$$S_{ew} = \left(D + G \cdot \left(\frac{N}{1 - NM}\right)\right) \tag{14.4}$$

- disturbance-input sensitivity function:

$$S_{uw} = \left(\frac{N}{1 - NM} \right) \tag{14.5}$$

- noise-input sensitivity function:

$$S_{u\eta} = \left(\frac{N}{1 - NM} \right) \tag{14.6}$$

The H_∞ control problem is to find a stabilizing feedforward compensator N which minimizes a scalar γ [1] such that:

$$\left\| \begin{pmatrix} W_1 \cdot S_{ve} \\ W_2 \cdot S_{ue} \\ W_3 \cdot S_{u\eta} \end{pmatrix} \right\|_\infty < \gamma \tag{14.7}$$

where W_1, W_2 and W_3 are corresponding weighting functions (which can be interpreted as analytic inverses of templates for the sensitivity functions). A similar formulation can be given for the H_2 control problem (see [2]).

The H_∞ approach has been applied on the test bench described in Chap. 2, see [3]. For an H_2 approach see [2] where the case of active suppression of vibrations in flexible structures has been considered.

Another approach can be considered by using an Youla–Kučera parametrization of the compensator N with a central stabilizing controller (R_0 and S_0) and a Q IIR filter as indicated in Fig. 14.2. In this case, using convex optimization one can try to find Q such that the difference between the transfer function of the primary path and the one of the compensated system be minimized in a frequency region of interest

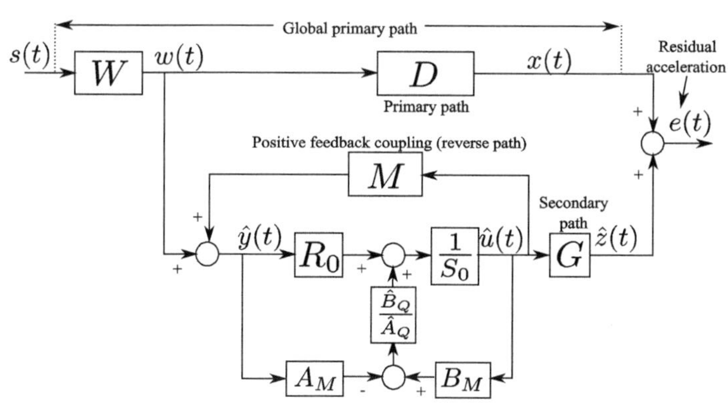

Fig. 14.2 Linear feedforward compensation scheme using Youla–Kučera parametrization of the feedforward compensator

where the spectrum of the disturbance is significant (of course H_2 and H_∞ can also be used in this configuration as well as the convex optimization procedure).

It is important to point out that in order to design a linear controller using a frequency weighted minimization one needs not only the models of the system but also the model of the disturbance. To get information about the disturbance and the system's models, it is required to have access to the system. In other terms, data acquisition under a protocol is mandatory for designing a linear feedforward compensator (in order to identify the models and the characteristics of the disturbance).

14.2 Indirect Approach for the Design of the Feedforward Compensator from Data

Assume that the secondary and the reverse path models have been already identified using the procedure described in Sect. 6.3. In order to design the feedforward compensator, one needs in addition to identify the model of the primary path (D). The primary path model has to be identified with the available input signal $w(t)$ which comes from the system (generated by the disturbance) and measuring the residual force or acceleration which in the absence of the compensator system is denoted $x(t)$ (it is the output of the primary path in the absence of the compensator system—see Fig. 14.1). The quality of the primary path identified model depends upon the richness of the disturbance signal $w(t)$. In fact the identified model will be relevant only in the frequency regions where $w(t)$ has enough energy.

To summarize:

1. one collects the input/output data ($w(t)$ and $x(t)$);
2. one identifies the primary path model from these data; and
3. one proceeds to the design of the linear feedforward compensator based on the primary, secondary and reverse path models and the PSD of $w(t)$ (the image of the disturbance).

The third step is equivalent of finding $\hat{N}$ in order to minimize $e(t)$ in the sense of a certain criterion for the given $w(t)$. This approach will be termed "indirect" since it requires several intermediate steps in order to design the feedforward compensator from data. As it will be shown in the next section, it is possible to formulate the estimation of $\hat{N}$ as the estimation of a reduced order controller.

14.3 Direct Approach for the Design of the Feedforward Compensator from Data

The interesting point is that the design of the linear compensator can be viewed as the estimation of a reduced order controller (see Chap. 9). Given, a set of relevant data collected on the upstream transducer ($w(t)$) and on the residual accelerometer or force transducer ($x(t)$) (in the absence of the compensator system), the problem can be formulated as estimating a reduced order filter N which minimizes the error between the measured acceleration (or force) and the predicted acceleration (force) given by the predictor of the compensation path.

In Fig. 14.3, if $G = 1$, this becomes an estimation of a reduced order controller in closed-loop operation which can be done using the techniques of Chap. 9 (Closed-Loop Input Matching). In the general case ($G \neq 1$) the problem can be reformulated as in Fig. 14.4 where one takes advantage of the hypothesis of perfect matching and of the linearity in steady state which allows to revert the order of the various blocks without changing the global transfer function.

Figure 14.5 makes further connections with the "controller estimation in closed-loop" techniques. What is added in fact is the filtering of the input through G and the filtering of the prediction error through V (allowing to further shape the frequencies characteristics of the estimated filter).

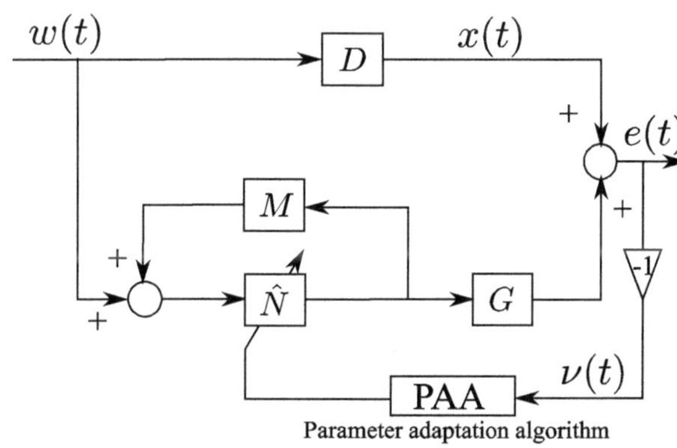

Fig. 14.3 Adaptive feedforward compensation scheme

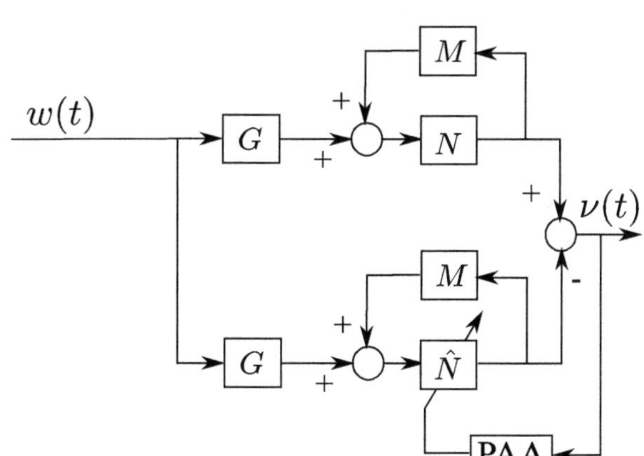

Fig. 14.4 Equivalent formulation of the estimation of the linear compensator as an estimation of the controller in closed-loop operation

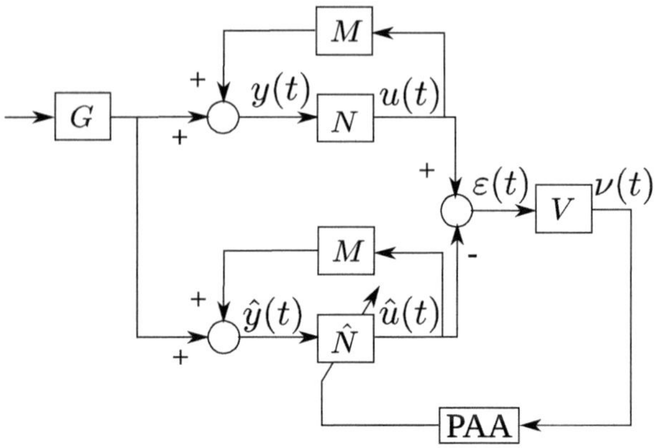

Fig. 14.5 Rapprochement between the estimation of the linear feedforward compensator and the techniques for reduced order controller estimation in closed-loop operation

The optimal feedforward compensator (unknown and of high order) is defined by:

$$N(q^{-1}) = \frac{R(q^{-1})}{S(q^{-1})} \tag{14.8}$$

where:

$$R(q^{-1}) = r_0 + r_1 q^{-1} + \cdots + r_{n_R} q^{-n_R} \tag{14.9}$$

$$S(q^{-1}) = 1 + s_1 q^{-1} + \cdots + s_{n_S} q^{-n_S} = 1 + q^{-1} S^*(q^{-1}) \tag{14.10}$$

and

$$\theta^T = [s_1, \ldots s_{n_S}, r_0, r_1, \ldots r_{n_R}] = [\theta_S^T, \theta_R^T] \tag{14.11}$$

is the vector of parameters of the optimal filter N assuring perfect matching.

The reduced order estimation of the optimal feedforward filter is defined by

$$\hat{N}(q^{-1}) = \frac{\hat{R}(q^{-1})}{\hat{S}(q^{-1})} \tag{14.12}$$

where:

$$\hat{R}(q^{-1}) = \hat{r}_0 + \hat{r}_1 q^{-1} + \cdots + \hat{r}_{n_R} q^{-n_R} \tag{14.13}$$

$$\hat{S}(q^{-1}) = 1 + \hat{s}_1 q^{-1} + \cdots + \hat{s}_{n_S} q^{-n_S} = 1 + q^{-1} \hat{S}^*(q^{-1}) \tag{14.14}$$

and

$$\hat{\theta}^T = [\hat{s}_1, \ldots \hat{s}_{n_S}, \hat{r}_0, \hat{r}_1, \ldots \hat{r}_{n_R}] = [\hat{\theta}_S^T, \hat{\theta}_R^T] \tag{14.15}$$

is the vector of constant estimated parameters of $\hat{N}$.

The input–output relationships for the estimated feedforward filter are given by:

$$\hat{u}(t+1) = -\hat{S}^*(q^{-1})\hat{u}(t) + \hat{R}(q^{-1})\hat{y}(t+1) = \hat{\theta}^T\phi(t) = [\hat{\theta}_S^T, \hat{\theta}_R^T]\begin{bmatrix}\phi_{\hat{u}}(t)\\ \phi_{\hat{y}}(t)\end{bmatrix} \tag{14.16}$$

where

$$\begin{aligned}\phi^T(t) &= [-\hat{u}(t), \ldots - \hat{u}(t-n_S+1), \hat{y}(t+1), \ldots \hat{y}(t-n_R+1)]\\ &= [\phi_{\hat{u}}^T(t), \phi_{\hat{y}}^T(t)]\end{aligned} \tag{14.17}$$

Going back to the system in Fig. 14.5, one has

$$\nu(t+1) = V(q^{-1})\frac{A_M(q^{-1})}{P(q^{-1})}[\theta - \hat{\theta}]^T\phi(t) \tag{14.18}$$

where P is the characteristic polynomial of the "positive" feedback loop.[2] Considering the model of the reverse path

$$M(q^{-1}) = \frac{B_M(q^{-1})}{A_M(q^{-1})} \tag{14.19}$$

P is given by

$$P(q^{-1}) = A_M(q^{-1})S(q^{-1}) - B_M(q^{-1})R(q^{-1}). \tag{14.20}$$

The identification of $\hat{N}$ can be viewed as an L_2 minimization problem which has a relevant interpretation in the frequency domain.

Using Parseval's relation, the asymptotic bias distribution of the estimated parameters in the frequency domain can be obtained starting from the expression of $\nu(t)$ by taking into account that the algorithm minimizes (almost) a criterion of the form

$$\lim_{N\to\infty}\frac{1}{N}\sum_{t=1}^{N}\nu^2(t).$$

This allows to reformulate the asymptotic behaviour of the estimated compensator (using the formulas given in Sect. 9.3.1) as shown next. Taking into account that the external excitation is filtered by G, and that the prediction error filtered by V, the estimated $\hat{N}$ (characterized by the parameter vector $\hat{\theta}$ given in (14.15)) will have the following asymptotic behaviour in the frequency domain (taking into account a perfect matching condition Eq. (14.1), see also Chap. 15):

---

[2]The term A_M/P comes from the expression of the prediction error, similar to that obtained in Sect. 8.2.1, for the CLOE configuration or in Sect. 9.3, for the CLIM algorithm, with the obvious change in notation (S is replaced by A_M).

$$\hat{\theta}^* = \arg\min_{\hat{\theta}} \int_{-\pi}^{\pi} \Big[|S_{NM}|^2 |N - \hat{N}|^2 |S_{\hat{N}M}|^2 |G|^2 |V|^2 \phi_w(\omega) + |V|^2 \phi_\eta(\omega) \Big] \mathrm{d}\omega \tag{14.21}$$

where ϕ_w and ϕ_η are the spectral densities of the disturbance $w(t)$ and of the measurement noise and S_{NM} and $S_{\hat{N}M}$ are the output sensitivity functions of the internal closed-loop for N and $\hat{N}$, respectively:

$$S_{NM} = \frac{1}{1 - NM}, \tag{14.22}$$

$$S_{\hat{N}M} = \frac{1}{1 - \hat{N}M}. \tag{14.23}$$

From (14.21), one concludes that a good approximation of N will be obtained in the frequency region where ϕ_w is significant and where G and V have a high gain (usually G should have high gain in the frequency region where ϕ_w is significant in order to counteract the effect of $w(t)$). The choice of V will clearly influence the estimated $\hat{N}$. The quality of the estimated $\hat{N}$ will be affected also by the output sensitivity functions of the internal closed-loop $N - M$.

One can also consider to use the adaptive algorithms which will be developed later on for the adaptation of N using the basic configurations indicated in Chap. 1 but in a self-tuning regime, i.e., the adaptation gain will tend asymptotically to zero. Both methods will be demonstrated in the next section.

The use of algorithms for estimation in closed-loop of the controller as indicated previously or the use of adaptive feedforward algorithms in a self-tuning regime is a one step design procedure since the intermediate step of identifying the primary path which is needed for design using H_∞, H_2 or convex optimization disappears. It is in fact another way of exploiting the available data.

14.4 Direct Estimation of the Feedforward Compensator and Real-Time Tests

In this section, it is shown how a linear feedforward controller for the AVC system described in Sect. 2.3 can be directly obtained from the data collected in the absence of the compensator system. The resulting compensator will be tested in real time on the test bench.

First, the use of closed-loop input matching algorithms (see Sect. 9.3) for estimating a reduced order feedforward compensator is demonstrated. As explained in Chap. 9, this problem is similar to the one of closed-loop model identification. Then self-tuning operation of an adaptive simulated feedforward compensation scheme in self-tuning operation will be used to obtain the feedforward compensator.

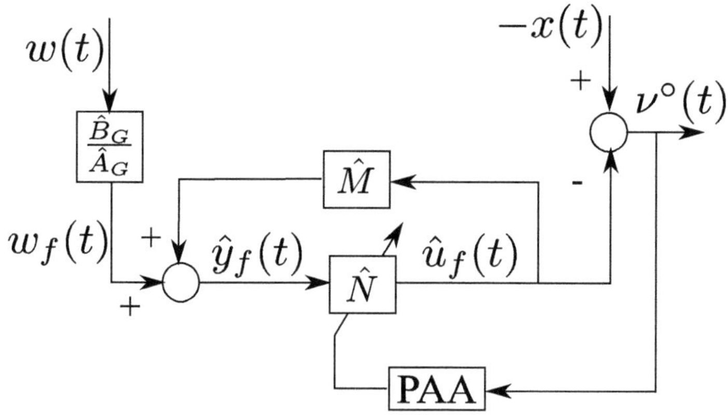

Fig. 14.6 Equivalent formulation for estimation of the feedforward compensator in closed-loop operation

While both approaches use the same data, the operation mode is different. The algorithms for identification in closed-loop take into account the fact that all data are available. The simulated adaptive feedforward compensation schemes in self-tuning regime operates like in real time, i.e., the algorithms ignore the availability of the data over the entire time horizon.

It is supposed that models for the secondary and the reverse paths are available. Section 6.3 gives details on how these models can be obtained. Measured data on the real system for $w(t)$ and $x(t)$ will be used instead of the identified model of the primary path. These data have been obtained using as excitation to the system a PRBS with shift register of length $N = 16$ and frequency divider $p = 1$. Around 82 s of real-time data have been obtained.[3]

Let us begin with the identification of the feedforward compensator using closed-loop input matching algorithms (see Sect. 9.3). The basic scheme is shown in Fig. 14.6, where the excitation signal is obtained by filtering the measured $w(t)$ through the estimated model of the secondary path

$$w_f(t) = \frac{\hat{B}_G}{\hat{A}_G} w(t). \tag{14.24}$$

Let us note that with respect to the closed-loop input matching algorithm given in Eqs. (9.26)–(9.28), one has the following changes in notations: $\hat{c}(t)$ becomes $\hat{y}_f(t)$ and $\hat{u}(t)$ becomes $\hat{u}_f(t)$.

The closed-loop input error is defined as the difference between the available measurement $x(t)$ and $\hat{u}_f(t)$ generated by the closed-loop predictor using $\hat{N}$. Two algorithms are considered. The first one, corresponds to the basic closed-loop controller identification algorithm with the objective to achieve closed-loop input match-

[3] See file *24-Sep-2015_19h0_data_BO_prim_82s_prim* on the book website.

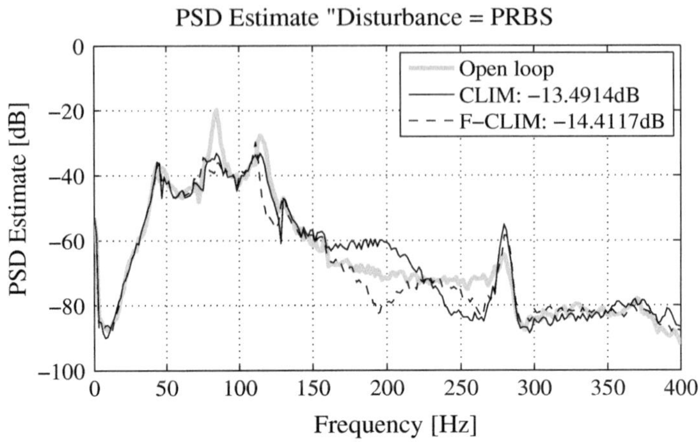

Fig. 14.7 Power spectral density estimates for closed-loop identified feedforward compensator with 20 parameters

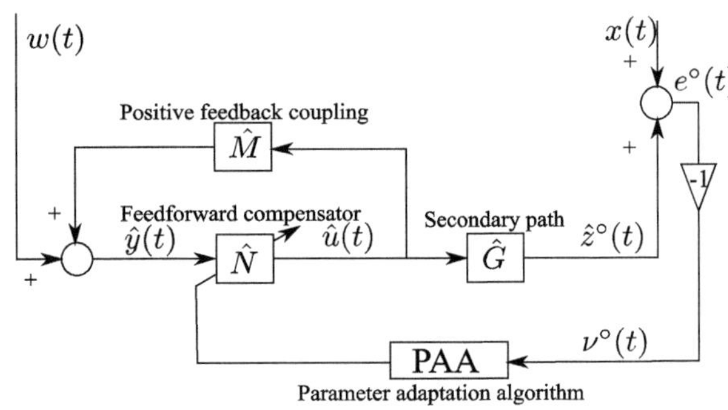

Fig. 14.8 Estimation of the feedforward compensator using a simulated adaptive feedforward compensation scheme in self-tuning regime

ing (CLIM) and uses directly as regressor vector $\phi(t)$ given in (14.17). The second one, corresponds to the filtered version of the same algorithm, F-CLIM as given in Sect. 9.3, where $\phi(t)$ is filtered through $\hat{A}_M/\hat{P}$. Real-time results obtained with the estimated compensators with $n_R = 9$ and $n_S = 10$ (20 parameters) are given in Fig. 14.7. One obtains an attenuation of -13.5 dB for the CLIM algorithm and -14.4 dB for the F-CLIM algorithm.

Based on the discussion in Sect. 14.3, for the self-tuning operation, two schemes are considered. In the first one, Fig. 14.8, the secondary path is considered at its true position downstream from the feedforward compensator. $\hat{M}(q^{-1})$ and $\hat{G}(q^{-1})$ in Fig. 14.8 represent the identified models for the reverse and secondary paths, respectively. $w(t)$ and $x(t)$ are real time measured signals in the absence of the controller (open-loop).

The second scheme is represented in Fig. 14.9. The main difference is that the model of the secondary path (without the inherent one step delay) $\hat{B}_G^*/\hat{A}_G$ is introduced upstream from the feedforward compensator $\hat{N}$ and the one sampling period

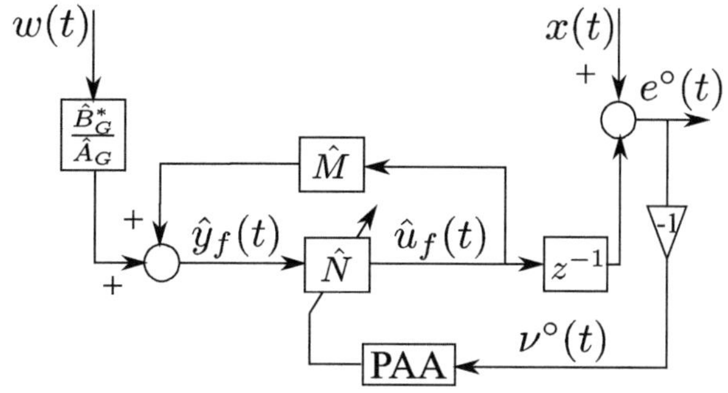

Fig. 14.9 A modified simulated feedforward compensation scheme in self-tuning regime for estimation of the feedforward compensator

delay of the secondary path is left at its usual position between feedforward compensator and measured residual acceleration (or force). The corresponding algorithms are similar to the ones obtained based on controller order reduction (see Fig. 14.6), with the exception of the one sampling period delay which appears between the feedforward compensator and the measurement of the residual acceleration.

The following PAA, derived from the algorithms which will be presented in Chap. 15, has been used for estimating the parameters of the feedforward compensator using a simulated adaptive feedforward compensation scheme:

$$\hat{\theta}(t+1) = \hat{\theta}(t) + F_I(t)\Phi(t)\nu(t+1) \tag{14.25a}$$

$$\nu(t+1) = \frac{\nu^0(t+1)}{1+\Phi^T(t)F_I(t)\Phi(t)} \tag{14.25b}$$

$$F_I(t+1) = F_I(t) - \frac{F_I(t)\Phi(t)\Phi^T(t)F_I(t)}{1+\Phi^T(t)F_I(t)\Phi(t)}, \quad F_I(0) > 0 \tag{14.25c}$$

where

$$\hat{\theta}^T(t) = \left[\hat{s}_1(t), \ \ldots \ \hat{s}_{n_S}(t), \ \hat{r}_0(t) \ \ldots \ \hat{r}_{n_R}(t)\right] = \left[\hat{\theta}_S^T(t), \ \hat{\theta}_R^T(t)\right] \tag{14.26}$$

is the vector of estimated parameters of $\hat{N}$. This algorithm is characterized by a decreasing adaptation gain which allows to obtain asymptotically a fixed value of the estimated parameters.

One defines $\phi(t)$ as the observation vector given by

$$\begin{aligned}\phi^T(t) &= \left[-\hat{u}(t), \ \ldots \ -\hat{u}(t-n_S+1), \ \hat{y}(t+1), \ \ldots \ \hat{y}(t-n_R+1)\right] \\ &= \left[\phi_{\hat{u}}^T(t), \ \phi_{\hat{y}}^T(t)\right],\end{aligned} \tag{14.27}$$

for the scheme presented in Fig. 14.8 and by

$$\begin{aligned}\phi^T(t) &= \left[-\hat{u}_f(t), \; \dots \; -\hat{u}_f(t-n_S+1), \; \hat{y}_f(t+1), \; \dots \; \hat{y}_f(t-n_R+1)\right] \\ &= \left[\phi^T_{\hat{u}_f}(t), \; \phi^T_{\hat{y}_f}(t)\right],\end{aligned} \tag{14.28}$$

for the one in Fig. 14.9.

$\Phi(t)$ is obtained by filtering $\phi(t)$ in order to satisfy a certain stability condition, which will be detailed in Chap. 15. Two types of filtering can be considered. The first type, labelled FUPLR (for Filtered-U Pseudo Linear Regression), uses only filtering by the estimated model of the secondary path. For the scheme in Fig. 14.8 this is achieved by filtering $\phi(t)$ through $L(q^{-1}) = \hat{G}(q^{-1})$, while for the scheme given in Fig. 14.9 the filtering is done through[4] $L(q^{-1}) = z^{-1}$. The stability condition is that

$$\frac{A_M G}{P\hat{G}} - \frac{1}{2} \tag{14.29}$$

should be SPR for the scheme given in Fig. 14.8 and

$$\frac{A_M}{P} - \frac{1}{2} \tag{14.30}$$

should be SPR for the scheme given in Fig. 14.9.

Feedforward compensators with orders $n_R = 9$, $n_S = 10$ (20 parameters) have been estimated. Experimental results are shown in Figs. 14.10 and 14.11 (black continuous lines). The global attenuation results are similar, −13.6 dB using the scheme given in Fig. 14.8 and −13.32 dB using the configuration given in Fig. 14.9.

The second type, labelled FUSBA (Filtered-U Stability Based Algorithm) corresponds to a filter chosen in order to satisfy a certain positive real condition (resulting from a stability analysis). In order to achieve this, one filters $\phi(t)$ by $L(q^{-1}) = \frac{\hat{A}_M\hat{G}}{\hat{P}}$, for the scheme given in Fig. 14.8, and by $L(q^{-1}) = \frac{z^{-1}\hat{A}_M}{\hat{P}}$, for the scheme given in Fig. 14.9 (see also Chaps. 9 and 15), where $\hat{A}_M$ is the denominator of the estimated reverse path model and $\hat{P}$ is the estimated characteristic polynomial of the internal positive feedback loop given by $\hat{P} = \hat{A}_M S - \hat{B}_M R$. In this case, the conditions of (14.29) and (14.30) become

$$\frac{A_M G\hat{P}}{P\hat{A}_M\hat{G}} - \frac{1}{2} \tag{14.31}$$

should be SPR for the scheme given in Fig. 14.8 and

$$\frac{A_M\hat{P}}{P\hat{A}_M} - \frac{1}{2} \tag{14.32}$$

[4]Note that the FUPLR filtering considered for the scheme in Fig. 14.8 is an exact algorithm for the configuration given in Fig. 14.4 which is equivalent to the configuration in Fig. 14.9.

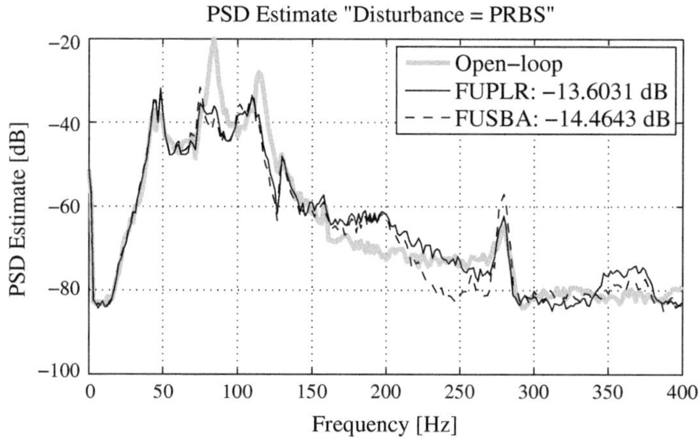

Fig. 14.10 Power spectral density estimates for closed-loop identified feedforward controller using the scheme given in Fig. 14.8 (G downstream)

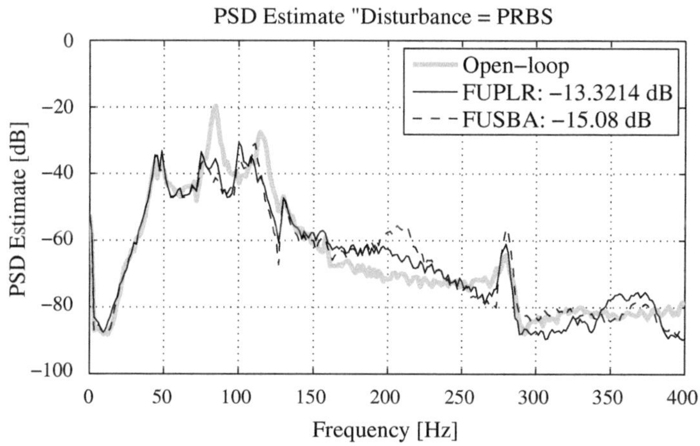

Fig. 14.11 Power spectral density estimates for closed-loop identified feedforward controller using the scheme given in Fig. 14.9 (G upstream)

should be SPR for the scheme given in Fig. 14.9. These conditions are much easier to satisfy if the estimated models of the system are good. Experimental results given in Figs. 14.10 and 14.11 (dashed lines) for the FUSBA algorithms show an improved performance. The global attenuation obtained is of -14.46 dB for the scheme given in Fig. 14.8 and -15.08 dB for the configuration given in Fig. 14.9. The results are very close to those obtained using CLIM algorithm for estimation of reduced order controllers (see Fig. 14.7).

From the above experimental results, it is clear that the stability-based algorithm FUSBA is more efficient than the FUPLR algorithm; however, an initial run with

FUPLR is necessary before using FUSBA in order to estimate the filter used in FUSBA.

Although the best compensator have been obtained with the simulated adaptive feedforward compensation scheme in self-tuning regime (Fig. 14.9), they are very close to the results obtained with the estimation in closed-loop of the reduced order controllers (Fig. 14.7). So both schemes can be used for direct estimation of a linear feedforward compensator from data collected in the absence of the feedforward compensation.

Note also that in [3] a reduced order H_∞ feedforward compensator with 40 parameters (instead of 20 parameters in this chapter) designed on the basis on an identified primary path model (indirect approach) was shown to provide a global attenuation of 14.7 dB, when operating with real data on the same system.

14.5 Concluding Remark

- The classical approach to the design of the feedforward compensator requires the knowledge of the model of the primary path and of the disturbance model in addition to the models of the secondary path and of the reverse path.
- Identification of a reliable model for the primary path rely on the frequency content of the available disturbance measurement.
- The indirect approach for designing the feedforward compensator includes:
  - Identification of the primary path.
  - Design of the feedforward compensator based on the knowledge of the primary, secondary and reverse path models and the PSD of the disturbance.
- The design of the linear feedforward compensator can be viewed as a special type of estimation in closed-loop of a reduced order controller, and therefore, one can directly obtain a feedforward compensator from collected data (measurement of the disturbance and of the residual acceleration or force).
- The major advantage of using a direct feedforward compensator identification approach for designing a linear compensator is that it short cuts the problem of identifying a reliable model for the primary path (quality of the model will depend on the frequency characteristics of the disturbance) and the problem of defining the appropriate weighting functions for design.

14.6 Notes and References

Linear feedforward designs based on LQG/H_2 methods are proposed in [2, 4–7]. Robust linear feedforward compensator based on H_∞ theory is presented in [3, 4, 8–14]. Mixed H_2/H_∞ techniques are used in [15, 16] and minimax LQG solutions in [11, 17, 18]. In [19], also a cautious Wiener filter is developed (see also [20]). Note that classical LQG/H_2 does not provide guaranteed robustness margins, while the classical H_∞ method is a worst-case design approach that does not necessarily provide good performance. The solutions mentioned previously present various mixed designs that try to efficiently combine the advantages of each method. Note that all these approaches assume the availability of the model of the primary path and of the disturbance in addition to the models of the secondary path and of the reverse path.

In [14] an interesting application of AVC to load reduction on the blades of a smart rotor using H_∞ feedback and fixed structure feedforward control is proposed. Note also the use of both *feedback* and *feedforward* controllers. Other examples of mixed controllers can be found in: [21–27]. An application to smart rotors vibration attenuation can be found in [28, 29].

References

1. Zhou K, Doyle J (1998) Essentials of robust control. Prentice-Hall International, Upper Saddle River
2. Rotunno M, de Callafon R (2003) Design of model-based feedforward compensators for vibration compensation in a flexible structure. Department of Mechanical and Aerospace Engineering. University of California, San Diego, Internal report
3. Alma M, Martinez J, Landau I, Buche G (2012) Design and tuning of reduced order H_∞ feedforward compensators for active vibration control. IEEE Trans Control Syst Technol 20(2):554–561. doi:10.1109/TCST.2011.2119485
4. Fraanje R (2004) Robust and fast schemes in broadband active noise and vibration control. Ph.D. thesis, University of Twente, Twente, The Netherlands
5. Bingham B, Atalla M, Hagood N (2001) Comparison of structural-acoustic control designs on an vibrating composite panel. J Sound Vib 244(5):761–778. doi:10.1006/jsvi.2000.3536
6. Doelman N (1993) Design of systems for active sound control. Ph.D. thesis, Delft University of Technology, Delft, The Netherlands
7. Mørkholt J, Elliott S (1998) Active vibration control using state space LQG and internal model control methods. In: Proceedings of fourth international conference on motion and vibration control. Institute of Robotics, pp 559–564
8. Bai M, Lin H (1997) Comparison of active noise control structures in the presence of acoustical feedback by using the H_∞ synthesis technique. J Sound Vib 206:453–471
9. Carmona J, Alvarado V (2000) Active noise control of a duct using robust control theory. IEEE Trans Control Syst Technol 8(6):930–938
10. Kaiser O (2001) Active control of sound transmission through a double wall structure. Ph.D. thesis, Swiss Federal Institute of Technology (ETH), Zürich, Switzerland
11. Petersen IR, Ugrinovskii VA, Savkin AV (2000) Robust control design using H_∞ methods. Communications and control engineering. Springer London, London. doi:10.1007/978-1-4471-0447-6
12. Prajna S, Kaiser O, Pietrzko S, Morari M (2000) Robust active control of a vibrating plate. In: National conference on noise control engineering. Newport Beach, USA. http://control.ee.ethz.ch/index.cgi?page=publications;action=details;id=283
13. Zhang BL, Tang GY (2013) Active vibration H_∞ control of offshore steel jacket platforms using delayed feedback. J Sound Vib 332(22):5662–5677. doi:10.1016/j.jsv.2013.06.029
14. van Wingerden J, Hulskamp A, Barlas T, Houtzager I, Bersee H, van Kuik G, Verhaegen M (2011) Two-degree-of-freedom active vibration control of a prototyped "smart" rotor. IEEE Trans Control Syst Technol 19(2):284–296. doi:10.1109/TCST.2010.2051810

15. Lin JY, Luo ZL (2000) Internal model-based LQG/H_∞ design of robust active noise controllers for an acoustic duct system. IEEE Trans Control Syst Technol 8(5):864–872. doi:10.1109/87.865860
16. Rafaely B, Elliott S (1999) H_2/H_∞ active control of sound in a headrest: design and implementation. IEEE Trans Control Syst Technol 7(1):79–84. doi:10.1109/87.736757
17. Petersen IR, Pota HR (2003) Minimax LQG optimal control of a flexible beam. Control Eng Pract 11(11):1273–1287. doi:10.1016/S0967-0661(02)00240-X
18. Petersen IR (2004) Multivariable control of noise in an acoustic duct. Eur J Control 10(6):557–572. doi:10.3166/ejc.10.557-572
19. Fraanje R, Verhaegen M, Doelman N (1999) Convergence analysis of the iltered-U LMS algorithm for active noise control in case perfect cancellation is not possible. Signal Process 73:255–266
20. Sternad M, Ahlén A (1993) Robust filtering and feedforward control based on probabilistic descriptions of model errors. Automatica 29(3):661–679. doi:10.1016/0005-1098(93)90062-X
21. Ma H, Tang GY, Hu W (2009) Feedforward and feedback optimal control with memory for offshore platforms under irregular wave forces. J Sound Vib 328(45):369–381. doi:10.1016/j.jsv.2009.08.025. http://www.sciencedirect.com/science/article/pii/S0022460X09006890
22. Rohlfing J, Gardonio P (2014) Ventilation duct with concurrent acoustic feed-forward and decentralised structural feedback active control. J Sound Vib 333(3):630–645. doi:10.1016/j.jsv.2013.09.022. http://www.sciencedirect.com/science/article/pii/S0022460X13007761
23. Luo J, Veres SM (2010) Frequency domain iterative feedforward/feedback tuning for MIMO ANVC. Automatica 46(4):735–742. doi:10.1016/j.automatica.2010.01.025. http://www.sciencedirect.com/science/article/pii/S0005109810000452
24. Su X, Jia Y (2015) Constrained adaptive tracking and command shaped vibration control of flexible hypersonic vehicles. IET Control Theory Appl 9(12):1857–1868. doi:10.1049/iet-cta.2014.0750
25. Wang J, Wang Y, Cao S (2011) Add-on feedforward compensation for vibration rejection in HDD. IEEE/ASME Trans Mechatron 16(6):1164–1170. doi:10.1109/TMECH.2010.2085008
26. Leang K, Zou Q, Devasia S (2009) Feedforward control of piezoactuators in atomic force microscope systems. IEEE Control Syst 29(1):70–82. doi:10.1109/MCS.2008.930922
27. Seki K, Tsuchimoto Y, Iwasaki M (2014) Feedforward compensation by specified step settling with frequency shaping of position reference. IEEE Trans Ind Electron 61(3):1552–1561. doi:10.1109/TIE.2013.2259778
28. Navalkar S, van Wingerden J, van Solingen E, Oomen T, Pasterkamp E, van Kuik G (2014) Subspace predictive repetitive control to mitigate periodic loads on large scale wind turbines. Mechatronics 24(8):916–925. doi:10.1016/j.mechatronics.2014.01.005
29. Dench M, Brennan M, Ferguson N (2013) On the control of vibrations using synchrophasing. J Sound Vib 332(20):4842–4855. doi:10.1016/j.jsv.2013.04.044. http://www.sciencedirect.com/science/article/pii/S0022460X1300401X

Chapter 15
Adaptive Feedforward Compensation of Disturbances

15.1 Introduction

In a number of application areas an image (a correlated measurement) of the disturbances acting upon the system can be made available. This information is very useful in active vibration control (AVC) and active noise control (ANC) for attenuating the disturbances using a feedforward compensation scheme (when the use of feedback is limited as a consequence of the Bode integral). Nevertheless, the feedforward compensator will depend not only upon the dynamics of the plant but also upon the characteristics of the disturbances. Since the characteristics (the model) of the disturbances are generally unknown and may be time-varying, adaptive feedforward compensation has to be considered. As indicated in Chap. 1, this solution has been proposed many years ago. Probably, the first references are [1–3].

Adaptive feedforward broad-band vibration (or noise) compensation is currently used in ANC and AVC when an image of the disturbance is available [4–9]. Nevertheless, at the end of the nineties it was pointed out that in most of these systems there is a physical "positive" feedback coupling between the compensator system and the measurement of the image of the disturbance (vibration or noise) [6, 7, 9, 10]. This is a very important issue and in Chap. 2 it has been shown on the considered test bench (Fig. 2.12) that effectively this internal positive feedback is significant and therefore cannot be ignored.

The corresponding block diagrams in open-loop operation and with the compensator system are shown in Fig. 15.1. The signal $w(t)$ is the image of the disturbance measured when the compensator system is not used (open-loop). The signal $\hat{y}(t)$ denotes the effective output provided by the measurement device when the compensator system is active and which will serve as input to the adaptive feedforward compensator $\hat{N}$. The output of this filter denoted by $\hat{u}(t)$ is applied to the actuator through an amplifier. The transfer function G (the secondary path) characterizes the dynamics from the output of the filter $\hat{N}$ to the residual acceleration

I.D. Landau et al., *Adaptive and Robust Active Vibration Control*,
Advances in Industrial Control, DOI 10.1007/978-3-319-41450-8_15

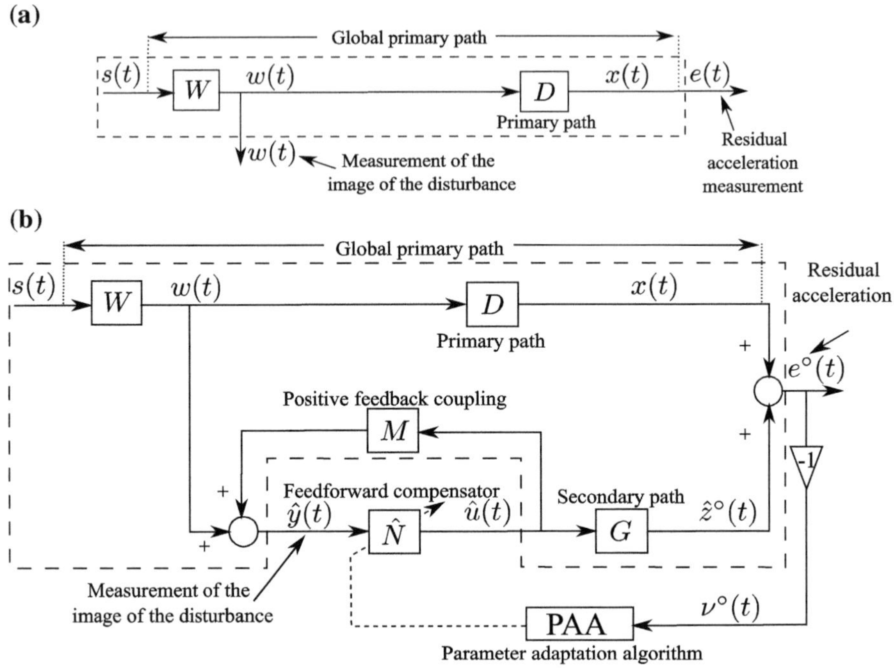

Fig. 15.1 Feedforward AVC: in open-loop (**a**) and with adaptive feedforward compensator (**b**)

measurement (amplifier + actuator + dynamics of the mechanical system). Subsequently, we will call the transfer function between $w(t)$ and the measurement of the residual acceleration (force) as the "primary path".

The coupling between the output of the feedforward compensator and the measurement $\hat{y}(t)$ through the compensator actuator is denoted by M. As indicated in Fig. 15.1, this coupling is a "positive" feedback. The positive feedback may destabilize the system.[1] The system is no longer a pure feedforward compensator.

In many cases, this unwanted coupling raises problems in practice and makes the analysis of adaptive (estimation) algorithms more difficult. The problem is to estimate and adapt the parameters of the feedforward compensator in the presence of this internal positive feedback.

There is also another reason why one should go towards using an adaptive approach. The linear design requires the availability of a reliable model for the primary path (if H_∞, H_2, or other model-based design techniques are used). Nevertheless, the signal $w(t)$ which is an upward measure of the disturbance does not necessarily have the appropriate PSD for correctly identifying the model of the primary

[1] Different solutions for reducing the effect of this internal positive feedback are reviewed in [8, 9].

path over a large frequency range. In other terms, the model which will be used for design will depend upon the characteristics of $w(t)$. Therefore, changes in the characteristics of the disturbance will affect also the model of the primary path used for design.[2]

It is important to make the following remarks, when the feedforward compensator is absent (open-loop operation):

1. One can identify very reliable models for the secondary path and the "positive" feedback path by applying appropriate excitation on the actuator (for example, a PRBS).
2. One can get an estimation of the primary path transfer function from the spectral densities of $w(t)$ and $e^{\circ}(t)$ when the compensator system is at rest (in this case $e^{\circ}(t) = x(t)$), but the quality of this model depends upon the spectral characteristics of $w(t)$.

It is also important to note that the estimation of the feedforward compensator in Fig. 15.1, as indicated in Chap. 14, can be interpreted as an identification-in-closed-loop operation or as an estimation in closed-loop of a (reduced-order) controller [11]. Therefore, to a certain extent, the methods given in Chaps. 8 and 9 as well as Chap. 14 are an inspiration source for solving this problem in real time.

The objective in this chapter, is to develop recursive algorithms for online estimation and adaptation of the parameters of the feedforward compensator N (which will be denoted as $\hat{N}$) for broad-band disturbances $w(t)$ (or $s(t)$) with unknown and variable spectral characteristics and in the presence of possible variations of the primary path model (D). The resulting algorithms, while minimizing the residual error (acceleration or force in AVC, noise in ANC), should assure the stability of the internal positive feedback loop created by the mechanical or acoustical coupling. Like for *adaptive regulation* (see Chap. 12) the *adaptive* operation and the *self-tuning* operation of the system should be considered.

In Sect. 15.2, the system representation and the feedforward filter structure will be given. The algorithms for adaptive feedforward compensation will be developed in Sect. 15.3 and analyzed in Sect. 15.4. Section 15.5 will present real-time results obtained on an AVC system. A modified adaptive algorithm using residual error filtering is presented in Sect. 15.6. Finally, in Sect. 15.7, algorithms for adaptive feedforward compensation in the presence of a fixed feedback controller are presented and experimental results are given in Sect. 15.8. The results of this chapter are also applicable to ANC systems.

[2]Design of adaptive AVC does not require either the model of the disturbance or the model of the primary path.

15.2 Basic Equations and Notations

The objective is to estimate (and to adapt) the parameters of the feedforward compensator $N(q^{-1})$ such that the measured residual error (acceleration or force in AVC, noise in ANC) be minimized in the sense of a certain criterion. The description of the various blocks will be made with respect to Fig. 15.1 (see also Sect. 6.3).

The primary path is characterized by the asymptotically stable transfer operator[3]:

$$D(q^{-1}) = \frac{B_D(q^{-1})}{A_D(q^{-1})} \tag{15.1}$$

where[4]

$$B_D(q^{-1}) = b_1^D q^{-1} + \cdots + b_{n_{B_D}}^D q^{-n_{B_D}} \tag{15.2}$$

$$A_D(q^{-1}) = 1 + a_1^D q^{-1} + \cdots + a_{n_{A_D}}^D q^{-n_{A_D}} \tag{15.3}$$

The unmeasurable value of the output of the primary path (when the compensation is active) is denoted as $x(t)$. The secondary path is characterized by the asymptotically stable transfer operator:

$$G(q^{-1}) = \frac{B_G(q^{-1})}{A_G(q^{-1})} \tag{15.4}$$

where

$$B_G(q^{-1}) = b_1^G q^{-1} + \cdots + b_{n_{B_G}}^G q^{-n_{B_G}} = q^{-1} B_G^*(q^{-1}) \tag{15.5}$$

$$A_G(q^{-1}) = 1 + a_1^G q^{-1} + \cdots + a_{n_{A_G}}^G q^{-n_{A_G}} \tag{15.6}$$

The positive feedback coupling is characterized by the asymptotically stable transfer operator:

$$M(q^{-1}) = \frac{B_M(q^{-1})}{A_M(q^{-1})} \tag{15.7}$$

where

$$B_M(q^{-1}) = b_1^M q^{-1} + \cdots + b_{n_{B_M}}^M q^{-n_{B_M}} = q^{-1} B_M^*(q^{-1}) \tag{15.8}$$

$$A_M(q^{-1}) = 1 + a_1^M q^{-1} + \cdots + a_{n_{A_M}}^M q^{-n_{A_M}} \tag{15.9}$$

---

[3] The complex variable z^{-1} will be used for characterizing the system's behaviour in the frequency domain and the delay operator q^{-1} will be used for describing the system's behaviour in the time domain.

[4] The following notation for polynomials is used: $A(q^{-1}) = a_0 + \sum_{i=1}^{n_A} a_i q^{-i} = a_0 + q^{-1} A^*(q^{-1})$.

Both B_G and B_M have a one-step discretization delay. The identified models of the secondary path and of the positive feedback coupling will be denoted as $\hat{G}$ and $\hat{M}$, respectively.

The optimal feedforward filter (unknown) is defined by

$$N(q^{-1}) = \frac{R(q^{-1})}{S(q^{-1})} \tag{15.10}$$

where

$$R(q^{-1}) = r_0 + r_1 q^{-1} + \cdots + r_{n_R} q^{-n_R} \tag{15.11}$$
$$S(q^{-1}) = 1 + S_1 q^{-1} + \cdots + S_{n_S} q^{-n_S} = 1 + q^{-1} S^*(q^{-1}) \tag{15.12}$$

The estimated compensator is denoted by $\hat{N}(q^{-1})$ or $\hat{N}(\hat{\theta}, q^{-1})$ when it is a linear filter with constant coefficients or $\hat{N}(t, q^{-1})$ during estimation (adaptation) of its parameters.

The input of the feedforward compensator is denoted by $\hat{y}(t)$ and it corresponds to the sum of the measurement provided by the primary transducer (force or acceleration transducer in AVC or a microphone in ANC) and of the output of the positive feedback path. In the absence of the compensation loop (open-loop operation) $\hat{y}(t) = w(t)$. The *a posteriori* output of the feedforward compensator (which is the control signal applied to the secondary path) is denoted by $\hat{u}(t+1) = \hat{u}(t+1|\hat{\theta}(t+1))$. The input–output relationship for the estimated feedforward compensator is given by the equation of the *a priori* output:

$$\begin{aligned}\hat{u}^\circ(t+1) &= \hat{u}(t+1|\hat{\theta}(t)) = -\hat{S}^*(t, q^{-1})\hat{u}(t) + \hat{R}(t, q^{-1})\hat{y}(t+1) \\ &= \hat{\theta}^T(t)\phi(t) = [\hat{\theta}_S^T(t), \hat{\theta}_R^T(t)] \begin{bmatrix} \phi_{\hat{u}}(t) \\ \phi_{\hat{y}}(t) \end{bmatrix}\end{aligned} \tag{15.13}$$

where

$$\hat{\theta}^T(t) = [\hat{s}_1(t), \ldots \hat{s}_{n_S}(t), \hat{r}_0(t), \ldots \hat{r}_{n_R}(t)] = [\hat{\theta}_S^T(t), \hat{\theta}_R^T(t)] \tag{15.14}$$
$$\begin{aligned}\phi^T(t) &= [-\hat{u}(t), -\hat{u}(t - n_S + 1), \hat{y}(t+1), \ldots \hat{y}(t - n_R + 1)] \\ &= [\phi_{\hat{u}}^T(t), \phi_{\hat{y}}^T(t)]\end{aligned} \tag{15.15}$$

and $\hat{u}(t)$, $\hat{u}(t-1)\ldots$, are the *a posteriori* outputs of the feedforward compensator generated by

$$\hat{u}(t+1) = \hat{u}(t+1|\hat{\theta}(t+1)) = \hat{\theta}^T(t+1)\phi(t) \tag{15.16}$$

while $\hat{y}(t+1)$, $\hat{y}(t)$, ... are the measurements provided by the primary transducer.[5]

[5] $\hat{y}(t+1)$ is available before adaptation of parameters starts at $t+1$.

The *a priori* output of the secondary path will be denoted as $\hat{z}^{\circ}(t+1)$.

$$\hat{z}^{\circ}(t+1) = \hat{z}(t+1|\hat{\theta}(t)) = \frac{B_G^*(q^{-1})}{A_G(q^{-1})}\hat{u}(t) \tag{15.17}$$

The *a posteriori* unmeasurable value of the output of the secondary path is denoted by

$$\hat{z}(t+1) = \hat{z}(t+1|\hat{\theta}(t+1)) \tag{15.18}$$

The measured primary signal (called also reference) satisfies the following equation:

$$\hat{y}(t+1) = w(t+1) + \frac{B_M^*(q^{-1})}{A_M(q^{-1})}\hat{u}(t). \tag{15.19}$$

The measured residual error satisfies the following equation:

$$e^{\circ}(t+1) = x(t+1) + \hat{z}^{\circ}(t+1). \tag{15.20}$$

The *a priori* adaptation error is defined as

$$\nu^{\circ}(t+1) = -e^{\circ}(t+1) = -x(t+1) - \hat{z}^{\circ}(t+1). \tag{15.21}$$

The *a posteriori* adaptation (residual) error (which is computed) will be given by

$$\nu(t+1) = \nu(t+1|\hat{\theta}(t+1)) = -x(t+1) - \hat{z}(t+1). \tag{15.22}$$

When using an estimated filter $\hat{N}$ with constant parameters: $\hat{u}^{\circ}(t) = \hat{u}(t)$, $\hat{z}^{\circ}(t) = \hat{z}(t)$ and $\nu^{\circ}(t) = \nu(t)$.

15.3 Development of the Algorithms

The algorithms for adaptive feedforward compensation will be developed under the following hypotheses:

(H1) The signal $w(t)$ is bounded, i.e.,

$$|w(t)| \leq \alpha \qquad \forall t \qquad (0 \leq \alpha \leq \infty) \tag{15.23}$$

or

$$\lim_{N\to\infty} \sum_{t=1}^{N} w^2(t) \leq N\varepsilon^2 + K_r \tag{15.24}$$
$$0 \leq \varepsilon^2 < \infty \qquad 0 < K_r < \infty$$

(which is equivalently to say that $s(t)$ is bounded and $W(q^{-1})$ in Fig. 15.1 is asymptotically stable).

(H2) (Perfect matching condition) There exists a filter $N(q^{-1})$ of finite dimension such that[6]:

$$\frac{N}{(1-NM)}G = -D \tag{15.25}$$

and the characteristic polynomial of the "internal" feedback loop:

$$P(z^{-1}) = A_M(z^{-1})S(z^{-1}) - B_M(z^{-1})R(z^{-1}) \tag{15.26}$$

is a Hurwitz polynomial.

(H3) The effect of the measurement noise upon the measured residual error is neglected (deterministic context).

Once the algorithms will be developed under these hypotheses, H2 and H3 can be removed and the algorithms can be analyzed in this modified context [12].

The key point in the development of the algorithms is to establish a relation between the errors on the estimation of the parameters of the feedforward compensator and the adaptation error (the measured residual acceleration or force with minus sign). Under hypotheses H1, H2 and H3, for the system described by Eqs. (15.1) and (15.22) using a feedforward compensator $\hat{N}$ with constant parameters, one has

$$\nu(t+1) = \frac{A_M(q^{-1})G(q^{-1})}{P(q^{-1})}[\theta - \hat{\theta}]^T\phi(t) \tag{15.27}$$

where

$$\theta^T = [s_1, \ldots s_{n_S}, r_0, r_1, \ldots r_{n_R}] = [\theta_S^T, \theta_R^T] \tag{15.28}$$

is the vector of parameters of the optimal filter N assuring perfect matching,

$$\hat{\theta}^T = [\hat{s}_1, \ldots \hat{s}_{n_S}, \hat{r}_0, \ldots \hat{r}_{n_R}] = [\hat{\theta}_S^T, \hat{\theta}_R^T] \tag{15.29}$$

is the vector of constant estimated parameters of $\hat{N}$,

$$\begin{aligned}\phi^T(t) &= [-\hat{u}(t), \ldots - \hat{u}(t-n_S+1), \hat{y}(t+1), \ldots \hat{y}(t-n_R+1)] \\ &= [\phi_{\hat{u}}^T(t), \phi_{\hat{y}}^T(t)]\end{aligned} \tag{15.30}$$

and $\hat{y}(t+1)$ is given by

$$\hat{y}(t+1) = w(t+1) + \frac{B_M^*(q^{-1})}{A_M(q^{-1})}\hat{u}(t) \tag{15.31}$$

The derivation of the expression (15.27) is given in Appendix D.1.

[6] In many cases, the argument q^{-1} or z^{-1} will be dropped out.

When $\hat{\theta}$ will be replaced with a time-varying estimation, (15.27) will take the form of the basic Eq. (4.125) shown in Chap. 4 and the basic adaptation algorithm given in Eqs. (4.121)–(4.123) can be used; however, a positive real (sufficient) condition will be imposed on $A_M G/P$ to assure stability. Therefore, filtering has to be introduced. One considers filtering the vector $\phi(t)$ through an asymptotically stable filter $L(q^{-1}) = B_L/A_L$. Equation (15.27) for $\hat{\theta}$ = constant becomes

$$\nu(t+1) = \frac{A_M(q^{-1})G(q^{-1})}{P(q^{-1})L(q^{-1})}[\theta - \hat{\theta}]^T \phi_f(t) \tag{15.32}$$

with

$$\phi_f(t) = L(q^{-1})\phi(t) \tag{15.33}$$

Equation (15.32) will be used to develop the adaptation algorithms neglecting the non-commutativity of the operators when $\hat{\theta}$ is time-varying (however, an exact algorithm can be derived in such cases—following the methodology given in [13, Sect. 5.5.3]).

Replacing the fixed estimated parameters by the current estimated parameters, (15.32) becomes the equation of the *a posteriori* adaptation error $\nu(t+1)$ (which is computed):

$$\nu(t+1) = \frac{A_M(q^{-1})G(q^{-1})}{P(q^{-1})L(q^{-1})}[\theta - \hat{\theta}(t+1)]^T \phi_f(t) \tag{15.34}$$

Equation (15.34) has the standard form for an *a posteriori* adaptation error given in Sect. 4.3, which immediately suggests to use the following PAA:

$$\hat{\theta}(t+1) = \hat{\theta}(t) + F(t)\Phi(t)\nu(t+1); \tag{15.35}$$

$$\nu(t+1) = \frac{\nu^{\circ}(t+1)}{1 + \Phi^T(t)F(t)\Phi(t)}; \tag{15.36}$$

$$F(t+1) = \frac{1}{\lambda_1(t)}\left[F(t) - \frac{F(t)\Phi(t)\Phi^T(t)F(t)}{\frac{\lambda_1(t)}{\lambda_2(t)} + \Phi^T(t)F(t)\Phi(t)}\right] \tag{15.37}$$

$$1 \geq \lambda_1(t) > 0; \quad 0 \leq \lambda_2(t) < 2; \; F(0) > 0 \tag{15.38}$$

$$\Phi(t) = \phi_f(t) \tag{15.39}$$

where $\lambda_1(t)$ and $\lambda_2(t)$ allow to obtain various profiles for the adaptation gain $F(t)$ (see Sects. 4.3.4 and 15.5) in order to operate in *adaptive* regime (the trace of the adaptation gain matrix has a strictly positive inferior minimum value) or in *self-tuning* regime (decreasing gain adaptation, the trace of the adaptation gain matrix goes to zero).

Three choices for the filter L will be considered, leading to three different algorithms:

Algorithm I: $L = G$
Algorithm II (FUPLR): $L = \hat{G}$
Algorithm III (FUSBA):

$$L = \frac{\hat{A}_M}{\hat{P}}\hat{G} \tag{15.40}$$

where

$$\hat{P} = \hat{A}_M\hat{S} - \hat{B}_M\hat{R} \tag{15.41}$$

is an estimation of the characteristic polynomial of the internal feedback loop computed on the basis of available estimates of the parameters of the filter $\hat{N}$.[7]

Algorithm I is a "theoretical" algorithm since in practice the true model G is not available.[8] So FUPLR can be viewed as an approximation of Algorithm I. FUSBA can be used after a short initialization horizon using FUPLR.

The following procedure is applied at each sampling time for *adaptive* operation:

1. Get the measured image of the disturbance $\hat{y}(t+1)$ and the measured residual error $e^{\circ}(t+1)$.
2. Compute $\phi(t)$ and $\phi_f(t)$ using (15.30) and (15.33).
3. Estimate the parameter vector $\hat{\theta}(t+1)$ using the parametric adaptation algorithm (15.35)–(15.39).
4. Compute the control (using (15.16)) and apply it:

$$\hat{u}(t+1) = -\hat{S}^*(t+1, q^{-1})\hat{u}(t) + \hat{R}(t+1, q^{-1})\hat{y}(t+1). \tag{15.42}$$

15.4 Analysis of the Algorithms

A detailed analysis of the algorithm can be found in [12]. In what follows, we will recall the main properties and their implications.

15.4.1 The Perfect Matching Case

Stability of the algorithms

For Algorithms I, II and III, the equation for the *a posteriori* adaptation error has the form:

$$\nu(t+1) = H(q^{-1})[\theta - \hat{\theta}(t+1)]^T\Phi(t) \tag{15.43}$$

---

[7]In the field of adaptive feedforward compensation names are associated to various adaptation algorithms. Algorithm II uses the same filtering of the regressor as FULMS algorithm but with a matrix adaptation gain which lead to a structure called "pseudolinear regression" [14]. So Algorithm II can be termed FUPLR. Algorithm III is obtained from a stability point of view and it can be termed FUSBA (stability based algorithm).

[8]See Appendix D, Sect. D.2 for further details.

where

$$H(q^{-1}) = \frac{A_M(q^{-1})G(q^{-1})}{P(q^{-1})L(q^{-1})}, \ \Phi = \phi_f. \tag{15.44}$$

Neglecting the non-commutativity of time-varying operators, one can straightforwardly use Theorem 4.1. Therefore, the sufficient stability condition for any initial conditions $\hat{\theta}(0)$, $\nu^\circ(0)$, $F(0)$ is that

$$H'(z^{-1}) = H(z^{-1}) - \frac{\lambda_2}{2}, \max_t [\lambda_2(t)] \leq \lambda_2 < 2 \tag{15.45}$$

is a strictly positive real (SPR) transfer function.

It is interesting to remark that for Algorithm III (FUSBA) taking into account (15.40), the stability condition is that

$$\frac{A_M}{\hat{A}_M}\frac{\hat{P}}{P}\frac{G}{\hat{G}} - \frac{\lambda_2}{2} \tag{15.46}$$

should be an SPR transfer function.

Remark 1 This condition can be rewritten for $\lambda_2 = 1$ as ([14]):

$$\left| \left(\frac{A_M}{\hat{A}_M} \cdot \frac{\hat{P}}{P} \cdot \frac{G}{\hat{G}} \right)^{-1} - 1 \right| < 1 \tag{15.47}$$

for all ω. This roughly means that it always holds provided that the estimates of A_M, P and G are close to the true values (i.e., $H(e^{-j\omega})$ in this case are close to a unit transfer function).

Remark 2 For constant adaptation gain $\lambda_2(t) \equiv 0$, the strict positive realness on $H'(z^{-1})$ implies at all the frequencies

$$-90^\circ < \angle \frac{A_M(e^{-j\omega})G(e^{-j\omega})}{P_0(e^{-j\omega})} - \angle \frac{\hat{A}_M(e^{-j\omega})\hat{G}(e^{-j\omega})}{\hat{P}_0(e^{-j\omega})} < 90^\circ.$$

Therefore, the interpretation of the SPR stability condition on the transfer function H' is that the angle between the direction of adaptation and the direction of the inverse of the true gradient (not computable) should be less than 90°. For time-varying adaptation gains the condition is sharper since in this case $Re\{H(e^{-j\omega})\}$ should be larger than $\frac{\lambda_2}{2}$ at all frequencies.

Remark 3 The poles of the internal positive closed-loop will be asymptotically inside the unit circle if the SPR condition is satisfied; however, transiently they may be outside the unit circle. It is possible to force these poles to remain inside of the unit circle during transient using adaptive algorithms with projection (see [13]); however, the SPR condition remains the same.

Effect of the Measurement Noise

There are two sources of measurement noise, one acting on the primary transducer which gives an image of the disturbance and the second acting on the measurement of the residual error (force, acceleration).

For the primary transducer, the effect of the measurement noise is negligible since the signal-to-noise ratio is very high. The situation is different for the residual error where the effect of the noise cannot be neglected. The analysis carried on in [12] using the averaging method ([13]) allows to conclude that under the same positive real condition that for deterministic stability, using a decreasing adaptation gain (self-tuning regime) one has

$$Prob\{\lim_{t\to\infty} \hat{\theta}(t) \in D_C\} = 1,$$

where

$$D_C = \{\hat{\theta} : \Phi^T(t,\hat{\theta})(\theta - \hat{\theta}) = 0\}.$$

If, furthermore,

$$\Phi^T(t,\hat{\theta})(\theta - \hat{\theta}) = 0$$

has a unique solution (richness condition), the condition that $H'(z^{-1})$ be strictly positive real implies that

$$Prob\{\lim_{t\to\infty} \hat{\theta}(t) = \theta\} = 1.$$

15.4.2 The Case of Non-perfect Matching

If $\hat{N}(t, q^{-1})$ does not have the appropriate dimension there is no chance to satisfy the perfect matching condition. Two questions are of interest in this case:

1. What are the additional hypotheses assuring the stability of the adaptation algorithm in this situation?
2. What are the approximation properties in the frequency domain for the reduced-order compensator estimated asymptotically by the adaptation algorithm?

Boundedness of the Residual Error

It has been shown in [12] that the residual error will remain bounded provided that

1. There exists a reduced-order filter $\hat{N}$ characterized by the unknown polynomials $\hat{S}$ (of order $n_{\hat{S}}$) and $\hat{R}$ (of order $n_{\hat{R}}$), for which the closed-loop formed by $\hat{N}$ and M is asymptotically stable, i.e., $A_M\hat{S} - B_M\hat{R}$ is a Hurwitz polynomial.

2. The output of the optimal filter satisfying the matching condition can be expressed as

$$\hat{u}(t+1) = -\hat{S}^*(q^{-1})\hat{u}(t) + \hat{R}(q^{-1})\hat{y}(t+1) + \upsilon(t+1) \tag{15.48}$$

where $\upsilon(t+1)$ is a norm bounded signal.

The first hypothesis simply says that the internal positive feedback loop can be stabilized by a feedforward compensator of the size used.

Equation (15.48) can be interpreted as a decomposition of the optimal filter into two parallel blocks, one is the reduced order filter and the other block with output $\upsilon(t)$ corresponds to the neglected dynamics (input additive uncertainty). The boundedness of $\upsilon(t)$ requires that the neglected dynamics in the feedforward compensator be stable.

Bias Distribution

The distribution of the matching error in the frequency domain (generically called "bias") is an important information upon the expected performances in the case of non-perfect matching. Using Parseval's relation, the asymptotic bias distribution of the matching error can be obtained starting from the expression of $\nu(t)$ by taking into account that the algorithm minimizes (almost) a criterion of the form[9]

$$\lim_{N\to\infty} \frac{1}{N} \sum_{t=1}^{N} \nu^2(t).$$

For details see [13].

The bias distribution (for Algorithm III) will be given by

$$\hat{\theta}^* = \arg\min_{\hat{\theta}} \int_{-\pi}^{\pi} \left[\left| D(e^{-j\omega}) - \frac{\hat{N}(e^{-j\omega})G(e^{-j\omega})}{1-\hat{N}(e^{-j\omega})M(e^{-j\omega})} \right|^2 \phi_w(\omega) \right. $$
$$\left. +\phi_\eta(\omega) \right] \mathrm{d}\omega, \tag{15.49}$$

where ϕ_w and ϕ_η are the spectral densities of the disturbance $w(t)$ and of the measurement noise, respectively. Taking into account (15.25), one obtains

$$\hat{\theta}^* = \arg\min_{\hat{\theta}} \int_{-\pi}^{\pi} \left[|S_{NM}|^2 |N - \hat{N}|^2 |S_{\hat{N}M}|^2 |G|^2 \phi_w(\omega) + \phi_\eta(\omega) \right] \mathrm{d}\omega, \tag{15.50}$$

where S_{NM} and $S_{\hat{N}M}$ are the output sensitivity functions of the internal closed-loop for N and, respectively, $\hat{N}$:

$$S_{NM} = \frac{1}{1-NM}; \quad S_{\hat{N}M} = \frac{1}{1-\hat{N}M}.$$

[9] The results are valid for the asymptotic behaviour obtained when using a decreasing adaptation gain.

From (15.49) and (15.50), one concludes that a good approximation of N will be obtained in the frequency region where ϕ_w is significant and where G has a high gain (usually G should have high gain in the frequency region where ϕ_w is significant in order to counteract the effect of $w(t)$); however, the quality of the estimated $\hat{N}$ will be affected also by the output sensitivity functions of the internal closed-loop $N - M$. With a decreasing adaptation gain, the measurement noise will not influence the asymptotic estimate of N.

15.4.3 Relaxing the Positive Real Condition

Averaging Approach

For the FUPLR algorithm, it is possible to relax the strictly positive real (SPR) condition taking into account that

1. The disturbance (input to the system) is a broad-band signal.[10]
2. Most of the adaptation algorithms work with a low adaptation gain.

Under these two assumptions, the behaviour of the algorithm can be well described by the "averaging theory" developed in [14, 15] (see also [13, Sect. 4.2]). When using the averaging approach, the basic assumption of a slow adaptation holds for small adaptation gains (constant and scalar in [15] with $\lambda_2(t) = 0, \lambda_1(t) = 1$; asymptotically decreasing matrix gain in [14] with $\lim_{t\to\infty} \lambda_1(t) = 1, \lambda_2(t) = \lambda_2 > 0$).

In the context of averaging, the basic condition for stability is that

$$\lim_{N\to\infty} \frac{1}{N} \sum_{t=1}^{N} \Phi(t) H'(q^{-1}) \Phi^T(t) = \frac{1}{2} \int_{-\pi}^{\pi} \Phi(e^{j\omega})[H'(e^{j\omega}) + H'(e^{-j\omega})]\Phi^T(e^{-j\omega}) \mathrm{d}\omega > 0 \tag{15.51}$$

be a positive definite matrix ($\Phi(e^{j\omega})$ is the Fourier transform of the regressor vector $\Phi(t)$).

One can view (15.51) as the weighted energy of the observation vector Φ. Of course the SPR sufficient condition upon $H'(z^{-1})$ (see 15.45) allows to satisfy this; however, in the averaging context it is only needed that (15.51) be true. *This allows that H' be non-positive real in a limited frequency band.* Expression (15.51) can be rewritten as follows:

[10]The fact that the disturbance is a broad-band signal will imply that one has persistence of excitation.

$$\int_{-\pi}^{\pi} \Phi(e^{j\omega})[H' + H'^{*}]\Phi^{T}(e^{-j\omega})\mathrm{d}\omega =$$
$$\sum_{i=1}^{r} \int_{\alpha_i}^{\alpha_i+\Delta_i} \Phi(e^{j\omega})[H' + H'^{*}]\Phi^{T}(e^{-j\omega})\mathrm{d}\omega -$$
$$\sum_{j=1}^{p} \int_{\beta_j}^{\beta_j+\Delta_j} \Phi(e^{j\omega})[\bar{H}' + \bar{H}'^{*}]\Phi^{T}(e^{-j\omega})\mathrm{d}\omega > 0 \tag{15.52}$$

where H' is strictly positive real in the frequency intervals $[\alpha_i, \alpha_i + \Delta_i]$ and $\bar{H}' = -H'$ is positive real in the frequencies intervals $[\beta_j, \beta_j + \Delta_j]$ (H'^{*} denotes the complex conjugate of H'). The conclusion is that H' does not need to be SPR. It is enough that the "positive" weighted energy exceeds the "negative" weighted energy. This explains why the algorithm FUPLR using low adaptation gains will work in practice in most of the cases even if the performance will be affected (particularly in the frequency regions where the SPR condition is violated). It is however important to remark that if the disturbance is a single sinusoid (which violates the hypothesis of broad-band disturbance) located in the frequency region where H' is not SPR, the algorithm may diverge (see [14, 15]).

Without doubt, the best approach for relaxing the SPR conditions, is to use algorithm FUSBA (see (15.40)) instead of FUPLR. This is motivated by (15.46) and (15.47). As it will be shown in the next section, this algorithm gives the best results both in simulations and in real-time experiments.

Use of "Integral + Proportional" Parameter Adaptation Algorithm

This approach is discussed in Appendix E.

15.5 Adaptive Attenuation of Broad-band Disturbances—Experimental Results

The active distributed flexible structure presented in Sect. 2.3, will be considered for experimental validation of the algorithms proposed. The structure of the system is described in Fig. 2.10. For a view of the system, see Fig. 2.9.

The incoming disturbance is the position of the mobile part of the inertial actuator on top of the structure (see Fig. 2.10).[11] The residual acceleration $e(t)$ and the input to the feedforward compensator $\hat{y}(t)$ are measured by accelerometers. The control input is the position of the mobile part of the inertial actuator located on the bottom of the structure.

[11]The inertial actuator is driven by an external source.

15.5.1 Broad-band Disturbance Rejection Using Matrix Adaptation Gain

The performance of the system for rejecting broad-band disturbances will be illustrated using the adaptive feedforward compensation scheme. The adaptive filter structure for most of the experiments has been $n_R = 9, n_S = 10$ (total of 20 parameters) and this complexity does not allow to verify the "perfect matching condition" (not enough parameters). The influence of the number of parameters upon the performance of the system has been also investigated (up to 40 parameters).

A PRBS excitation on the global primary path will be considered as the disturbance. The corresponding spectral densities of $w(t)$ in open-loop and of $\hat{y}(t)$ when feedforward compensation is active are shown in Fig. 15.2. The effect of the mechanical feedback coupling is significant.

Two modes of operation can be considered, depending on the particular choices taken in (15.39):

- For adaptive operation, Algorithms II and III have been used with *decreasing adaptation gain* ($\lambda_1(t) = 1, \lambda_2(t) = 1$) combined with a *constant trace* adaptation gain. When the trace of the adaptation matrix is below a given value, the constant trace gain updating modifies the values of $\lambda_1(t)$ and $\lambda_2(t)$ so that the trace of F is kept constant. The corresponding formula is

$$tr F(t+1) = \frac{1}{\lambda_1(t)} tr \left[F(t) - \frac{F(t)\Phi(t)\Phi(t)^T F(t)}{\alpha + \Phi(t)^T F(t)\Phi(t)} \right] = tr F(t) \qquad (15.53)$$

This assures the evolution of the PAA in the optimal direction but the adaptation step size does not go to zero, therefore maintaining adaptation capabilities for

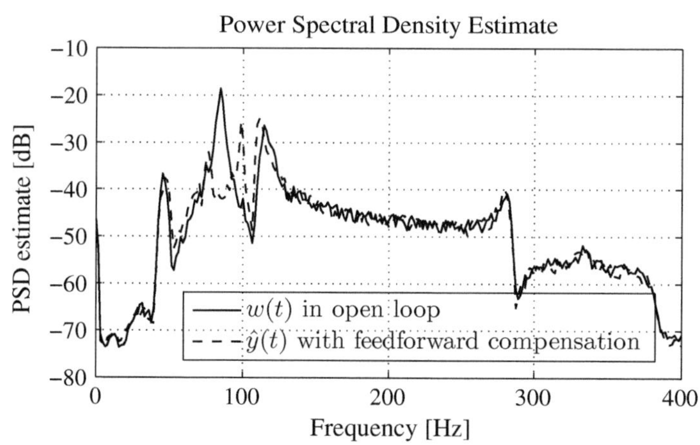

Fig. 15.2 Spectral densities of the image of the disturbance in open-loop $w(t)$ and with the feedforward compensation scheme $\hat{y}(t)$ (experimental)

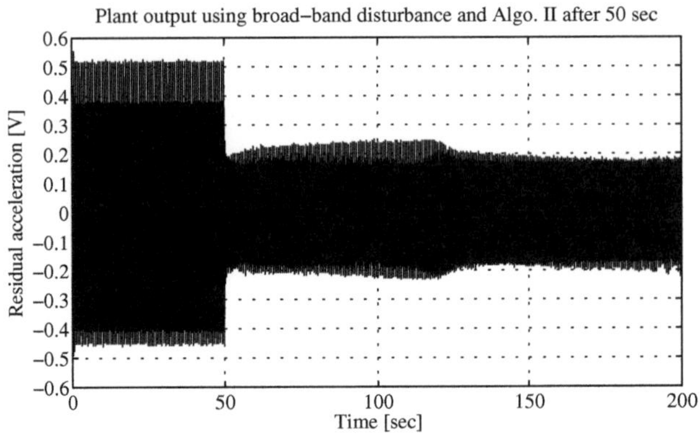

Fig. 15.3 Real-time results obtained with Algorithm II (FUPLR) using matrix adaptation gain

possible changes in disturbance or variations of the primary path model. For details, see [13, 16].

- In self-tuning operation, a decreasing adaptation gain $F(t)$ is used and the adaptation step size goes to zero. Then, if a degradation of the performance is observed, as a consequence of a change of the disturbance characteristics, the PAA is restarted.

The parametric adaptation algorithms have been implemented using the UD factorization [13] (see Appendix B).[12] The experiments have been carried out by first applying the disturbance and then starting the adaptive feedforward compensation after 50 s. Time domain results obtained in open-loop and with adaptive feedforward compensation using Algorithms II (FUPLR) and III (FUBSA) on the AVC system are shown in Figs. 15.3 and 15.4, respectively. The filter for the Algorithm III has been computed based on the parameter estimates obtained with Algorithm II at t = 3600 s (almost same results are obtained if the initialization horizon is of the order of 200 s). The initial trace of the matrix adaptation gain for 20 parameters was 10 and the constant trace has been fixed at 0.2.

As it can be seen, the transient duration for Algorithm II (FUPLR) is approximately 75 s, while for Algorithm III (FUBSA) is approximately 12 s.

Time domain comparison between Algorithms II and III can be found in Table 15.1, where the two algorithms are compared in terms of closed-loop variances and global attenuation. Also the open-loop variance is given as reference value. It can be seen that the performance of Algorithm III (FUSBA) is better than performance of Algorithm II (FUPLR).

Figure 15.5 shows the time evolution of the trace of the adaptation gain matrix. As it can be seen after 2.5 s the trace of the matrix gain remains constant assuring the real-time adaptation capabilities. Figure 15.6 shows the power spectral densities of the residual acceleration measured on the AVC in open-loop (without compensator) and using adaptive feedforward compensation (the measurements are made between 175

[12] An array implementation as in [17] can be also considered.

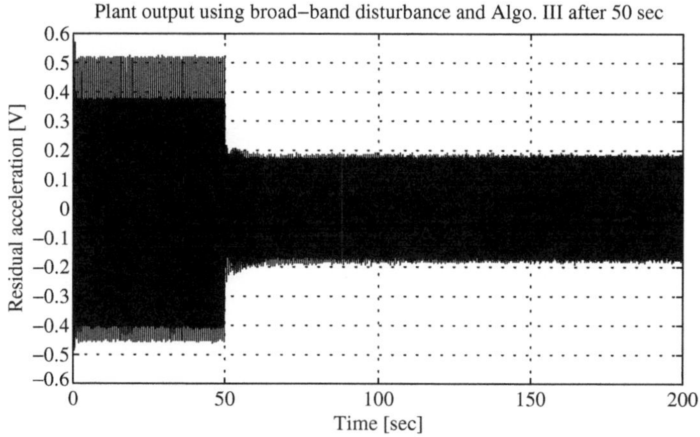

Fig. 15.4 Real-time results obtained with Algorithm III (FUSBA) using matrix adaptation gain

Table 15.1 Performance of Algorithms II (FUPLR) and III (FUSBA)

	Open-loop var.	Closed-loop var.	Global att. (dB)
Algo. II (FUPLR)	0.0354	0.0058	15.68
Algo. III (FUSBA)	0.0354	0.0054	16.23

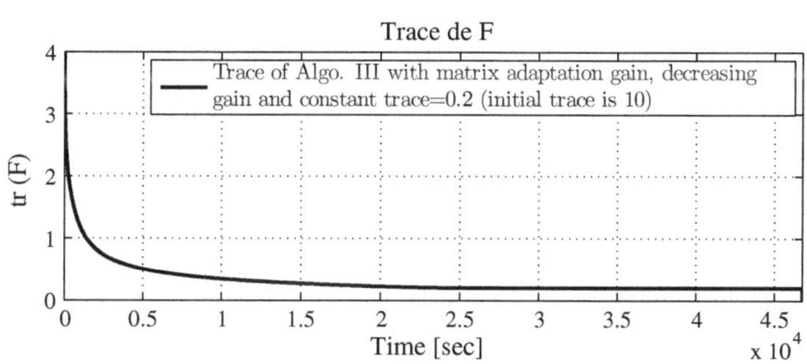

Fig. 15.5 Evolution of the trace of the matrix adaptation gain for Algorithm III (experimental)

and 200 s—the adaptation transient is finished well before 175 s). The corresponding global attenuations are also given. Algorithm III (FUSBA) performs slightly better than Algorithm II (FUPLR).

The H_∞ design ([18]) provides less good performance than Algorithm II (FUPLR) and III (FUSBA) with matrix adaptation gain (Fig. 15.6) despite that it has a double number of parameters (40 instead of 20). In addition, the H_∞ compensator does not have adaptation capabilities as it will be shown in Sect. 15.5.1.1.

To better understand the differences between Algorithm II (FUPLR) and Algorithm III (FUSBA), in Fig. 15.7 the Bode diagram of the estimated A_M/P transfer

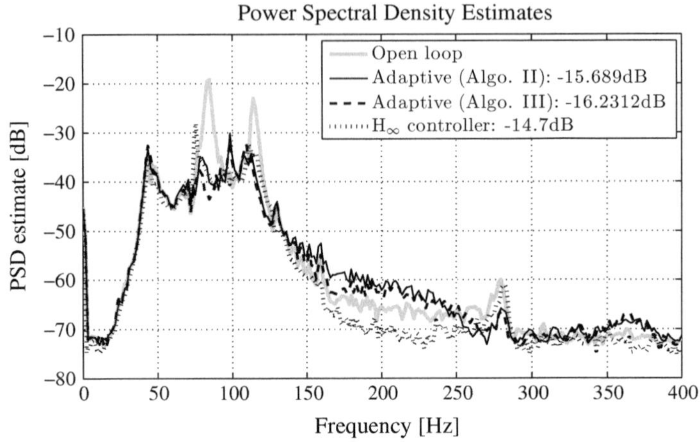

Fig. 15.6 Power spectral densities of the residual acceleration in open-loop, with adaptive feedforward compensation (20 parameters), and with H_∞ controller (40 parameters) (Disturbance = PRBS)

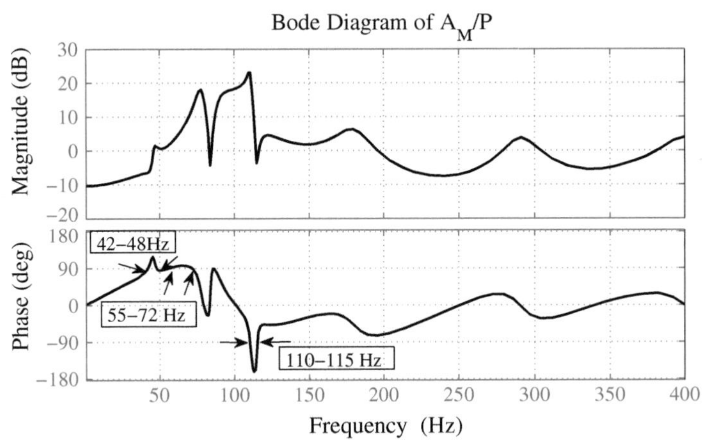

Fig. 15.7 Bode diagram of the estimated transfer function A_M/P

function is shown. Assuming that $\hat{G} = G$, using Algorithm II with constant adaptation gain, $\frac{A_M}{P}$ should be SPR. It can be seen that A_M/P it is not strictly positive real (phase outside of the interval $[-90, \ +90]$ degrees) in the frequency intervals $[42, \ 48]$, $[55, \ 72]$, and $[110, \ 115]$ Hz (while for the Algorithm III with constant adaptation gain the estimated transfer function which should be SPR is equal to 1).

The influence of the number of parameters upon the performance of the system is summarized in Table 15.2 for the case of Algorithm III. The global attenuation is slightly improved when the number of parameters of the compensator is augmented over 20 (the PSD are almost the same).

Table 15.2 Influence of the number of parameters upon the global attenuation

Number of parameters	20	32	40
Global attenuation (dB)	16.23	16.49	16.89

15.5.1.1 Testing Adaptation Capabilities When Disturbance Characteristics Change

Adaptation capabilities with respect to the characteristic of the disturbance is a key issue. This has been tested by adding a sinusoidal disturbance at 1500 s (adaptation Algorithm III (FUSBA) with constant trace set at 1). Figure 15.8 shows the time domain results in the case when the adaptation is stopped *prior* to the application of the additional sinusoidal disturbance (upper diagram) and when the adaptation is active (lower diagram). The duration of the transient is approximately 25 s.

Figure 15.9 shows the evolution of the parameters when the sinusoidal disturbance is applied. The power spectral densities when adaptation is stopped *prior* to the application of the sinusoidal disturbance, when adaptation is active, and when the H_∞ compensator (not designed for this supplementary disturbance) is used are shown in Fig. 15.10. One can remark a strong attenuation of the sinusoidal disturbance (larger than 35 dB) without affecting other frequencies when the adaptation is active (similar results are obtained with Algorithm II). The H_∞ compensator [18] does a very little attenuation of the sinusoidal disturbance (2.6 dB). It does not have "adaptation capabilities". The linear compensators considered in Chap. 14 will not be able to cope with the new disturbance. Other results can be found in [13].

15.5.2 Broad-band Disturbance Rejection Using Scalar Adaptation Gain

Table 15.3 gives a summary of the proposed algorithms with matrix adaption gain (column 1) and with scalar adaptation gain (column 2). Column 3 gives the algorithms of Jacobson–Johnson [6] and column 4 gives the FULMS algorithm [19].

The algorithm of Jacobson–Johnson (column 3) was unstable even for very low adaptation gain. The explanation is clear. It does not use filtering at least by $\hat{G}$ and since G is not positive real (in particular, in the frequency zone where most of the energy of the disturbance is concentrated) the instability is not surprising.

To make a fair comparison of the algorithms given in columns 2 and 4 of Table 15.3, the same adaptation gain has been used. Since the FULMS is very sensitive to the value of the adaptation gain (becomes easily unstable and the transients are very bad) a value of 0.001 has been chosen for the scalar adaptation gain (for a higher value FULMS is unstable).

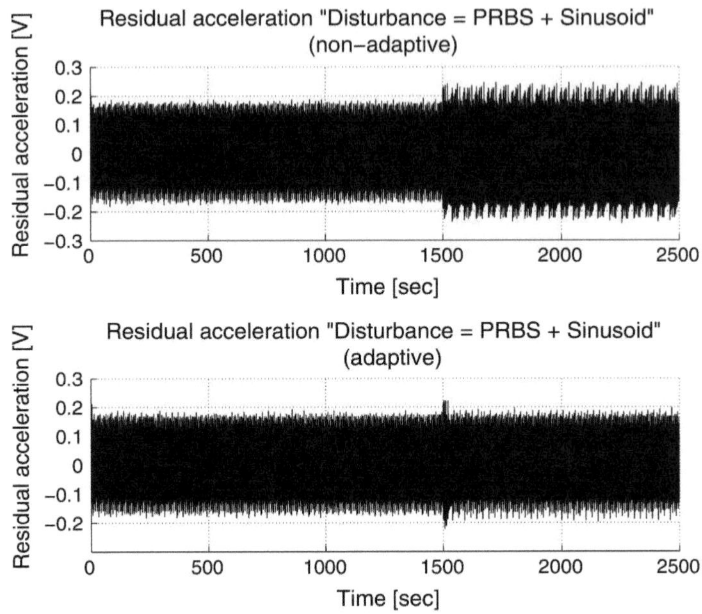

Fig. 15.8 Real-time results for rejection of an additional sinusoidal disturbance. *Upper diagram* adaptation stopped *prior* application of the disturbance. *Lower diagram* adaptation is active

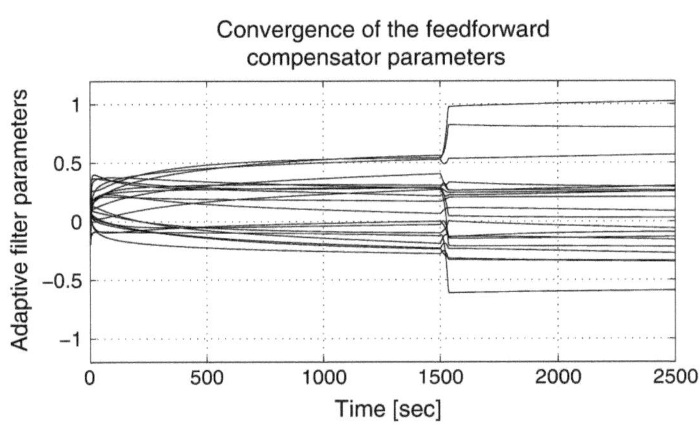

Fig. 15.9 Evolution of the compensator parameters when a sinusoidal disturbance is added using the FUSBA algorithm (experimental)

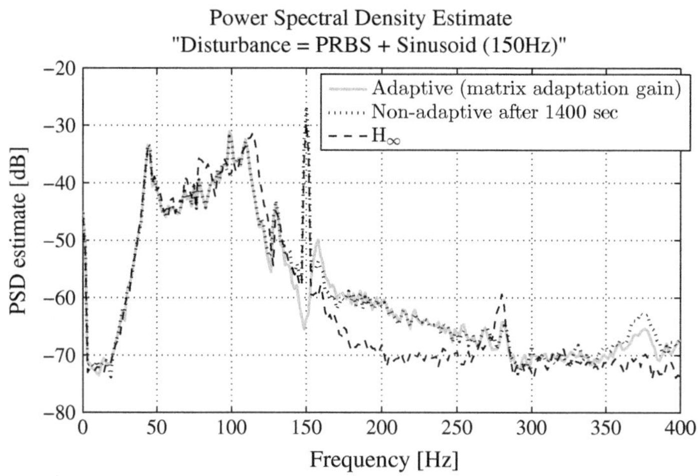

Fig. 15.10 Power spectral densities of the residual acceleration when an additional sinusoidal disturbance is added (Disturbance = PRBS + sinusoid)

The FULMS algorithm and the Algorithm II from column 2 use the same filtering of the regressor. The difference comes from the fact that the FULMS uses the *a priori* adaptation error while the Algorithm II of column 2 uses the *a posteriori* value of the adaptation error. The difference between these two algorithms can be also interpreted in terms of adaptation gains. The FULMS uses an unnormalized adaptation gain γ while Algorithm II, column 2, uses a normalized[13] adaptation gain $\gamma/(1+\gamma\Phi^T(t)\Phi(t))$.

Figure 15.11 shows the adaptation transient for the FULMS algorithm. The maximum value is unacceptable in practice (one cannot tolerate an overshoot over 50 % of the uncompensated residual acceleration). Figure 15.12 shows the adaptation transient for the scalar version of the Algorithm III. It is surprisingly good. Almost same transient behaviour is obtained with the scalar version of Algorithm II. Figures 15.13 and 15.14 show the evolution of the parameters for the FULMS algorithm and the scalar version of Algorithm III, respectively. One can see jumps in the evolution of the parameters for the FULMS algorithms and instabilities occur on a long run. For the Algorithm III evolution of the parameters is smooth and no instabilities occur in a long run (12 h).

The performances in the frequency domain are summarized in Fig. 15.15 where the power spectral densities and the global attenuation provided by the algorithms with scalar adaptation gain are shown.

[13]The scalar adaptation gain algorithms presented in this book can be denoted as NFULMS (normalized FULMS) for Algorithm II and SFUSBA (scalar FUSBA) for Algorithm III.

Table 15.3 Comparison of algorithms for adaptive feedforward compensation in AVC with mechanical coupling

	Book (Matrix gain)	Book (Scalar gain)	Jacobson–Johnson (Scalar gain)	FULMS (Scalar gain)
$\hat{\theta}(t+1) =$	$\hat{\theta}(t) + F(t)\Phi(t)\frac{\nu^{\circ}(t+1)}{1+\Phi^T(t)F(t)\Phi(t)}$	$\hat{\theta}(t) + \gamma(t)\Phi(t)\frac{\nu^{\circ}(t+1)}{1+\gamma(t)\Phi^T(t)\Phi(t)}$	$\hat{\theta}(t) + \mu\Phi(t)\frac{\nu^{\circ}(t+1)}{1+\gamma\psi^T(t)\Phi(t)}$	$\hat{\theta}(t) + \gamma(t)\Phi(t-1)\nu^{\circ}(t)$
Adapt. gain	$F(t+1)^{-1} = \lambda_1(t)F(t) +$ $+\lambda_2(t)\Phi(t)\Phi^T(t)$ $0 \leq \lambda_1(t) < 1, 0 \leq \lambda_2(t) < 2$ $F(0) > 0$	$\gamma(t) > 0$	$\gamma > 0, \quad 0 < \mu \leq 1$	$\gamma(t) > 0$
Adaptive	Decr. gain and const. trace	$\gamma(t) = \gamma = const$	$\gamma > 0$	$\gamma(t) = \gamma = const$
Self-tuning	$\lambda_2 = const. \lim_{t\to\infty} \lambda_1(t) = 1$	$\sum_{t=1}^{\infty} \gamma(t) = \infty, \quad \lim_{t\to\infty} \gamma(t) = 0$	Does not apply	$\sum_{t=1}^{\infty} \gamma(t) = \infty, \quad \lim_{t\to\infty} \gamma(t) = 0$
$\phi^T(t) =$	$[-\hat{y}(t), \ldots, \hat{u}(t+1), \ldots]$	$[-\hat{y}(t), \ldots, \hat{u}(t+1), \ldots]$	$[-\hat{y}(t), \ldots, \hat{u}(t+1), \ldots]$	$[-\hat{y}(t), \ldots, \hat{u}(t+1), \ldots]$
$\Phi(t) =$	$L\phi(t)$ FUPLR: $L_2 = \hat{G}$ FUSBA: $L_3 = \frac{\hat{A}_M}{\hat{P}}\hat{G}$ $\hat{P} = \hat{A}_M\hat{S} - \hat{B}_M\hat{R}$	$L\phi(t)$ NFULMS: $L_2 = \hat{G}$ SFUSBA: $L_3 = \frac{\hat{A}_M}{\hat{P}}\hat{G}$ $\hat{P} = \hat{A}_M\hat{S} - \hat{B}_M\hat{R}$	$\phi(t)$	$L\phi(t)$ $L = \hat{G}$
$G = \frac{B_G}{A_G}$	$B_G = b_{1_G}z^{-1} + b_{2_G}z^{-2} + \cdots$ $A_G = 1 + a_{1_G}z^{-1} + a_{2_G}z^{-2} + \cdots$	$B_G = b_{1_G}z^{-1} + b_{2_G}z^{-2} + \cdots$ $A_G = 1 + a_{1_G}z^{-1} + \cdots$	$B_G = 1, \quad A_G = 1$ or $G = SPR$	$B_G = b_{1_G}z^{-1} + b_{2_G}z^{-2} + \cdots$ $A_G = 1 + a_{1_G}z^{-1} + \cdots$
$M = \frac{B_M}{A_M}$	$B_M = b_{1_M}z^{-1} + b_{2_M}z^{-2} + \cdots$ $A_M = 1 + a_{1_M}z^{-1} + a_{2_M}z^{-2} + \cdots$	$B_M = b_{1_M}z^{-1} + b_{2_M}z^{-2} + \cdots$ $A_M = 1 + a_{1_M}z^{-1} + \cdots$	$B_M = b_{1_M}z^{-1} + b_{2_M}z^{-2} + \cdots$ $A_M = 1$	$B_M = b_{1_M}z^{-1} + b_{2_M}z^{-2} + \cdots$ $A_M = 1$
$D = \frac{B_D}{A_D}$	$B_D = b_{1_D}z^{-1} + b_{2_D}z^{-2} + \cdots$ $A_D = 1 + a_{1_D}z^{-1} + a_{2_D}z^{-2} + \cdots$	$B_D = b_{1_D}z^{-1} + b_{2_D}z^{-2} + \cdots$ $A_D = 1 + a_{1_D}z^{-1} + \cdots$	$B_D = b_{1_D}z^{-1} + b_{2_D}z^{-2} + \cdots$ $A_D = 1$	$B_D = b_{1_D}z^{-1} + b_{2_D}z^{-2} + \cdots$ $A_D = 1 + a_{1_D}z^{-1} + \cdots$
Stability condition	$\frac{A_M G}{PL} - \frac{\lambda}{2} = SPR$, $\lambda = \max \lambda_2(t)$	$\frac{A_M G}{PL} = SPR$	$G = SPR$	Unknown
Conv. condition	$\frac{A_M G}{PL} - \frac{\lambda}{2} = SPR, \lambda = \lambda_2$	$\frac{A_M G}{PL} = SPR$	Does not apply	$\frac{G}{P\hat{G}} = SPR$

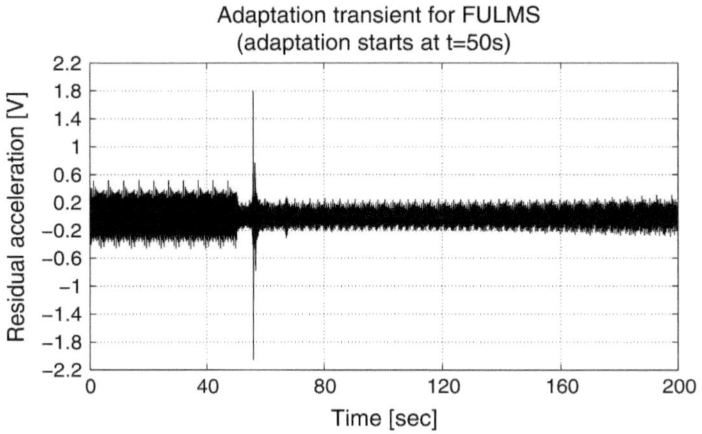

Fig. 15.11 Real-time results obtained with FULMS algorithm

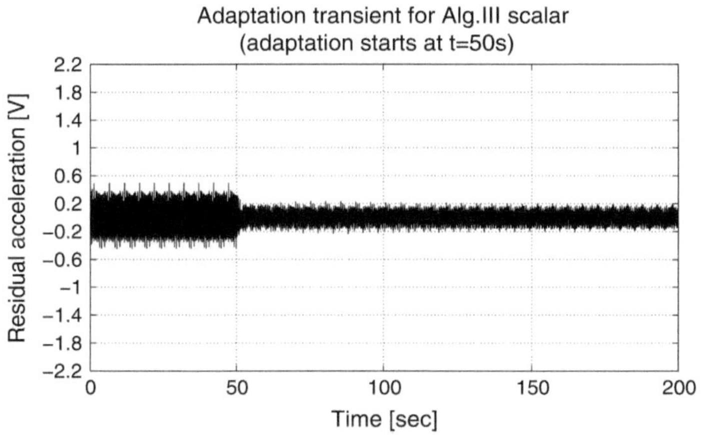

Fig. 15.12 Real-time results obtained with Algorithm III using scalar adaptation gain

15.5.2.1 Testing Adaptation Capabilities When Disturbance Characteristics Change

Adaptation capabilities with respect to the characteristic of the disturbance have been tested by adding a sinusoidal disturbance like for the case of matrix adaptation gain. The FULMS has been destabilized by the application of the sinusoidal disturbance. Figure 15.16 shows the PSD of the residual acceleration when the adaptation is stopped before the sinusoidal disturbance is applied and when the adaptation is active. The performance of the adaptation Algorithm III with scalar gain is less good than in the case of matrix adaptation gain (see Fig. 15.10). The sinusoidal disturbance is attenuated in the scalar case by 20 dB while the attenuation is over 35 dB with

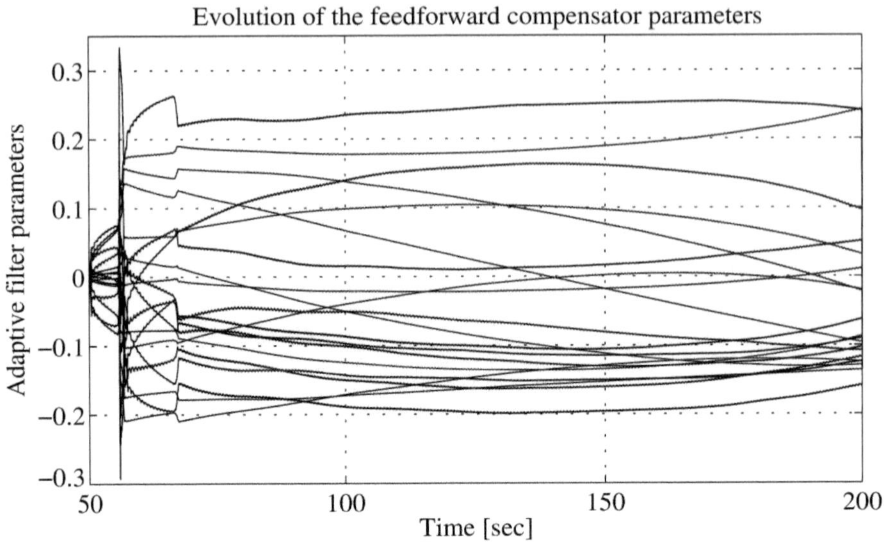

Fig. 15.13 Evolution of the feedforward compensator parameters (experimental)—Algorithm FULMS

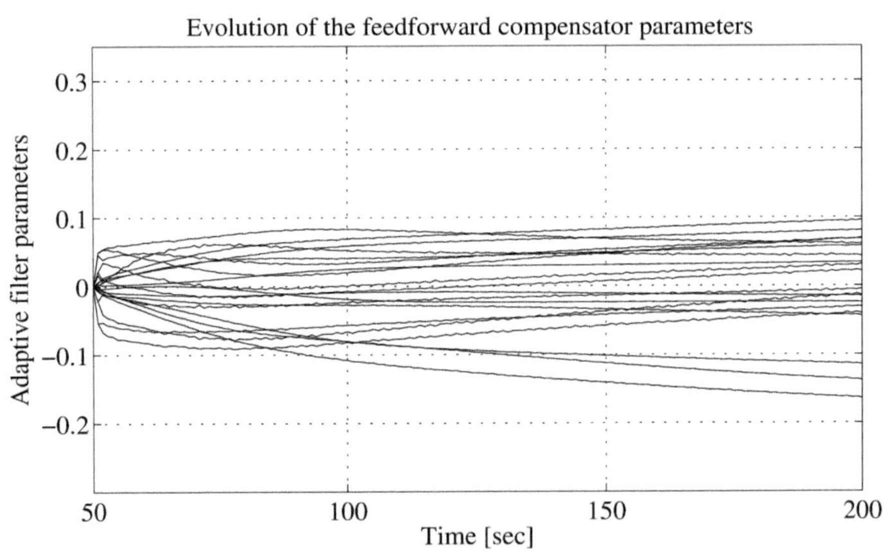

Fig. 15.14 Evolution of the feedforward compensator parameters (experimental)—Algorithm III using scalar adaptation gain

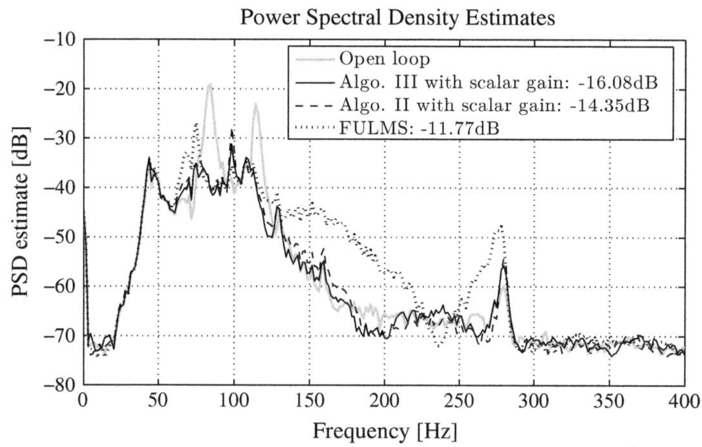

Fig. 15.15 Power spectral densities of the residual acceleration in open-loop and with adaptive feedforward compensation using scalar adaptation gain (Disturbance = PRBS)

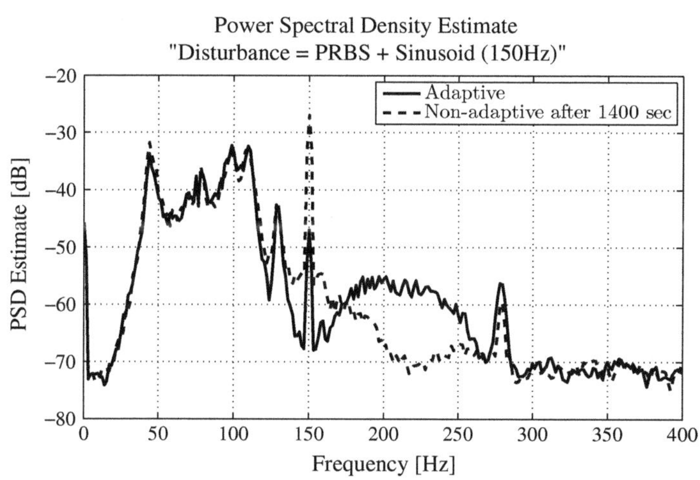

Fig. 15.16 Power spectral densities of the residual acceleration using scalar adaptation gain when a sinusoidal disturbance is added (Disturbance = PRBS + sinusoid) (experimental)

a matrix adaptation gain. In addition the performance is degraded in the frequency region 170–270 Hz which does not occur when using a matrix adaptation gain.

15.6 Adaptive Feedforward Compensation with Filtering of the Residual Error

Another solution to fulfil the strictly positive real condition (popular in adaptive control [13]) is to introduce a filter on the residual error in order to generate the adaptation error. Some of the references considering the use of the filter on the adaptation error are [17, 20–23]. As it will be shown, the filtering of the residual error will affect its power spectral density. There are a number of situations where shaping the residual error in the frequency domain is very useful.

Recall from Sect. 15.2 that the measured residual acceleration (or force) satisfies the following equation:

$$e^{\circ}(t+1) = x(t+1) + \hat{z}^{\circ}(t+1). \tag{15.54}$$

Then the filtered *a priori* adaptation error is defined as

$$\begin{aligned} \nu^{\circ}(t+1) &= \nu(t+1|\hat{\theta}(t)) \\ &= \varepsilon^{\circ}(t+1) + \sum_{i=1}^{n_1} \upsilon_i^B \varepsilon(t+1-i) - \sum_{i=1}^{n_2} \upsilon_i^A \nu^{\circ}(t+1-i), \end{aligned} \tag{15.55}$$

where

$$\varepsilon^{\circ}(t+1) = -e^{\circ}(t+1) = -x(t+1) - \hat{z}^{\circ}(t+1) \tag{15.56}$$

and

$$\varepsilon(t+1) = -e(t+1) = -x(t+1) - \hat{z}(t+1) \tag{15.57}$$

are also called, respectively, the *a priori* and the *a posteriori* unfiltered adaptation errors.

The coefficients $\upsilon_i^X,\ X \in \{B,\ A\}$, are the coefficients of an IIR filter, with all poles and zeros inside the unit circle, acting on the adaptation error

$$V(q^{-1}) = \frac{B_V(q^{-1})}{A_V(q^{-1})}, \tag{15.58}$$

where

$$X_V(q^{-1}) = 1 + q^{-1}X_V^*(q^{-1}) = 1 + \sum_{i=1}^{n_j} \upsilon_i^X q^{-i},\ X \in \{B,\ A\}. \tag{15.59}$$

The filtered *a posteriori* unmeasurable (but computable) adaptation error is given by

$$\nu(t+1) = \nu(t+1|\hat{\theta}(t+1)) \tag{15.60}$$

$$= \varepsilon(t+1) + \sum_{i=1}^{n_1} \upsilon_i^B \varepsilon(t+1-i) - \sum_{i=1}^{n_2} \upsilon_i^A \nu(t+1-i), \tag{15.61}$$

with $\varepsilon(t+1)$ given in (15.57).

The PAA given in Eqs. (15.35) through (15.39) is transformed as follows[14]

$$\hat{\theta}(t+1) = \hat{\theta}(t) + F(t)\Phi(t)\nu(t+1); \tag{15.62}$$

$$\varepsilon(t+1) = \frac{\varepsilon^\circ(t+1)}{1+\Phi^T(t)F(t)\Phi(t)}; \tag{15.63}$$

$$\nu(t+1) = \varepsilon(t+1) + \sum_{i=1}^{n_1} \upsilon_i^B \varepsilon(t+1-i) - \sum_{i=1}^{n_2} \upsilon_i^A \nu(t+1-i), \tag{15.64}$$

$$F(t+1) = \frac{1}{\lambda_1(t)}\left[F(t) - \frac{F(t)\Phi(t)\Phi^T(t)F(t)}{\frac{\lambda_1(t)}{\lambda_2(t)} + \Phi^T(t)F(t)\Phi(t)}\right] \tag{15.65}$$

$$1 \geq \lambda_1(t) > 0; \quad 0 \leq \lambda_2(t) < 2; \; F(0) > 0 \tag{15.66}$$

$$\Phi(t) = \phi_f(t) = L\phi(t) \tag{15.67}$$

Equation (15.27) becomes

$$\nu(t+1) = \frac{A_M(q^{-1})G(q^{-1})V(q^{-1})}{P(q^{-1})L(q^{-1})}[\theta - \hat{\theta}]^T \phi_f(t) \tag{15.68}$$

For the stability of the system, the selection of L and V should be done such that $\frac{A_M(q^{-1})G(q^{-1})V(q^{-1})}{P(q^{-1})L(q^{-1})}$ be SPR (for $\lambda_2 = 0$). Nevertheless, in practice one uses Algorithms II (FUPLR) or III (FUSBA) and V is added mainly for shaping the PSD of the residual error in the frequency domain. The new algorithms are termed FUePLR and FUeSBA, respectively, to denote the filtering of the residual error in addition to the filtering of the observation vector.

Using FUSBA presented in Sect. 15.3 with the prediction error filtered by $V(q^{-1})$, the estimated $\hat{N}$ feedforward compensator will minimize the following criterion in the frequency domain (taking into account (15.25)):

$$\hat{\theta}^* = \arg\min_{\hat{\theta}} \int_{-\pi}^{\pi} [|S_{NM}|^2|N-\hat{N}|^2|S_{\hat{N}M}|^2|G|^2|V|^2\phi_w(\omega)+|V|^2\phi_\eta(\omega)]d\omega \tag{15.69}$$

where ϕ_w and ϕ_η are the spectral densities of the disturbance $w(t)$ and of the measurement noise and S_{NM} and $S_{\hat{N}M}$ are the output sensitivity functions of the internal closed-loop for N and respectively, $\hat{N}$: $S_{NM} = \frac{1}{1-NM}$, $S_{\hat{N}M} = \frac{1}{1-\hat{N}M}$.

[14]This algorithm can be termed FUeSBA since both the input and the error are filtered.

Comparison of (15.69) with (15.50) allows to conclude that V will further shape the power spectral density of the residual error.

A number of experimental tests have been done to compare the adaptation capability of the algorithms with residual error filtering in addition to observation vector filtering. As broad-band disturbance, a PRBS generated by a 15-bit register and filtered through a Butterworth band-pass filter between 20 and 380 Hz has been used. A sinusoidal signal has been added at 250 Hz.

The residual error filter has been chosen as $V(q^{-1}) = 1 - 0.9q^{-1}$. Using an adaptive feedforward compensator with 20 parameters ($n_R = 9, n_S = 10$) the global attenuation achieved is 15.8 dB with FUPLR algorithm and 16.24 dB with FUePLR algorithm.

15.7 Adaptive Feedforward + Fixed Feedback Compensation of Broad-band Disturbances

As already mentioned throughout the book, feedforward compensation has to be considered when the performance/robustness compromise cannot be achieved by feedback only. Nevertheless, nothing prevents the use of (adaptive) feedforward compensation on top of a feedback controller. Defining jointly the control objective of the feedback controller and feedforward compensation is a problem-dependent issue. One can assign to the feedback controller an active damping task and the feedforward compensation will enhance the performances. Alternatively, one can design a stabilizing controller which attenuates certain type of disturbances under robustness constraints and the performance of the system will be enhanced by the feedforward compensation. The combination of feedback and feedforward compensation is often termed "hybrid" compensation.

Figure 15.17 gives the block diagram of such a system. The transfer operators characterizing the primary path (D), the secondary path (G) and the reverse path (M) have been defined in Sect. 15.2, Eqs. (15.1), (15.4) and (15.7), as well as the optimal feedforward compensator N and the estimated feedforward compensator $\hat{N}$. The vector of the estimated feedforward parameters has been defined in (15.29).

The fixed feedback RS controller K, computed on the basis of the model $\hat{G}$ which stabilizes the system and attenuates disturbances on the output $e(t)$, is characterized by the asymptotically stable transfer function

$$K(q^{-1}) = \frac{B_K(q^{-1})}{A_K(q^{-1})}, \tag{15.70}$$

where

$$B_K(q^{-1}) = b_0^K + b_1^K q^{-1} + \cdots + b_{n_{B_K}}^K q^{-n_{B_K}}, \tag{15.71}$$

$$A_K(q^{-1}) = 1 + a_1^K q^{-1} + \cdots + a_{n_{A_K}}^K q^{-n_{A_K}}. \tag{15.72}$$

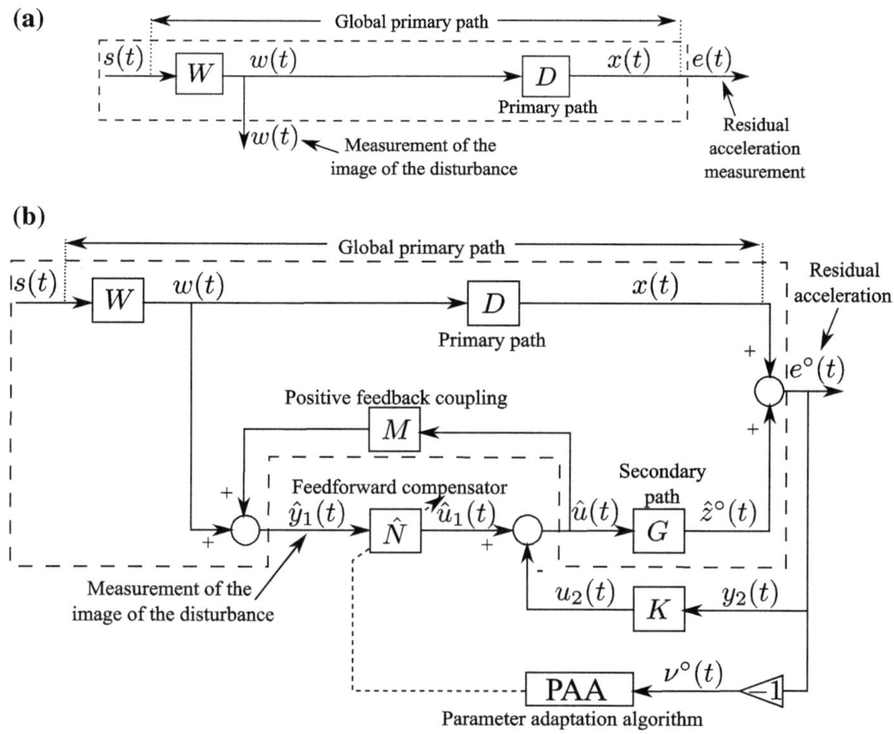

Fig. 15.17 Feedforward–feedback AVC—the control scheme: **a** in open-loop and **b** with adaptive feedforward + fixed feedback compensator

The input of the feedforward compensator (called also reference) is denoted by $\hat{y}_1(t)$. The output of the feedforward compensator is denoted by $\hat{u}_1(t+1) = \hat{u}_1(t+1|\hat{\theta}(t+1))$ (*a posteriori* output). The measured input applied to the feedforward compensator can be written as

$$\hat{y}_1(t+1) = w(t+1) + \frac{B_M^*(q^{-1})}{A_M(q^{-1})}\hat{u}(t), \tag{15.73}$$

where

$$\hat{u} = \hat{u}_1(t) - u_2(t), \tag{15.74}$$

$\hat{u}_1(t)$ and $u_2(t)$ are the outputs given by the adaptive feedforward and the fixed feedback compensators, respectively. $\hat{u}$ is the effective input sent to the control actuator.

The *a priori* output of the estimated feedforward compensator is given by

$$\begin{aligned}\hat{u}_1^\circ(t+1) &= \hat{u}_1(t+1|\hat{\theta}(t)) = -\hat{S}^*(t,q^{-1})\hat{u}_1(t) + \hat{R}(t,q^{-1})\hat{y}_1(t+1)\\ &= \hat{\theta}^T(t)\phi(t) = \left[\hat{\theta}_S^T(t), \hat{\theta}_R^T(t)\right]\begin{bmatrix}\phi_{\hat{u}_1}(t)\\ \phi_{\hat{y}_1}(t)\end{bmatrix}\end{aligned} \tag{15.75}$$

where $\hat{\theta}(t)$ has been given in (15.29) and

$$\phi^T(t) = [-\hat{u}_1(t), \ldots - \hat{u}_1(t - n_S + 1), \hat{y}_1(t + 1), \ldots \hat{y}_1(t - n_R + 1)] \\ = [\phi^T_{\hat{u}_1}(t), \ \phi^T_{\hat{y}_1}(t)] \quad (15.76)$$

The input to the feedback (fixed) compensator is given by the performance variable, therefore $y_2(t) = e(t)$. Its output will be $u_2(t) = K(q^{-1})y_2(t)$. The unmeasurable value of the output of the primary path (when the compensation is active) is denoted as $x(t)$. The *a priori* output of the secondary path is denoted as $\hat{z}^\circ(t + 1) = \hat{z}(t + 1|\hat{\theta}(t))$ while its input is $\hat{u}(t)$. One has

$$\hat{z}^\circ(t + 1) = \frac{B^*_G(q^{-1})}{A_G(q^{-1})}\hat{u}(t) = \frac{B^*_G(q^{-1})}{A_G(q^{-1})}\hat{u}(t|\hat{\theta}(t)). \quad (15.77)$$

The measured residual acceleration (or force) satisfies the following equation:

$$e^\circ(t + 1) = x(t + 1) + \hat{z}^\circ(t + 1). \quad (15.78)$$

The *a priori* and *a posteriori* adaptation errors are defined as

$$\nu^\circ(t + 1) = \nu(t + 1|\hat{\theta}(t)) = -e^\circ(t + 1) \quad (15.79)$$

and

$$\nu(t + 1) = \nu(t + 1|\hat{\theta}(t + 1)) = -e(t + 1) = -x(t + 1) - \hat{z}(t + 1) \quad (15.80)$$

where the *a posteriori* value of the output of the secondary path $\hat{z}(t + 1)$ (dummy variable) is given by

$$\hat{z}(t + 1) = \hat{z}(t + 1|\hat{\theta}(t + 1)) = \frac{B^*_G(q^{-1})}{A_G(q^{-1})}\hat{u}(t|\hat{\theta}(t + 1)). \quad (15.81)$$

For compensators with constant parameters $\nu^\circ(t) = \nu(t), e^\circ(t) = e(t), \hat{z}^\circ(t) = \hat{z}(t)$, $\hat{u}^\circ(t) = \hat{u}(t)$.

15.7.1 Development of the Algorithms

The algorithms for adaptive feedforward compensation in the presence of feedback controller will be developed under the same hypotheses as in Sect. 15.3 except hypothesis H2 which is replaced by [24]:

(H2′) (Perfect matching condition) There exists a filter $N(q^{-1})$ of finite dimension such that

$$\frac{N(z^{-1})}{1 - N(z^{-1})M(z^{-1})} G(z^{-1}) = -D(z^{-1}) \tag{15.82}$$

and the characteristic polynomials

- of the "internal" positive coupling loop

$$P(z^{-1}) = A_M(z^{-1})S(z^{-1}) - B_M(z^{-1})R(z^{-1}), \tag{15.83}$$

- of the closed-loop (G-K)

$$P_{cl}(z^{-1}) = A_G(z^{-1})A_K(z^{-1}) + B_G(z^{-1})B_K(z^{-1}) \tag{15.84}$$

- and of the coupled feedforward–feedback loop

$$P_{fb-ff} = A_M S[A_G A_K + B_G B_K] - B_M R A_K A_G \tag{15.85}$$

are Hurwitz polynomials.

Like for the previous feedforward compensation configurations, the key point in the development of the algorithm is to establish a relation between the error in the estimation of the parameters of the feedforward compensator and the measured residual acceleration or force. Under the hypotheses H1, H3 and the new hypothesis H2′, for the system described in Sect. 15.2 using a feedforward compensator $\hat{N}$ with constant parameters and a feedback controller K, the equation of the adaptation error (the measured residual acceleration or force with minus sign) for constant estimated parameters is given by [24]:

$$\nu(t+1) = \frac{A_M A_G A_K G}{P_{fb-ff}} [\theta - \hat{\theta}]^T \phi(t) \tag{15.86}$$

where

$$\theta^T = [s_1, \ldots s_{n_S}, r_0, r_1, \ldots r_{n_R}] = [\theta_S^T, \theta_R^T] \tag{15.87}$$

is the vector of parameters of the optimal filter N assuring perfect matching,

$$\hat{\theta}^T = [\hat{s}_1, \ldots \hat{s}_{n_S}, \hat{r}_0, \ldots \hat{r}_{n_R}] = [\hat{\theta}_S^T, \hat{\theta}_R^T] \tag{15.88}$$

is the vector of constant estimated parameters of $\hat{N}$,

$$\begin{aligned}\phi^T(t) = [-\hat{u}_1(t), \ldots - \hat{u}_1(t - n_S + 1), \hat{y}_1(t+1), \ldots \hat{y}_1(t - n_R + 1)] \\ = [\phi_{\hat{u}_1}^T(t), \ \phi_{\hat{y}_1}^T(t)]\end{aligned} \tag{15.89}$$

and $\hat{y}_1(t+1)$ is given by (15.73).

The derivation of the expression (15.86) is given in Appendix D.

Of course this expression can be particularized for the case without internal positive coupling ($B_M = 0$ and $A_M = 1$) and for the case of the absence of feedback ($K = 0$). Details are given in [24].

Filtering the vector $\phi(t)$ through an asymptotically stable filter $L(q^{-1}) = \frac{B_L}{A_L}$, Eq. (15.86) for $\hat{\theta}$ = constant becomes

$$\nu(t+1) = \frac{A_M A_G A_K G}{P_{fb-ff} L}[\theta - \hat{\theta}]^T \phi_f(t) \tag{15.90}$$

$$\phi_f(t) = L(q^{-1})\phi(t). \tag{15.91}$$

Equation (15.90) has been used to develop the adaptation algorithms neglecting the non-commutativity of the operators when $\hat{\theta}$ is time-varying (however, an exact algorithm can be derived in such cases—see [13]).

Replacing the fixed estimated parameters by the current estimated parameters, Eq. (15.90) becomes the equation of the *a posteriori* residual (adaptation) error $\nu(t+1)$ (which is computed):

$$\nu(t+1|\hat{\theta}(t+1)) = \frac{A_M A_G A_K}{P_{fb-ff} L} G[\theta - \hat{\theta}(t+1)]^T \phi_f(t). \tag{15.92}$$

Equation (15.92) has the standard form for an *a posteriori* adaptation error ([13]), which immediately suggests to use the same parametric adaptation algorithm given in equations (15.35) through (15.39). The stability of the algorithm has been analyzed in [24] and the main results are recalled next.

15.7.2 Analysis of the Algorithms

Stability of the Algorithms

The equation for the *a posteriori* adaptation error has the form

$$\nu(t+1) = H(q^{-1})[\theta - \hat{\theta}(t+1)]^T \Phi(t) \tag{15.93}$$

where

$$H(q^{-1}) = \frac{A_M A_G A_K}{P_{fb-ff} L} G, \ \Phi = \phi_f. \tag{15.94}$$

Neglecting the non-commutativity of time-varying operators, one can straightforwardly use Theorem 4.1. Therefore, the sufficient stability condition for any initial conditions $\hat{\theta}(0)$, $\nu^{\circ}(0)$, $F(0)$ is that

$$H'(z^{-1}) = H(z^{-1}) - \frac{\lambda_2}{2}, \max_t [\lambda_2(t)] \leq \lambda_2 < 2 \tag{15.95}$$

Table 15.4 Global attenuations for various configurations

	Feedback only	Feedforward only (H_∞)	Adaptive feedforward only	Feedforward (H_∞) feedback	Feedback and adaptive feedforward
Att. (dB)	–14.40	–14.70	–16.23	–18.42	–20.53

be an SPR transfer function.

Various choices can be considered for the filter $L(q^{-1})$ in order to satisfy the positive real condition (see [12, 24]). It is important to remark that the positive real condition is strongly influenced by the presence of the feedback controller and its design. The best performances are in general obtained by taking $L(q^{-1})$ as an estimation of $\frac{A_M A_G A_K}{P_{fb-ff}} G$ (see (15.94)).

Relaxation of the positive real condition by averaging arguments is discussed in [12] (same procedure and conclusions as in Sect. 15.4) and by adding proportional adaptation in [25]. Filtering of the residual error can also be considered for satisfying the positive real condition, but this will modify the criterion which is minimized [21, 25].

Analysis of the algorithms when hypotheses H2′ and H3 are violated can be found in [12]. The conclusions of this analysis are similar those given in Sect. 15.4.

15.8 Adaptive Feedforward + Fixed Feedback Attenuation of Broad-band Disturbances—Experimental Results

A summary of various results obtained on the system described in Sect. 2.3 will be presented next. The adaptive feedforward compensator structure for all the experiments has been $n_R = 9$, $n_S = 10$ (total of 20 parameters) and this complexity does not allow to verify the "perfect matching condition" (which requires more than 40 parameters). A feedback RS controller has been also introduced to test the potential improvement in performance.

Table 15.4 summarizes the global attenuation results for various configurations. Clearly, the hybrid adaptive feedforward—fixed feedback scheme brings a significant improvement in performance with respect to adaptive feedforward compensation alone. This can be seen on the power spectral densities shown in Fig. 15.18.[15] A pseudorandom binary sequence (PRBS) excitation on the global primary path has been considered as the disturbance.

It is important to point out that the design of a linear feedforward + feedback requires not only the perfect knowledge of the disturbance characteristics but also of the model of the primary path, while an adaptive approach does not require these informations. To illustrate the adaptation capabilities of the algorithms presented, a sinusoidal disturbance of 150 Hz has been added to the PRBS disturbance.

[15] For the adaptive schemes, the PSD is evaluated after the adaptation transient has settled.

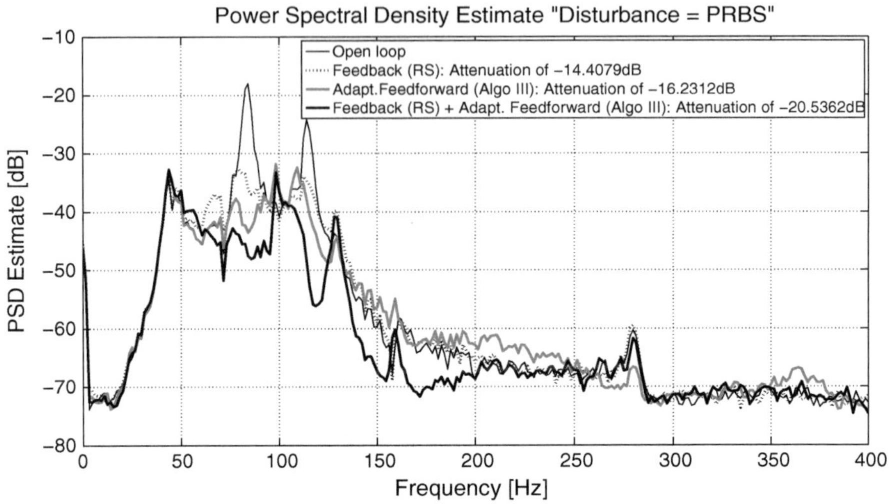

Fig. 15.18 Power spectral densities of the residual acceleration for various control configurations (Disturbance = PRBS)

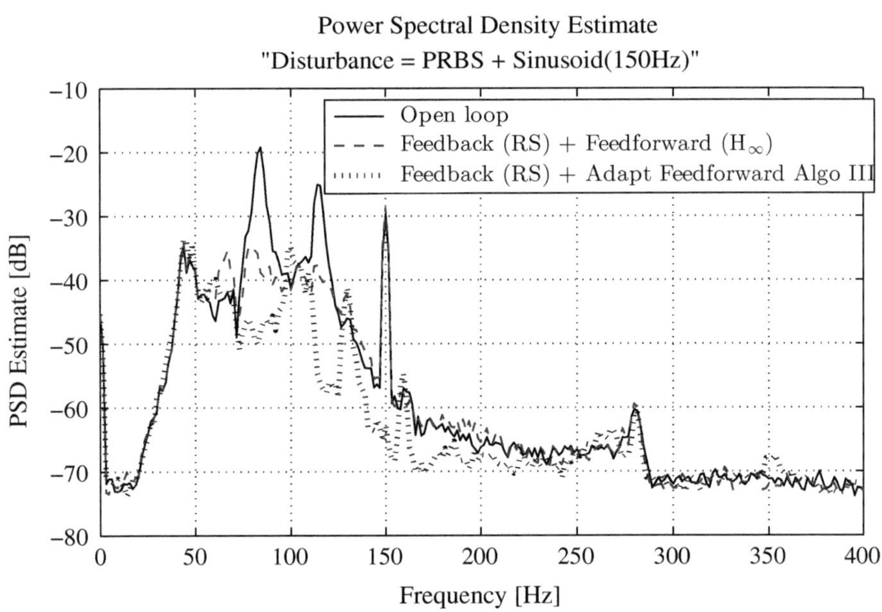

Fig. 15.19 Power spectral densities when an additional sinusoidal disturbance is added (Disturbance = PRBS + sinusoid)

Figure 15.19 shows the power spectral densities in open-loop, when using an adaptive feedforward compensation algorithm and when the H_∞ feedforward compensator which is not designed for this additional disturbance is used. One can remark that the hybrid adaptive feedforward–feedback scheme introduces a strong attenuation of the sinusoidal disturbance (larger than 30 dB) without affecting other frequencies (compare with Fig. 15.18) while the model-based H_∞ feedforward compensator + feedback controller has not been able to attenuate the sinusoidal disturbance.

15.9 Concluding Remarks

- If a measurement correlated with the disturbance is available, an adaptive feedforward compensation scheme can be built.
- This approach is currently used for active vibration control and active noise control when broad-band disturbances should be attenuated.
- It is important to emphasize the existence of an inherent positive feedback coupling between the actuator and the measure of the image of the disturbance which has a strong impact on the stability of the adaptive feedforward compensation system.
- Stable adaptation algorithms preserving the stability of the inner positive feedback loop has been developed.
- To assure the stability of the adaptive feedforward compensation schemes, the regressor vector should be appropriately filtered.
- Parameter adaptation algorithms with matrix adaptation gain and scalar adaptation gain can be used.
- Adaptive feedforward compensation can be used on top a feedback loop.

15.10 Notes and References

The first attempts in the literature of adaptive feedforward active vibration and noise compensation have been done neglecting the positive feedback coupling. Most of the initial work was centred around the use of the **Least Mean Squares (LMS)** gradient search algorithm introduced in [26, 27] (see also Chap. 4). Applications of LMS-type algorithm in active control can be found in [28–33]. Further references include [34–36].

A powerful approach for stability analysis of adaptive feedforward compensation algorithms is the **hyperstability** theory [37–40] which prompted out the importance of the strict positive realness of some transfer functions in order to assure stability. The initial framework for studying adaptive systems using hyperstability was established in [41–43] and a complete theoretical analysis can be found in [13]. Application of this approach in the context of adaptive feedforward compensation is considered in [12, 17, 44, 45]. Related problems are discussed in [20, 46–48].

Improved numerical efficiency for adaptation algorithms is discussed in [17, 21] (limited to the case without positive feedback coupling). FIR adaptive feedforward compensators using orthonormal basis functions are discussed in [49, 50].

In [51], a mixed adaptive feedback with GPC feedforward controller using online identification of the system is applied to acoustic noise control, structural vibration control and optical jitter control.

There has been also an important development in materials used for AVC systems. Among them, piezoelectric sensors and actuator are widely used for structural vibration cancellation (see book [52] and some of the applications in [53–57]).

Numerous applications of AVC concern hard disk drives or DVD/CDD [31, 58]. Also force tracking with feedforward motion estimation for beating heart surgery is presented in [59].

Various AVC problems in passenger vehicles are discussed in [60–62]. In the field of aerial vehicles some interesting applications are [63, 64]. Vibration control on flexible structures is discussed in [65, 66]. Multichannel adaptive algorithms have extensively been used in adaptive optics applications [67, 68].

References

1. Burgess J (1981) Active adaptive sound control in a duct: a computer simulation. J Acoust Soc Am 70:715–726
2. Widrow B, Shur D, Shaffer S (1981) On adaptive inverse control. In: Proceedings of the 15th asilomar conference circuits, systems and computers. Pacific Grove, CA, USA
3. Widrow B, Stearns S (1985) Adaptive signal processing. Prentice-Hall, Englewood Cliffs, New Jersey
4. Elliott S, Nelson P (1994) Active noise control. Noise/news international, pp 75–98
5. Elliott S, Sutton T (1996) Performance of feedforward and feedback systems for active control. IEEE Trans Speech Audio Process 4(3):214–223. doi:10.1109/89.496217
6. Jacobson CA, Johnson CR, Jr, McCormick DC, Sethares WA (2001) Stability of active noise control algorithms. IEEE Signal Process Lett 8(3):74–76. doi:10.1109/97.905944
7. Zeng J, de Callafon R (2006) Recursive filter estimation for feedforward noise cancellation with acoustic coupling. J Sound Vib 291(3–5):1061–1079. doi:10.1016/j.jsv.2005.07.016
8. Kuo M, Morgan D (1996) Active noise control systems-algorithms and DSP implementation. Wiley, New York
9. Kuo S, Morgan D (1999) Active noise control: a tutorial review. Proc IEEE 87(6):943–973. doi:10.1109/5.763310
10. Hu J, Linn J (2000) Feedforward active noise controller design in ducts without independent noise source measurements. IEEE Trans Control Syst Technol 8(3):443–455
11. Landau I, Karimi A, Constantinescu A (2001) Direct controller order reduction by identification in closed loop. Automatica 37:1689–1702
12. Landau I, Alma M, Airimitoaie T (2011) Adaptive feedforward compensation algorithms for active vibration control with mechanical coupling. Automatica 47(10):2185–2196. doi:10.1016/j.automatica.2011.08.015
13. Landau ID, Lozano R, M'Saad M, Karimi A (2011) Adaptive control, 2nd edn. Springer, London
14. Ljung L, Söderström T (1983) Theory and practice of recursive identification. MIT Press, Cambridge
15. Anderson B, Bitmead R, Johnson C, Kokotovic P, Kosut R, Mareels I, Praly L, Riedle B (1986) Stability of adaptive systems. MIT Press, Cambridge
16. Landau I, Zito G (2005) Digital control systems—design, identification and implementation. Springer, London

17. Montazeri P, Poshtan J (2010) A computationally efficient adaptive IIR solution to active noise and vibration control systems. IEEE Trans Autom Control AC-55:2671–2676
18. Alma M, Martinez J, Landau I, Buche G (2012) Design and tuning of reduced order H_∞ feedforward compensators for active vibration control. IEEE Trans Control Syst Technol 20(2):554–561. doi:10.1109/TCST.2011.2119485
19. Eriksson L (1991) Development of the filtered-U LMS algorithm for active noise control. J Acoust Soc Am 89(1):257–261
20. Larimore MG, Treichler JR, Johnson CR (1980) SHARF: An algorithm for adapting IIR digital filters. IEEE Trans Acoust Speech Signal Process 28(4):428–440. doi:10.1109/TASSP.1980.1163428
21. Montazeri A, Poshtan J (2011) A new adaptive recursive RLS-based fast-array IIR filter for active noise and vibration control systems. Signal Process 91(1):98–113. doi:10.1016/j.sigpro.2010.06.013
22. Sun X, Chen DS (2002) A new infinte impulse response filter-based adaptive algorithm for active noise control. J Sound Vib 258(2):385–397. doi:10.1006/jsvi.2002.5105
23. Sun X, Meng G (2004) Steiglitz-Mcbride type adaptive IIR algorithm for active noise control. J Sound Vib 273(1–2):441–450. doi:10.1016/j.jsv.2003.07.023
24. Alma M, Landau ID, Airimitoaie TB (2012) Adaptive feedforward compensation algorithms for AVC systems in the presence of a feedback controller. Automatica 48(10):982–985
25. Airimitoaie TB, Landau ID (2013) Improving adaptive feedforward vibration compensation by using integral+proportional adaptation. Automatica 49(5):1501–1505. doi:10.1016/j.automatica.2013.01.025
26. Widrow B, Hoff M (1960) Adaptive swithching circuits. Oric IRE WESCON Convention Record 4(Session 16):96–104
27. Widrow B (1971) Adaptive filters. In: Kalman R, DeClaris H (eds) Aspects of network and system theory. Holt, Rinehart and Winston
28. Billoud DG (2001) LL-6508 active control at lord corporation—A reality
29. Tang Y, Zhu ZC, Shen G, Li X (2015) Improved feedforward inverse control with adaptive refinement for acceleration tracking of electro-hydraulic shake table. J Vib Control. doi:10.1177/1077546314567725
30. Xiang M, Wei T (2014) Autobalancing of high-speed rotors suspended by magnetic bearings using lms adaptive feedforward compensation. J Vib Control 20(9):1428–1436. doi:10.1177/1077546313479990. http://jvc.sagepub.com/content/20/9/1428.abstract
31. Pan MC, Wei WT (2006) Adaptive focusing control of dvd drives in vehicle systems. J Vib Control 12(11):1239–1250. doi:10.1177/1077546306069037. http://jvc.sagepub.com/content/12/11/1239.abstract
32. Akraminia M, Mahjoob MJ, Niazi AH (2015) Feedforward active noise control using wavelet frames: simulation and experimental results. J Vib Control. doi:10.1177/1077546315581939. http://jvc.sagepub.com/content/early/2015/06/04/1077546315581939.abstract
33. Erdogan G, Alexander L, Rajamani R (2010) Adaptive vibration cancellation for tire-road friction coefficient estimation on winter maintenance vehicles. IEEE Trans Control Syst Technol 18(5):1023–1032. doi:10.1109/TCST.2009.2031326
34. Hassibi B, Sayed A, Kailath T (1996) H_∞ optimality of the LMS algorithm. IEEE Trans Signal Process 44(2):267–280. doi:10.1109/78.485923
35. Sayyarrodsari B, How J, Hassibi B, Carrier A (2001) Estimation-based synthesis of H_∞-optimal adaptive FIR filters for filtered-LMS problems. IEEE Trans Signal Process 49(1):164–178. doi:10.1109/78.890358
36. Fraanje R (2004) Robust and fast schemes in broadband active noise and vibration control. PhD thesis, University of Twente, Twente, The Netherlands
37. Popov V (1960) Criterii de stabilitate pentru sistemele automate conţinând elemente neunivoce, probleme de automatizare. Publishing House of the Romanian Academy, pp 143–151
38. Popov V (1966) Hiperstabilitatea Sistemelor Automate. Editura Academiei Republicii Socialiste România

39. Popov V (1963) Solution of a new stability problem for controlled systems. Autom Remote Control 24(1):1–23
40. Popov V (1973) Hyperstability of control systems, trans. edn. Springer, Heidelberg
41. Landau I, Silveira H (1979) A stability theorem with applications to adaptive control. IEEE Trans Autom Control 24(2):305–312. doi:10.1109/TAC.1979.1102009
42. Landau I (1979) Adaptive control: the model reference approach. Marcel Dekker, New York
43. Landau I (1980) An extension of a stability theorem applicable to adaptive control. IEEE Trans Autom Control 25(4):814–817. doi:10.1109/TAC.1980.1102440
44. Landau ID, Airimitoaie TB, Alma M (2012) A Youla-Kučera parametrized adaptive feedforward compensator for active vibration control with mechanical coupling. Automatica 48(9):2152–2158. doi:10.1016/j.automatica.2012.05.066. http://www.sciencedirect.com/science/article/pii/S0005109812002397
45. Landau ID, Airimitoaie TB, Alma M (2013) IIR Youla-Kučera parameterized adaptive feedforward compensators for active vibration control with mechanical coupling. IEEE Trans Control Syst Technol 21(3):765–779
46. Treichler J, Larimore M, Johnson C, Jr (1978) Simple adaptive IIR filtering. In: IEEE international conference on acoustics, speech, and signal processing, ICASSP '78, vol 3, pp 118–122. doi:10.1109/ICASSP.1978.1170389
47. Mosquera C, Gomez J, Perez F, Sobreira M (1999) Adaptive IIR filters for active noise control. In: Sixth international congress on sound and vibration, ICSV '99, pp 1571–1582
48. Snyder S (1994) Active control using IIR filters—a second look. In: 1994 IEEE international conference on acoustics, speech, and signal processing, 1994, ICASSP-94, vol 2, pp II/241 –II/244. doi:10.1109/ICASSP.1994.389675
49. de Callafon R, Zeng J, Kinney C (2010) Active noise control in a forced-air cooling system. Control Eng Pract 18(9):1045–1052. doi:10.1016/j.conengprac.2010.05.007
50. Yuan J (2007) Adaptive laguerre filters for active noise control. Appl Acoust 68(1):86–96. doi:10.1016/j.apacoust.2006.01.009. http://www.sciencedirect.com/science/article/pii/S0003682X06000260
51. Moon SM, Cole DG, Clark RL (2006) Real-time implementation of adaptive feedback and feedforward generalized predictive control algorithm. J Sound Vib 294(12):82–96. doi:10.1016/j.jsv.2005.10.017. http://www.sciencedirect.com/science/article/pii/S0022460X05007030
52. Moheimani SR, Fleming AJ (2006) Piezoelectric transducers for vibration control and damping. Springer, London
53. Trindade MA Jr, Pagani CC, Oliveira, LPR (2015) Semi-modal active vibration control of plates using discrete piezoelectric modal filters. J Sound Vib 351:17–28. doi:10.1016/j.jsv.2015.04.034. http://www.sciencedirect.com/science/article/pii/S0022460X15003867
54. Reza Moheimani S, Pota H, Petersen I (1997) Broadband disturbance attenuation over an entire beam. In: Control Conference (ECC), 1997 European, pp 3896–3901
55. Lin CY, Huang YH (2014) Enhancing vibration suppression in a periodically excited flexible beam by using a repetitive model predictive control strategy. J Vib Control. doi:10.1177/1077546314564451
56. Lee Y, Halim D (2004) Vibration control experiments on a piezoelectric laminate plate using spatial feedforward control approach. In: 43rd IEEE conference on decision and control, 2004. CDC, vol 3, pp 2403–2408. doi:10.1109/CDC.2004.1428764
57. Kuhnen K, Krejci P (2009) Compensation of complex hysteresis and creep effects in piezoelectrically actuated systems. A new Preisach modeling approach. IEEE Trans Autom Control, 54(3), 537–550. doi:10.1109/TAC.2009.2012984
58. Wang J, Wang Y, Cao S (2011) Add-on feedforward compensation for vibration rejection in HDD. IEEE/ASME Trans Mechatron 16(6):1164–1170. doi:10.1109/TMECH.2010.2085008
59. Yuen S, Perrin D, Vasilyev N, del Nido P, Howe R (2010) Force tracking with feed-forward motion estimation for beating heart surgery. IEEE Trans Robot 26(5):888–896. doi:10.1109/TRO.2010.2053734
60. Zhou Z, Chen X, Zhou B (2015) Feedforward compensation in vibration isolation system subject to base disturbance. J Vib Control 21(6):1201–1209. doi:10.1177/1077546313493311

61. Gan Z, Hillis AJ, Darling J (2015) Adaptive control of an active seat for occupant vibration reduction. J Sound Vib 349:39–55. doi:10.1016/j.jsv.2015.03.050. http://www.sciencedirect.com/science/article/pii/S0022460X15002898
62. Lee BH, Lee CW (2009) Model based feed-forward control of electromagnetic type active control engine-mount system. J Sound Vib 323(35):574–593. doi:10.1016/j.jsv.2009.01.033. http://www.sciencedirect.com/science/article/pii/S0022460X09000819
63. Wildschek A, Bartosiewicz Z, Mozyrska D (2014) A multi-input multi-output adaptive feed-forward controller for vibration alleviation on a large blended wing body airliner. J Sound Vib 333(17):3859–3880. doi:10.1016/j.jsv.2014.04.021. http://www.sciencedirect.com/science/article/pii/S0022460X14002867
64. Prakash S, Kumar TR, Raja S, Dwarakanathan D, Subramani H, Karthikeyan C (2016) Active vibration control of a full scale aircraft wing using a reconfigurable controller. J Sound Vib 361:32–49. doi:10.1016/j.jsv.2015.09.010. http://www.sciencedirect.com/science/article/pii/S0022460X15007130
65. Vipperman J, Burdisso R (1995) Adaptive feedforward control of non-minimum phase structural systems. J Sound Vib 183(3):369–382. doi:10.1006/jsvi.1995.0260. http://www.sciencedirect.com/science/article/pii/S0022460X8570260X
66. Hansen C, Snyder S, Qiu X, Brooks L, Moreau D (2012) Active Control of Noise and Vibration. CRC Press, Taylor and Francis Group, Milton Park
67. Hoagg J, Lacy S, Bernstein D (2005) Broadband adaptive disturbance rejection for a deployable optical telescope testbed. In: Proceedings of the American control conference, 2005, pp 4953–4958, vol 7. doi:10.1109/ACC.2005.1470791
68. Ruppel T, Dong S, Rooms F, Osten W, Sawodny O (2013) Feedforward control of deformable membrane mirrors for adaptive optics. IEEE Trans Control Syst Technol 21(3):579–589. doi:10.1109/TCST.2012.2186813

Chapter 16
Youla–Kučera Parametrized Adaptive Feedforward Compensators

16.1 Introduction

Since most of the adaptive feedforward vibration (or noise) compensation systems feature an internal "positive feedback" coupling between the compensator system and the correlated disturbance measurement which serves as reference, one may think building a stabilizing controller for this internal loop to which an additional filter will be added with the objective to enhance the disturbance attenuation capabilities while preserving the stabilization properties of the controller.

In order to achieve this, instead of a standard IIR feedforward compensator, one can use an Youla–Kučera parametrization of the adaptive feedforward compensator. The central compensator will assure the stability of the internal positive feedback loop and its performance are enhanced in real-time by the direct adaptation of the parameters of the Youla–Kučera Q-filter.

A block diagram of such adaptive feedforward compensator is shown in Fig. 16.1. FIR and IIR Q-filters can be used. Details of the specific algorithms can be found in [1, 2]. Comparisons between IIR, FIR YK and IIR YK adaptive feedforward have been done. The main conclusions are:

- For the same level of performance IIR YK requires the lower number of adjustable parameters.
- IIR YK and FIR YK allow easily the incorporation of an initial stabilizing controller of any dimension while for IIR feedforward compensator this is more difficult.

These facts justify the use of this approach for adaptive feedforward compensation in the presence of an internal positive feedback.

I.D. Landau et al., *Adaptive and Robust Active Vibration Control*,
Advances in Industrial Control, DOI 10.1007/978-3-319-41450-8_16

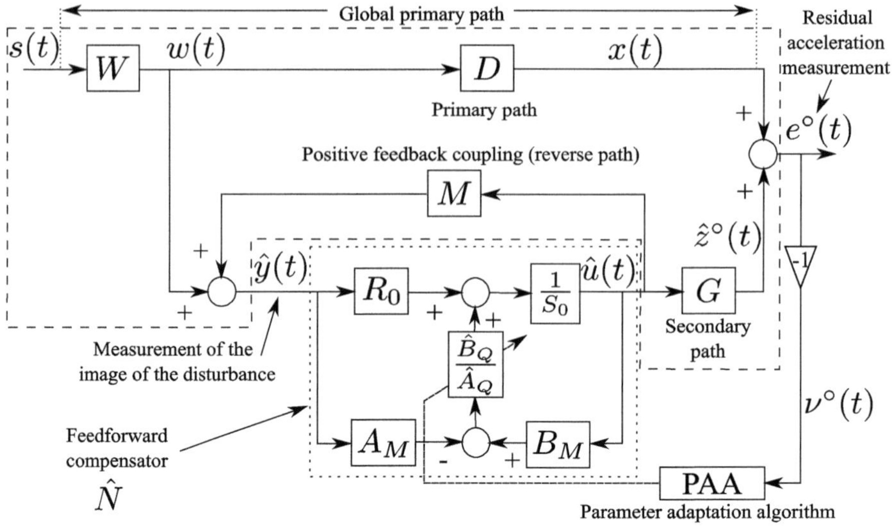

Fig. 16.1 Adaptive feedforward disturbance compensation using Youla–Kučera parametrization

16.2 Basic Equations and Notations

The block diagrams associated with an AVC system when an IIR (Infinite Impulse Response) Youla–Kučera compensator is active is shown in Fig. 16.1. The transfer operators of the various paths of the AVC system have been given in Sect. 15.2.

The optimal IIR feedforward compensator which will minimize the residual acceleration can be written, using the Youla–Kučera parametrization, as

$$N(q^{-1}) = \frac{R(q^{-1})}{S(q^{-1})} = \frac{A_Q(q^{-1})R_0(q^{-1}) - B_Q(q^{-1})A_M(q^{-1})}{A_Q(q^{-1})S_0(q^{-1}) - B_Q(q^{-1})B_M(q^{-1})} \tag{16.1}$$

where the optimal polynomial $Q(q^{-1})$ has an IIR structure

$$Q(q^{-1}) = \frac{B_Q(q^{-1})}{A_Q(q^{-1})} = \frac{b_0^Q + b_1^Q q^{-1} + \cdots + b_{n_{B_Q}}^Q q^{-n_{B_Q}}}{1 + a_1^Q q^{-1} + \cdots + a_{n_{A_Q}}^Q q^{-n_{A_Q}}} \tag{16.2}$$

and $R_0(q^{-1})$, $S_0(q^{-1}) = 1 + q^{-1}S_0^*(q^{-1})$ are the polynomials of the central (stabilizing) filter and $A_M(q^{-1})$, $B_M(q^{-1})$ are given in (15.7).

The estimated QIIR filter is denoted by $\hat{Q}(q^{-1})$ or $\hat{Q}(\hat{\theta}, q^{-1})$ when it is a linear filter with constant coefficients or $\hat{Q}(t, q^{-1})$ during estimation (adaptation). The vector of parameters of the optimal QIIR filter assuring perfect matching will be denoted by

$$\theta^T = [b_0^Q, \ldots, b_{n_{B_Q}}^Q, a_1^Q, \ldots, a_{n_{A_Q}}^Q] = [\theta_{B_Q}^T, \theta_{A_Q}^T]. \tag{16.3}$$

The vector of parameters for the estimated $\hat{Q}$IIR filter

$$\hat{Q}(q^{-1}) = \frac{\hat{B}_Q(q^{-1})}{\hat{A}_Q(q^{-1})} = \frac{\hat{b}_0^Q + \hat{b}_1^Q q^{-1} + \cdots + \hat{b}_{n_{B_Q}}^Q q^{-n_{B_Q}}}{1 + \hat{a}_1^Q q^{-1} + \cdots + \hat{a}_{n_{A_Q}}^Q q^{-n_{A_Q}}} \tag{16.4}$$

is denoted by

$$\hat{\theta}^T = [\hat{b}_0^Q, \ldots, \hat{b}_{n_{B_Q}}^Q, \hat{a}_1^Q, \ldots, \hat{a}_{n_{A_Q}}^Q] = [\hat{\theta}_{B_Q}^T, \hat{\theta}_{A_Q}^T]. \tag{16.5}$$

The input of the feedforward compensator (called also reference) is denoted by $\hat{y}(t)$. In the absence of the compensation loop (open-loop operation) $\hat{y}(t) = w(t)$. In the presence of the compensation this signal is the sum of $w(t$ and of the output of the reverse path M. The output of the feedforward compensator (which is the control signal applied to the secondary path) is denoted by $\hat{u}(t+1) = \hat{u}(t+1/\hat{\theta}(t+1))$ (*a posteriori* output).[1]

The *a priori* output of the estimated feedforward compensator using an YKIIR parametrization for the case of time-varying parameter estimates is given by (using (16.1))

$$\begin{aligned}\hat{u}^{\circ}(t+1) = \hat{u}(t+1/\hat{\theta}(t)) &= -\hat{S}^*(t,q^{-1})\hat{u}(t) + \hat{R}(t,q^{-1})\hat{y}(t+1)\\ &= -(\hat{A}_Q(t,q^{-1})S_0)^*\hat{u}(t) + \hat{A}_Q(t,q^{-1})R_0\hat{y}(t+1)\\ &\quad + \hat{B}_Q(t,q^{-1})\left(B_M^*\hat{u}(t) - A_M\hat{y}(t+1)\right),\end{aligned} \tag{16.6}$$

The *a posteriori* output is given by:

$$\begin{aligned}\hat{u}(t+1) &= -(\hat{A}_Q(t+1,q^{-1})S_0)^*\hat{u}(t) + \hat{A}_Q(t+1,q^{-1})R_0\hat{y}(t+1)\\ &\quad + \hat{B}_Q(t+1,q^{-1})\left(B_M^*\hat{u}(t) - A_M\hat{y}(t+1)\right).\end{aligned} \tag{16.7}$$

It should be observed that Eqs. (16.1), (16.2), (16.6) and (16.7) can be easily particularized for the case of a FIR Youla–Kučera parametrization by taking $\hat{A}_Q(t,q^{-1}) \equiv 1$.

The measured input of the feedforward compensator can also be written as

$$\hat{y}(t+1) = w(t+1) + \frac{B_M^*(q^{-1})}{A_M(q^{-1})}\hat{u}(t). \tag{16.8}$$

The unmeasurable value of the output of the primary path (when the compensation is active) is denoted $x(t)$. The *a priori* output of the secondary path will be denoted $\hat{z}^{\circ}(t+1) = \hat{z}(t+1|\hat{\theta}(t))$ while its input is $\hat{u}(t)$. One has

---

[1] In adaptive control and estimation the predicted output at $t+1$ can be computed either on the basis of the previous parameter estimates (*a priori*) or on the basis of the current parameter estimates (*a posteriori*).

$$\hat{z}^{\circ}(t+1) = \frac{B_G^*(q^{-1})}{A_G(q^{-1})}\hat{u}(t) = \frac{B_G^*(q^{-1})}{A_G(q^{-1})}\hat{u}(t|\hat{\theta}(t)), \tag{16.9}$$

where $\hat{\theta}(t)$ is the vector of estimated parameters given in (16.5). The measured residual acceleration (or force) satisfies the following equation

$$e^{\circ}(t+1) = x(t+1) + \hat{z}^{\circ}(t+1). \tag{16.10}$$

The *a priori* adaptation error is defined as

$$\nu^{\circ}(t+1) = \nu(t+1|\hat{\theta}(t)) = -e^{\circ}(t+1) = -x(t+1) - \hat{z}^{\circ}(t+1). \tag{16.11}$$

The *a posteriori* unmeasurable (but computable) adaptation error is given by

$$\nu(t+1) = \nu(t+1|\hat{\theta}(t+1)) = -e(t+1) = -x(t+1) - \hat{z}(t+1). \tag{16.12}$$

where the *a posteriori* value of the output of the secondary path $\hat{z}(t+1)$ (dummy variable) is given by

$$\hat{z}(t+1) = \hat{z}(t+1|\hat{\theta}(t+1)) = \frac{B_G^*(q^{-1})}{A_G(q^{-1})}\hat{u}(t|\hat{\theta}(t+1)). \tag{16.13}$$

For compensators with constant parameters $\nu^{\circ}(t) = \nu(t)$, $e^{\circ}(t) = e(t)$, $\hat{z}^{\circ}(t) = \hat{z}(t)$, $\hat{u}^{\circ}(t) = \hat{u}(t)$.

The objective is to develop stable recursive algorithms for adaptation of the parameters of the Q filter such that the measured residual error (acceleration or force in AVC, noise in ANC) be minimized in the sense of a certain criterion. This has to be done for broad-band disturbances $w(t)$ (or $s(t)$) with unknown and variable spectral characteristics and an unknown primary path model.

16.3 Development of the Algorithms

The algorithm for adaptive feedforward YKIIR compensators will be developed under the same hypotheses as in Sect. 15.3 except H2 which is modified as:

(H2″) (Perfect matching condition) There exists a value of the Q parameters such that[2]

$$\frac{G \cdot A_M(R_0 A_Q - A_M B_Q)}{A_Q(A_M S_0 - B_M R_0)} = -D \tag{16.14}$$

[2]The parenthesis (q^{-1}) or (z^{-1}) will be omitted in some of the following equations to make them more compact.

and there exists a central feedforward compensator N_0 (R_0, S_0) which stabilizes the inner positive feedback loop formed by N_0 and M and the characteristic polynomial of the closed-loop

$$P_0(z^{-1}) = A_M(z^{-1})S_0(z^{-1}) - B_M(z^{-1})R_0(z^{-1}) \tag{16.15}$$

is a Hurwitz polynomial.

Like for the standard IIR feedforward compensator, the algorithm will be developed under these hypotheses. Afterwards, hypotheses H2″ and H3 can be removed and the algorithm can be analyzed in this modified context.

A first step in the development of the algorithms is to establish for a fixed estimated compensator a relation between the error on the Q-parameters (with respect to the optimal values) and the adaptation error ν.

Under the hypotheses H1 and H3 from Chap. 15 for the system described by Eqs. (15.1)–(15.9) and the new hypothesis H2″ for the system described by (16.1) to (16.13) using an estimated IIR Youla–Kučera parameterized feedforward compensator with constant parameters one has:

$$\nu(t+1|\hat{\theta}) = \frac{A_M(q^{-1})G(q^{-1})}{A_Q(q^{-1})P_0(q^{-1})}[\theta - \hat{\theta}]^T\phi(t), \tag{16.16}$$

with $\phi(t)$ given by:

$$\begin{aligned}\phi^T(t) = [&\alpha(t+1), \alpha(t), \ldots, \alpha(t - n_{B_Q} + 1),\\ &-\beta(t), -\beta(t-1), \ldots, -\beta(t - n_{A_Q})].\end{aligned} \tag{16.17}$$

where:

$$\alpha(t+1) = B_M\hat{u}(t+1) - A_M\hat{y}(t+1) = B_M^*\hat{u}(t) - A_M\hat{y}(t+1) \tag{16.18a}$$

$$\beta(t) = S_0\hat{u}(t) - R_0\hat{y}(t). \tag{16.18b}$$

The derivation of expression (16.16) is given in Appendix D.

Throughout the remainder of this section and the next one, unless stated differently, the Youla–Kučera parametrization having a QIIR filter will be discussed. It should be observed that, in most of the cases, results for QFIR polynomials can be obtained by imposing $A_Q(q^{-1}) = 1$ and $\hat{A}_Q(q^{-1}) = 1$.

As it will be shown later on (like for the IIR feedforward compensator), it is convenient for assuring the stability of the system to filter the observation vector $\phi(t)$. Filtering the vector $\phi(t)$ through an asymptotically stable filter $L(q^{-1}) = \frac{B_L}{A_L}$, Eq. (16.16) for $\hat{\theta} = constant$ becomes

$$\nu(t+1|\hat{\theta}) = \frac{A_M(q^{-1})G(q^{-1})}{A_Q(q^{-1})P_0(q^{-1})L(q^{-1})}[\theta - \hat{\theta}]^T\phi_f(t) \tag{16.19}$$

with

$$\phi_f(t) = L(q^{-1})\phi(t) = [\alpha_f(t+1), \ldots, \alpha_f(t-n_{B_Q}+1), \\ \beta_f(t), \beta_f(t-1), \ldots, \beta_f(t-n_{A_Q})] \tag{16.20}$$

where $\alpha_f(t+1) = L(q^{-1})\alpha(t+1)$ and $\beta_f(t) = L(q^{-1})\beta(t)$.

Equation (16.19) will be used to develop the adaptation algorithms. When the parameters of $\hat{Q}$ evolve over time and neglecting the non-commutativity of the time-varying operators (16.19) transforms into[3]

$$\nu(t+1|\hat{\theta}(t+1)) = \frac{A_M(q^{-1})G(q^{-1})}{A_Q(q^{-1})P_0(q^{-1})L(q^{-1})}[\theta - \hat{\theta}(t+1)]^T \phi_f(t). \tag{16.21}$$

Equation (16.21) has the standard form for an *a posteriori* adaptation error given in Chap. 3 [3], which immediately suggests to use the following PAA:

$$\hat{\theta}(t+1) = \hat{\theta}(t) + F(t)\Phi(t)\nu(t+1); \tag{16.22}$$

$$\nu(t+1) = \frac{\nu^\circ(t+1)}{1+\Phi^T(t)F(t)\Phi(t)}; \tag{16.23}$$

$$F(t+1) = \frac{1}{\lambda_1(t)}\left[F(t) - \frac{F(t)\Phi(t)\Phi^T(t)F(t)}{\frac{\lambda_1(t)}{\lambda_2(t)} + \Phi^T(t)F(t)\Phi(t)}\right] \tag{16.24}$$

$$1 \geq \lambda_1(t) > 0;\ 0 \leq \lambda_2(t) < 2;\ F(0) > 0 \tag{16.25}$$

$$\Phi(t) = \phi_f(t), \tag{16.26}$$

where $\lambda_1(t)$ and $\lambda_2(t)$ allow to obtain various profiles for the adaptation gain $F(t)$ (see Sect. 4.3.4) in order to operate in *adaptive* regime (the trace of the adaptation gain matrix has a strictly positive inferior minimum value) or *self-tuning* regime (decreasing gain adaptation, the trace of the adaptation gain matrix goes asymptotically to zero). By taking $\lambda_2(t) \equiv 0$ and $\lambda_1(t) \equiv 1$, one gets a constant adaptation gain matrix and choosing $F = \gamma I, \gamma > 0$ one gets a scalar adaptation gain.

Several choices for the filter L will be considered, leading to different algorithms[4]:

Algorithm I: $L = G$
Algorithm IIa (FUPLR): $L = \hat{G}$
Algorithm IIb: $L = \frac{\hat{A}_M}{\hat{P}_0}\hat{G}$ with

$$\hat{P}_0 = \hat{A}_M S_0 - \hat{B}_M R_0 \tag{16.27}$$

[3]Nevertheless, exact algorithms can be developed taking into account the non-commutativity of the time-varying operators—see [3].

[4]One cannot use in practice Algorithm I as the true model of the secondary path is not known. Instead one can use Algorithm II with an estimation of the secondary path model.

Algorithm III (FUSBA):

$$L = \frac{\hat{A}_M}{\hat{P}}\hat{G} \tag{16.28}$$

with

$$\hat{P} = \hat{A}_Q(\hat{A}_M S_0 - \hat{B}_M R_0) = \hat{A}_Q \hat{P}_0, \tag{16.29}$$

where $\hat{A}_Q$ is an estimation of the denominator of the ideal QIIR filter computed on the basis of available estimates of the parameters of the filter $\hat{Q}$. For the Algorithm III several options for updating $\hat{A}_Q$ can be considered:

- Run Algorithm IIa or IIb for a certain time to get an estimate of $\hat{A}_Q$
- Run a simulation (using the identified models)
- Update $\hat{A}_Q$ at each sampling instant or from time to time using Algorithm III (after a short initialization horizon using Algorithm IIa or IIb)

When using a YKFIR structure $\hat{A}_Q \equiv 1$ and the implementation of Algorithm III is much simpler since $\hat{P} = \hat{P}_0$ is constant and known once the central controller is designed.

The following procedure is applied at each sampling time for *adaptive* or *self-tuning* operation:

1. Get the measured image of the disturbance $\hat{y}(t+1)$, the measured residual error $e^{\circ}(t+1)$ and compute $\nu^{\circ}(t+1) = -e^{\circ}(t+1)$.
2. Compute $\phi(t)$ and $\phi_f(t)$ using (16.17) and (16.20).
3. Estimate the parameter vector $\hat{\theta}(t+1)$ using the parametric adaptation algorithm (16.22) through (16.26).
4. Compute (using (16.7)) and apply the control:

$$\begin{aligned}\hat{u}(t+1) = &-(\hat{A}_Q(t+1,q^{-1})S_0)^*\hat{u}(t) + \hat{A}_Q(t+1,q^{-1})R_0\hat{y}(t+1)\\ &+ \hat{B}_Q(t+1,q^{-1})\left(B_M^*\hat{u}(t) - A_M\hat{y}(t+1)\right).\end{aligned} \tag{16.30}$$

16.4 Analysis of the Algorithms

16.4.1 The Perfect Matching Case

Stability of the Algorithms

For Algorithms I, IIa, IIb and III, the equation for the *a posteriori* adaptation error has the form:

$$\nu(t+1) = H(q^{-1})[\theta - \hat{\theta}(t+1)]^T \Phi(t), \tag{16.31}$$

where

$$H(q^{-1}) = \frac{A_M(q^{-1})G(q^{-1})}{A_Q(q^{-1})P_0(q^{-1})L(q^{-1})}, \quad \Phi = \phi_f. \tag{16.32}$$

Equation (16.31) has the standard form considered in Chap. 4 and therefore neglecting the non-commutativity of time-varying operators, one can conclude that the system is asymptotically stable for all initial conditions $\hat{\theta}(0)$, $\nu^\circ(0)$, $F(0)$, provided that

$$H'(z^{-1}) = H(z^{-1}) - \frac{\lambda_2}{2}, \max_t [\lambda_2(t)] \leq \lambda_2 < 2 \tag{16.33}$$

is a SPR transfer function.

This result can be particularized for the case of FIR Youla–Kučera adaptive compensators by taking in account that in this case $A_Q = 1$ in (16.32).

Remark 1 Using Algorithm III and taking into account (16.28), the stability condition for $\lambda_2 = 1$ can be transformed into [4, 5]:

$$\left| \left(\frac{A_M}{\hat{A}_M} \cdot \frac{\hat{A}_Q}{A_Q} \cdot \frac{\hat{P}_0}{P_0} \cdot \frac{G}{\hat{G}} \right)^{-1} - 1 \right| < 1 \tag{16.34}$$

for all ω. This roughly means that it always holds provided that the estimates of A_M, A_Q, P_0, and G are close to the true values (i.e., $H(e^{-j\omega})$ in this case is close to a unit transfer function).

Effect of the Measurement Noise

The situation is similar to that encountered for the standard IIR adaptive feedforward compensator and the results are similar. The parameters of the estimated feedforward compensator will converge to the same value as for the case without noise.

16.4.2 The Case of Non-perfect Matching

If $\hat{Q}(t, q^{-1})$ does not have the appropriate dimension there is no chance to satisfy the perfect matching condition. Two questions are of interest in this case:

1. The boundedness of the residual error.
2. The bias distribution in the frequency domain.

for the first point the answer is the same as for the IIR adaptive feedforward compensator (see Chap. 15), i.e., that the residual error will be bounded under similar conditions given in Sect. 15.4.2.

Bias Distribution

Following the same pathway as in Sect. 15.4.2 and using (16.14), the bias distribution (for Algorithm III) will be given by

$$\hat{\theta}^* = \arg\min_{\hat{\theta}} \int_{-\pi}^{\pi} \left[\left| D(e^{-j\omega}) + \frac{\hat{N}(e^{-j\omega})G(e^{-j\omega})}{1 - \hat{N}(e^{-j\omega})M(e^{-j\omega})} \right|^2 \phi_w(\omega) + \phi_\eta(\omega) \right] d\omega \tag{16.35}$$

where ϕ_w and ϕ_η are the spectral densities of the disturbance $w(t)$ and of the measurement noise. Taking into account Eq. (16.14), one obtains

$$\hat{\theta}^* = \arg\min_{\hat{\theta}} \int_{-\pi}^{\pi} \left[\left| \frac{GA_M^2}{P_0} \right|^2 \left| \frac{B_Q}{A_Q} - \frac{\hat{B}_Q}{\hat{A}_Q} \right|^2 \phi_w(\omega) + \phi_\eta(\omega) \right] d\omega. \tag{16.36}$$

From (16.36) one concludes that a good approximation of the Q filter will be obtained in the frequency region where ϕ_w is significant and where G has a high gain (usually G should have high gain in the frequency region where ϕ_w is significant in order to counteract the effect of $w(t)$). Nevertheless, the quality of the estimated $\hat{Q}$ filter will be affected also by the transfer function $\frac{GA_M^2}{P_0}$.

16.4.3 Relaxing the Positive Real Condition

Like for the IIR adaptive feedforward compensator the strictly positive real condition for stability (and convergence) can be relaxed if relatively small adaptation gains are used (slow adaptation). The algorithms will work in general provided that the weighted energy associated to the observation vector is in average positive, which allows in fact that the SPR condition be violated in some limited frequency regions. See analysis given in Sect. 15.4.3.

It was observed that despite satisfaction of condition (15.52) which will assure the stability of the system, attenuation is not very good in the frequency regions where the positive real condition (16.33) is violated.

Without doubt, the best approach for relaxing the SPR conditions is to use Algorithm III (given in (16.28)) instead of Algorithm IIa or IIb. This is motivated by (16.34). As it will be shown experimentally, this algorithm gives the best results.

16.4.4 Summary of the Algorithms

Table 16.1 summarizes the structure of the algorithms and the stability and convergence conditions for the algorithms presented in Chap. 15 and this chapter with matrix

Table 16.1 Comparison of algorithms for adaptive feedforward compensation in AVC with mechanical coupling

	YKIIR (in this chapter)	YKFIR (Chap. 16, [1])	IIR (Chap. 15, [7])	YKIIR	YKFIR	IIR (FUSBA)
	Matrix adaptation gain			Scalar adaptation gain		
$\hat{\theta}(t+1) =$	$\hat{\theta}(t) + F(t)\psi(t)\frac{\nu^{\circ}(t+1)}{1+\psi^T(t)F(t)\psi(t)}$			$\hat{\theta}(t) + \gamma(t)\psi(t)\frac{\nu^{\circ}(t+1)}{1+\gamma(t)\psi^T(t)\psi(t)}$		
Adapt. gain	$F(t+1)^{-1} = \lambda_1(t)F(t) + \lambda_2(t)\psi(t)\psi^T(t)$ $0 \le \lambda_1(t) < 1,\ 0 \le \lambda_2(t) < 2,\ F(0) > 0$			$\gamma(t) > 0$		
Adaptive	Decr. gain and const. trace			$\gamma(t) = \gamma = const$		
Self-tuning	$\lambda_2 = const.,\quad \lim_{t\to\infty} \lambda_1(t) = 1$			$\sum_{t=1}^{\infty} \gamma(t) = \infty,\quad \lim_{t\to\infty} \gamma(t) = 0$		
$\hat{\theta}(t) =$	$[\hat{b}_0^Q, \ldots, \hat{a}_1^Q, \ldots]$	$[\hat{b}_0^Q, \ldots]$	$[-\hat{s}_1(t), \ldots, \hat{r}_0(t), \ldots]$	$[\hat{b}_0^Q, \ldots, \hat{a}_1^Q, \ldots]$	$[\hat{b}_0^Q, \ldots]$	$[-\hat{s}_1(t), \ldots, \hat{r}_0(t), \ldots]$
$\phi^T(t) =$	$[\alpha(t+1), \ldots, \beta(t), \ldots]$ $\alpha(t) = B_M\hat{u}(t) - A_M\hat{y}(t)$ $\beta(t) = R_0\hat{y}(t) - S_0\hat{u}(t)$	$[\alpha(t+1), \ldots]$ $\alpha(t) = B_M\hat{u}(t) - A_M\hat{y}(t)$	$[-\hat{u}(t), \ldots$ $\hat{y}(t+1), \ldots]$	$[\alpha(t+1), \ldots, \beta(t), \ldots]$ $\alpha(t) = B_M\hat{u}(t) - A_M\hat{y}(t)$ $\beta(t) = R_0\hat{y}(t) - S_0\hat{u}(t)$	$[\alpha(t+1), \ldots]$ $\alpha(t) = B_M\hat{u}(t) - A_M\hat{y}(t)$	$[-\hat{u}(t), \ldots,$ $\hat{y}(t+1), \ldots]$
$\hat{P} =$	$\hat{A}_Q(\hat{A}_M S_0 - \hat{B}_M R_0)$	$\hat{A}_M S_0 - \hat{B}_M R_0$	$\hat{A}_M \hat{S} - \hat{B}_M \hat{R}$	$\hat{A}_Q(\hat{A}_M S_0 - \hat{B}_M R_0)$	$\hat{A}_M S_0 - \hat{B}_M R_0$	$\hat{A}_M \hat{S} - \hat{B}_M \hat{R}$
$P =$	$A_Q(A_M S_0 - B_M R_0)$	$A_M S_0 - B_M R_0$	$A_M \hat{S} - B_M \hat{R}$	$A_Q(A_M S_0 - B_M R_0)$	$A_M S_0 - B_M R_0$	$A_M \hat{S} - B_M \hat{R}$
$\psi(t) =$	$L\phi(t);\quad L_2 = \hat{G};\ L_3 = \frac{\hat{A}_M}{\hat{P}}\hat{G}$			$L\phi(t);\quad L_2 = \hat{G};\ L_3 = \frac{\hat{A}_M}{\hat{P}}\hat{G}$		
Stability condition	$\frac{A_M G}{PL} - \frac{\lambda}{2} = SPR\quad (\lambda = \max \lambda_2(t))$			$\frac{A_M G}{PL} = SPR$		
Conv. condition	$\frac{A_M G}{PL} - \frac{\lambda}{2} = SPR\quad (\lambda = \lambda_2)$			$\frac{A_M G}{PL} = SPR$		

and scalar adaptation gain for IIR Youla–Kučera feedforward compensators, for FIR Youla–Kučera feedforward compensators and for IIR adaptive feedforward compensators. The original references for these algorithms are [2, 6, 7]. These algorithms take also into account the internal positive feedback.

It was not possible to give in Table 16.1 all the options for the adaptation gain. Nevertheless, basic characteristics for adaptive operation (non-vanishing adaptation gain) and self-tuning operation (vanishing adaptation gain) have been provided.[5]

The parametric adaptation algorithms can be implemented using the UD factorization (see Appendix B) [3].[6]

16.5 Experimental Results

The same active distributed flexible mechanical structure as in Chap. 15 has been used for experimental validation. Identification of the system has been detailed in Chap. 6.

16.5.1 The Central Controllers and Comparison Objectives

Two central controllers have been used to test IIRYK adaptive feedforward compensators. The first (PP) has been designed using a pole placement method tuned for the case of positive feedback systems. Its main objective is to stabilize the internal positive feedback loop. The end result was a controller of orders $n_{R_0} = 15$ and $n_{S_0} = 17$. The second (H_∞) is a reduced order H_∞ controller with $n_{R_0} = 19$ and $n_{S_0} = 20$ from [9].[7] For the design of the H_∞ controller, the knowledge of the primary path and of the PSD of the disturbance is mandatory (which is not necessary for the design of PP controller).

The H_∞ controller assures a global attenuation of 14.70 dB while the PP controller achieves only 4.61 dB.

16.5.2 Broad-band Disturbance Rejection Using Matrix Adaptation Gain

Broad-band disturbance rejection capabilities using the two Youla–Kučera parameterizations with IIR and FIR filters described in column 2 and 3 of Table 16.1 are

---

[5]Convergence analysis in a stochastic environment can be applied only for vanishing adaptation gains.

[6]An array implementation as in [8] can be also considered.

[7]The orders of the initial H_∞ controller were: $n_{R_{H\infty}} = 70$ and $n_{S_{H\infty}} = 70$.

evaluated in this subsection. A comparison with the algorithm given in column 4 is made (see also [7]). For most of the experiments, the complexity of the IIRYK filter was $n_{B_Q} = 3$ and $n_{A_Q} = 8$, therefore having 12 parameters in the adaptation algorithm according to Eq. (16.2). For the FIRYK parametrization, an adaptive filter of order $n_{B_Q} = 31$ and $n_{A_Q} = 0$ (32 parameters) has been used. These values do not allow to verify the "perfect matching condition".

Two modes of operation can be considered: *adaptive* operation and *self-tuning* operation.

For reason of space only the experimental results in adaptive operation will be presented. Algorithms IIa and III have been used with *decreasing adaptation gain* ($\lambda_1(t) = 1$, $\lambda_2(t) = 1$) combined with a *constant trace* adaptation gain. For IIRYK the adaptation has been done starting with an initial gain of 0.02 (initial trace = initial gain × number of adjustable parameters, thus 0.24) and using a constant trace of 0.02. For FIRYK an initial gain of 0.05 (initial trace $0.05 \times 32 = 1.6$) and constant trace 0.1 have been used.

The experiments have been carried out by first applying the disturbance and then starting the adaptive feedforward compensation after 50 s. If not otherwise specified, the results which will be presented have been obtained with the H_∞ central controller. In the case of the IIRYK parametrization using Algorithm III, the filtering of the regressor is done adaptively. The last stable estimation of $A_Q(q^{-1})$ is used in (16.29).

Time domain results using IIRYK with Algorithms IIa and III are shown in Figs. 16.2 and 16.3 respectively. It can be seen that Algorithm III provides better performance than Algorithm IIa and this can be explained by a better approximation of the positive real condition (see discussion in Sect. 16.4.3). Figure 16.4 shows the evolution of the residual acceleration with the FIRYK adaptive compensator using Algorithm III [6].

The power spectral density of the residual acceleration (after adaptation transient is finished) for the considered algorithms are given in Fig. 16.5. The final attenuation

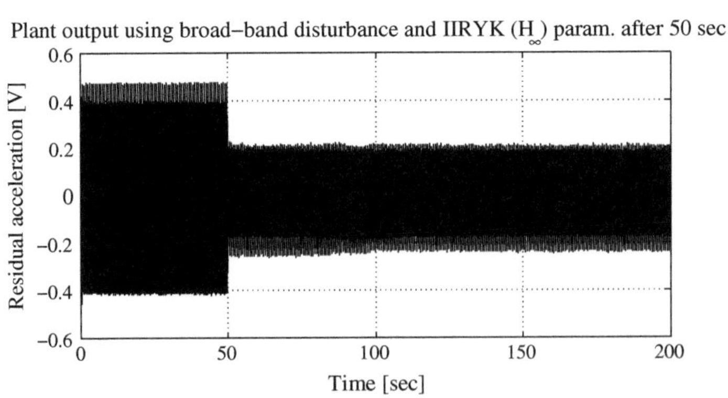

Fig. 16.2 Real-time residual acceleration obtained with the IIR Youla–Kučera parametrization ($n_{B_Q} = 3, n_{A_Q} = 8$) using Algorithm IIa with matrix adaptation gain and the H_∞ central controller

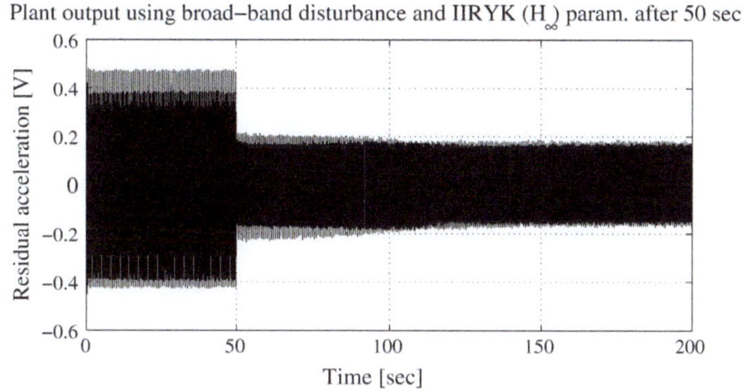

Fig. 16.3 Real-time residual acceleration obtained with the IIR Youla–Kučera parametrization ($n_{B_Q} = 3, n_{A_Q} = 8$) using Algorithm III with matrix adaptation gain and the H_∞ central controller

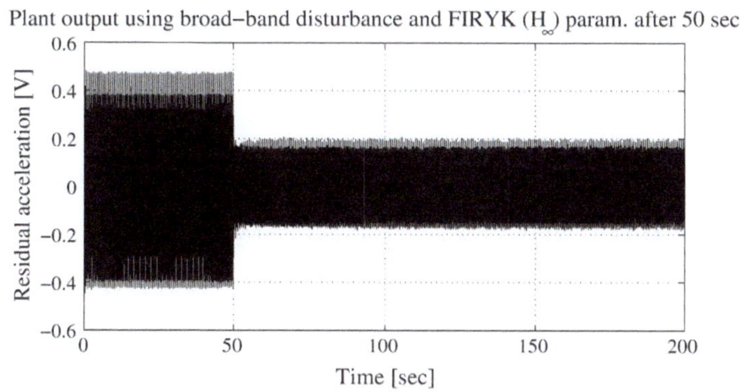

Fig. 16.4 Real-time results obtained with the FIR Youla–Kučera parametrization ($n_Q = 31$) using Algorithm III with matrix adaptation gain and the H_∞ central controller

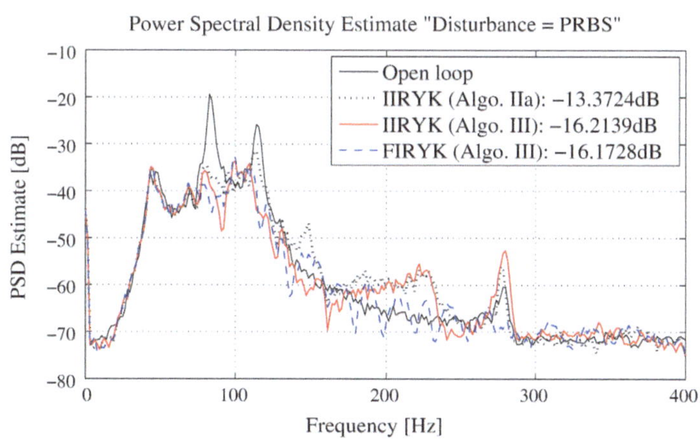

Fig. 16.5 Power spectral densities of the residual acceleration in open-loop, with IIRYK ($n_{B_Q} = 3$, $n_{A_Q} = 8$) and with FIRYK ($n_Q = 31$) using the H_∞ central controller (experimental)

Table 16.2 Influence of the number of the parameters upon the global attenuation

Total no. param.	0	8	16	32	40
IIR (dB)	–			16.49	16.89
FIRYK/H_∞ (dB)	14.70	15.40	15.60	16.52	16.03
FIRYK/PP (dB)	4.61	14.69	15.89	15.7	15.33
IIRYK/H_∞ (dB)	14.70	16.53	16.47		
IIRYK/PP (dB)	4.61	15.53	16.21		

provided by IIRYK using Algorithm III (16.21 dB) is better than that provided by IIRYK using Algorithm IIa (13.37 dB) and slightly better than that provided by using FIRYK with Algorithm III (16.17 dB) which uses significantly more adjustable parameters (32 instead of 12). Nevertheless, the adaptation transient is slightly more rapid for FIRYK.

An evaluation of the influence of the number of parameters upon the global attenuation for each algorithm is shown in Table 16.2. Each line gives global attenuation results for a certain algorithm (IIR/FIRYK/IIRYK). The central controller is also specified for the case of Youla–Kučera parametrized filters. The values for global attenuation are given in dB. The column headers give the number of coefficients.

One can say that a reduction of the number of adjustable parameters by a factor of (at least) 2 is obtained in the case of IIRYK/H_∞ with respect to FIRYK/H_∞ and IIR adaptive feedforward compensators for approximately the same level of performance (compare IIRYK/H_∞ with 8 parameters with the FIRYK/H_∞ with 16 parameters and the IIRYK/H_∞ with 16 parameters with the FIRYK/H_∞ with 32 parameters and with the IIR with 32 parameters). It can be noticed that the IIRYK/H_∞ is less sensitive than FIRYK/H_∞ with respect to the performances of the model-based central controller.

To verify the adaptive capabilities of the two parameterizations (FIRYK and IIRYK) with respect to changes in the characteristics of the disturbance, a narrow-band disturbance has been added after 1400 s of experimentation. This has been realized by using a sinusoidal signal of 150 Hz. Power spectral density estimates are shown in Fig. 16.6 for the IIRYK parametrization and in Fig. 16.7 for the FIRYK parametrization. Better results are obtained with the IIRYK parametrization and they are comparable with those obtained for IIR adaptive feedforward compensators (see Chap. 15, Fig. 15.10).

16.5.3 *Broad-band Disturbance Rejection Using Scalar Adaptation Gain*

The scalar adaptation gain algorithms given in Table 16.1, columns 5 and 6 have also been tested on the AVC system.

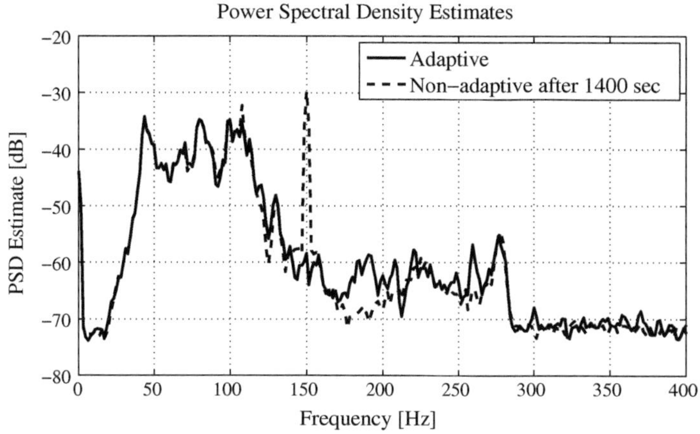

Fig. 16.6 Power spectral densities of the residual acceleration when an additional sinusoidal disturbance is added (Disturbance = PRBS + sinusoid) and the IIRYK parametrization is used

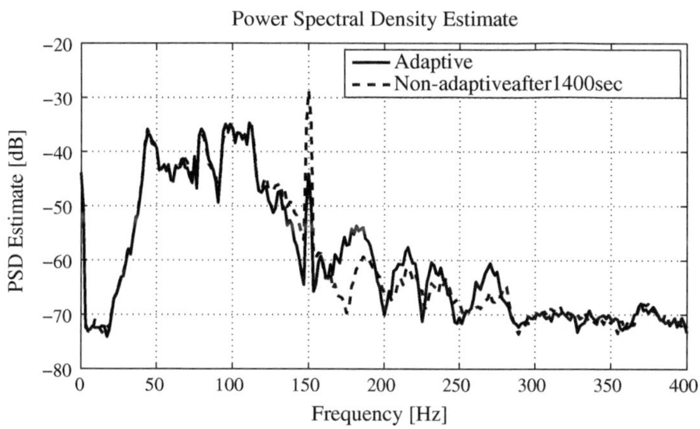

Fig. 16.7 Power spectral densities of the residual acceleration when an additional sinusoidal disturbance is added (Disturbance = PRBS + sinusoid) and the FIRYK parametrization is used

In the *adaptation* regime, a constant adaptation gain of 0.001 has been used for both parameterizations, as in [7] (see also Table 16.1). This corresponds to a constant trace of 0.012 for the IIRYK and 0.032 for the FIRYK (taking into account the number of adapted parameters). Figure 16.8 shows the adaptation transient for the scalar version of the IIRYK parametrization using Algorithm III. Surprisingly, the performances are close to those obtained with a matrix adaptation gain (a similar observation has been made in [7, Fig. 14]). Figure 16.9 shows the adaptation transient for the FIRYK parametrization using a scalar adaptation gain. It can be seen that the transient performances are slightly better for the IIRYK.

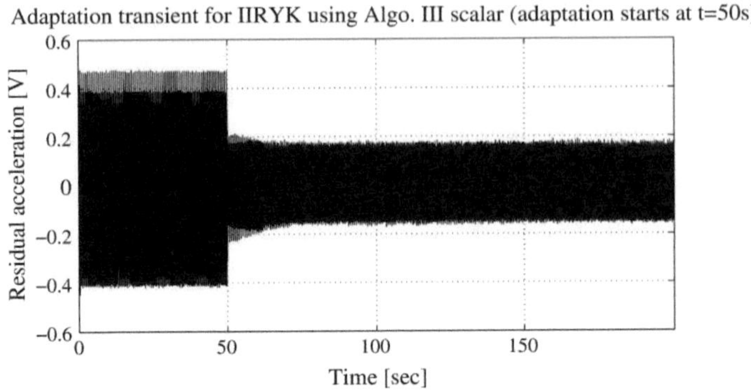

Fig. 16.8 Real-time residual acceleration obtained with the IIR Youla–Kučera parametrization ($n_{B_Q} = 3, n_{A_Q} = 8$) using Algorithm III with scalar adaptation gain and the H_∞ central controller

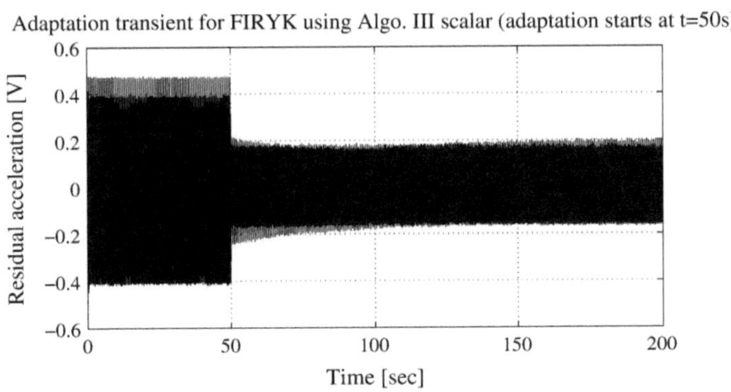

Fig. 16.9 Real-time residual acceleration obtained with the FIR Youla–Kučera parametrization ($n_Q = 31$) using Algorithm III with scalar adaptation gain and the H_∞ central controller

In terms of global attenuation, an IIRYK feedforward compensator with 12 parameters ($n_{B_Q} = 3, n_{A_Q} = 8$) gives a global attenuation of 16.45 dB and a FIRYK feedforward compensator with 32 parameters ($n_Q = 31$) achieves a global attenuation of 15.92 dB. This significant reduction in the number of adjustable parameters for the same level of performance when using IIRYK feedforward compensators, holds also if one compare IIRYK feedforward compensators with IIR feedforward compensators (with scalar adaptation gain). See Chap. 15 and [2, 7].

16.6 Comparison of the Algorithms

The Number of Adjustable Parameters

The main advantage of the IIRYK adaptive feedforward compensators compared with FIRYK adaptive compensators is that they require a significantly lower number of adjustable parameters for a given level of performance (a reduction by a factor of 2 in the application presented). This is without doubt a major practical advantage in terms of implementation complexity. A slight reduction of the number of adjustable parameters is also obtained with respect to IIR adaptive feedforward compensators.

The Poles of the Internal Positive Closed-Loop

For IIR adaptive feedforward compensators provided that the SPR condition for stability is satisfied, the poles of the internal "positive" loop will be asymptotically stable but they can be very close to the unit circle. For FIRYK, the poles of the internal positive feedback loop are assigned by the central stabilizing controller and they remain unchanged under the effect of adaptation. For IIRYK, part of the poles of the internal positive feedback loop are assigned by the central stabilizing controller but there are additional poles corresponding to $\hat{A}_Q$. These poles will be asymptotically inside the unit circle if the positive real condition for stability is satisfied but they can be very close to the unit circle (at least theoretically). Nevertheless, if one likes to impose that these poles lie inside a circle of a certain radius, this can be easily achieved by using parameter adaptation algorithms with "projections" [3, 10].

Implementation of the Filter for Algorithm III

For IIRYK adaptive compensators, one has to run first Algorithm IIa or IIb over a short-time horizon in order to get an estimate of $\hat{A}_Q$ for implementing the appropriate filter. A similar procedure has to be used also for IIR adaptive compensators (see Chap. 15 and [7]). For the IIRYK structure, the filter can be continuously improved by updating at each step the estimation of $\hat{A}_Q$ in the filter. Such a procedure is more difficult to apply to the IIR structure since the estimated closed-loop poles have to be computed at each step based on current estimates of the feedforward compensator's parameters and the knowledge of the reverse path $M(q^{-1})$. For FIRYK this initialization procedure is not necessary since the poles of the internal positive feedback loop remain unchanged under the effect of adaptation and a good estimation is provided by the knowledge of the central stabilizing compensator and of the model of the reverse path.

Initial Model-Based Design Compensator

Since the system as well as the initial characteristics of the disturbance can be identified, a model-based design of an initial feedforward compensator can be done. For a FIRYK or an IIRYK adaptive feedforward compensator, any model-based designed compensator can be used as the central controller (no matter what is its dimension). Its performances will be enhanced by the adaptation of the Q-parameters. Nevertheless, for IIR adaptive feedforward compensators the initial model-based designed compensator should have the same structure (number of parameters) as the adaptive structure.

Influence of the Initial Stabilizing Controller

The performances of IIRYK adaptive compensator are less sensitive than those of FIRYK adaptive compensator with respect to the performances of the initial model-based stabilizing controller.

16.7 Concluding Remarks

- Use of the Youla–Kučera parametrization for adaptive feedforward compensation in the presence of the internal positive feedback is justified by the separation of the stabilization of the internal positive feedback loop from the optimization of the feedforward compensator parameters.
- IIR or FIR Youla–Kučera structures can be used for the feedforward compensator.
- IIR Youla–Kučera compensator structure leads to the minimal number of adjustable parameters with respect to FIR Youla–Kučera feedforward compensator and IIR feedforward compensator.
- Youla–Kučera structure for the feedforward compensator allows to use a central controller of any order independently of the number of adjustable parameters.

16.8 Notes and References

The basis of the Youla–Kučera parametrization is discussed in [11]. Linear feedforward compensators using Youla–Kučera parametrization are presented in [12]. Youla–Kučera based adaptive feedforward compensator using orthonormal basis functions is considered in [13]. The orthonormal basis functions are presented in [14].

References

1. Landau ID, Airimitoaie TB, Alma M (2012) A Youla–Kučera parametrized adaptive feedforward compensator for active vibration control with mechanical coupling. Automatica 48(9):2152–2158. doi:10.1016/j.automatica.2012.05.066. http://www.sciencedirect.com/science/article/pii/S0005109812002397
2. Landau ID, Airimitoaie TB, Alma M (2013) IIR Youla–Kučera parameterized adaptive feedforward compensators for active vibration control with mechanical coupling. IEEE Trans Control Syst Technol 21(3):765–779
3. Landau ID, Lozano R, M'Saad M, Karimi A (2011) Adaptive control, 2nd edn. Springer, London
4. Ljung L, Söderström T (1983) Theory and practice of recursive identification. The MIT Press, Cambridge
5. Ljung L (1977) On positive real transfer functions and the convergence of some recursive schemes. IEEE Trans Autom Control AC-22:539–551

6. Landau I, Airimitoaie T, Alma M (2011) A Youla–Kučera parametrized adaptive feedforward compensator for active vibration control. In: Proceedings of the 18th IFAC world congress, Milano, Italy, pp 3427–3432
7. Landau I, Alma M, Airimitoaie T (2011) Adaptive feedforward compensation algorithms for active vibration control with mechanical coupling. Automatica 47(10):2185–2196. doi:10.1016/j.automatica.2011.08.015
8. Montazeri P, Poshtan J (2010) A computationally efficient adaptive IIR solution to active noise and vibration control systems. IEEE Trans Autom Control AC-55:2671–2676
9. Alma M, Martinez J, Landau I, Buche G (2012) Design and tuning of reduced order H_∞ feedforward compensators for active vibration control. IEEE Trans Control Syst Technol 20(2):554–561. doi:10.1109/TCST.2011.2119485
10. Goodwin G, Sin K (1984) Adaptive filtering prediction and control. Prentice Hall, New Jersy
11. Anderson B (1998) From Youla–Kučera to identification, adaptive and nonlinear control. Automatica 34(12):1485–1506. doi:10.1016/S0005-1098(98)80002-2. http://www.sciencedirect.com/science/article/pii/S0005109898800022
12. Tay TT, Mareels IMY, Moore JB (1997) High performance control. Birkhäuser, Boston
13. Zeng J, de Callafon R (2006) Recursive filter estimation for feedforward noise cancellation with acoustic coupling. J Sound Vib 291(3–5):1061–1079. doi:10.1016/j.jsv.2005.07.016
14. Heuberger P, Van den Hof P, Bosgra O (1995) A generalized orthonormal basis for linear dynamical systems. IEEE Trans Autom Control 40(3):451–465. doi:10.1109/9.376057

Appendix A
Generalized Stability Margin and Normalized Distance Between Two Transfer Functions

A.1 Generalized Stability Margin

In Sect. 7.2.4, the *modulus margin* has been introduced. It corresponds to the minimum distance between the Nyquist plot of the open-loop transfer function and the critical point $[-1,\ j0]$. The modulus margin has the expression

$$\Delta M = \left(\left| S_{yp}\left(e^{-j\omega}\right) \right|_{\max_{\omega}} \right)^{-1} = \| S_{yp}\left(e^{-j\omega}\right) \|_{\infty}^{-1},\ \forall\, 0 \leq \omega \leq \pi f_s \tag{A.1}$$

Stability of the closed-loop system requires that all the sensitivity functions be asymptotically stable. Furthermore it was shown in Sect. 7.2.5 that the uncertainties tolerated on the plant model depend upon the sensitivity functions. More specifically, the admissible uncertainties will be smaller as the maximum of the modulus of the various sensitivity functions grows.

One can ask if it is not possible to give a global characterization of the stability margin of a closed-loop system and its robustness, taking simultaneously into account all the four sensitivity functions. This problem can be viewed as the generalization of the *modulus margin*.

Denoting the controller by

$$K = \frac{R(z^{-1})}{S(z^{-1})} \tag{A.2}$$

and the transfer function of the plant model by

$$G = \frac{z^{-d}B(z^{-1})}{A(z^{-1})} \tag{A.3}$$

one defines for the closed-loop system (K, G) the matrix of sensitivity functions $(z = e^{j\omega})$

$$T(j\omega) = \begin{vmatrix} S_{yr}(e^{-j\omega}) & S_{yv}(e^{-j\omega}) \\ -S_{up}(e^{-j\omega}) & S_{yp}(e^{-j\omega}) \end{vmatrix} \tag{A.4}$$

I.D. Landau et al., *Adaptive and Robust Active Vibration Control*,
Advances in Industrial Control, DOI 10.1007/978-3-319-41450-8

where S_{yr}, S_{yv}, S_{up}, and S_{yp} represent

- the complementary sensitivity function

$$S_{yr}(z^{-1}) = \frac{KG}{1+KG} = \frac{z^{-d}B(z^{-1})R(z^{-1})}{P(z^{-1})},$$

- the output sensitivity function with respect to an input disturbance

$$S_{yv}(z^{-1}) = \frac{G}{1+KG} = \frac{z^{-d}B(z^{-1})S(z^{-1})}{P(z^{-1})};$$

- the output sensitivity function

$$S_{yp}(z^{-1}) = \frac{1}{1+KG} = \frac{A(z^{-1})S(z^{-1})}{P(z^{-1})};$$

- the input sensitivity function

$$S_{up}(z^{-1}) = -\frac{K}{1+KG} = -\frac{A(z^{-1})R(z^{-1})}{P(z^{-1})};$$

where

$$P(z^{-1}) = A(z^{-1})S(z^{-1}) + z^{-d}B(z^{-1})R(z^{-1}). \tag{A.5}$$

defines the poles of the closed-loop.

Similarly to the modulus margin, one defines the *generalized stability margin* as

$$b(K,G) = \begin{cases} \left(\left|T(e^{-j\omega})\right|_{\max\limits_{\omega}}\right)^{-1} = \|T(e^{-j\omega})\|_{\infty}^{-1} & \text{if } (K,G) \text{ is stable} \\ 0 & \text{if } (K,G) \text{ is unstable} \end{cases} \tag{A.6}$$

where:

$$\left|T(e^{-j\omega})\right|_{\max\limits_{\omega}} = \left|\bar{\sigma}(e^{-j\omega})\right|_{\max\limits_{\omega}} = \|T(e^{-j\omega})\|_{\infty},\ \forall\, \omega \in [0,\ \pi f_s] \tag{A.7}$$

In Eq. (A.7), $\bar{\sigma}(e^{-j\omega})$ is the largest singular value of $T(e^{-j\omega})$ computed using singular value decomposition [1, 2].

The generalized stability margin can be computed with the function *smarg.m* from the toolbox REDUC® [3].[1]

As the value of $b(K,G)$ decreases, the closed-loop system will be close to instability and it will be less robust with respect to the variations (or uncertainties) of the plant nominal transfer function.

---

[1] To be downloaded from the book website.

A.2 Normalized Distance Between Two Transfer Functions

Consider a transfer function G. Let denotes the number of unstable zeros by n_{z_i} and the number of unstable poles by n_{p_i}. The number of encirclements of the origin of the complex plane is given by

$$wno(G) = n_{z_i}(G) - n_{p_i}(G) \tag{A.8}$$

(positive value = counter clockwise encirclements; negative value = clockwise encirclements). It is possible to compare two transfer functions G_1 and G_2 only if they satisfy the following property:

$$wno(1 + G_2^* G_1) + n_{p_i}(G_1) - n_{p_i}(G_2) - n_{p_1}(G_2) = 0 \tag{A.9}$$

where G_2^* is the complex conjugate of G_2 and $n_{p_1}(G_2)$ is the number of poles of G_2 located on the unit circle.[2]

The *normalized distance* between two transfer functions satisfying the property of Eq. A.9 is called the *Vinnicombe distance* or ν-gap [4].

Let define the normalized difference between two transfer functions $G_1(e^{-j\omega})$ and $G_2(e^{-j\omega})$ as

$$\Psi\left(G_1(e^{-j\omega}), G_2(e^{-j\omega})\right) = \frac{G_1(e^{-j\omega}) - G_2(e^{-j\omega})}{\left(1 + \left|G_1(e^{-j\omega})\right|^2\right)^{\frac{1}{2}} \left(1 + \left|G_2(e^{-j\omega})\right|^2\right)^{\frac{1}{2}}} \tag{A.10}$$

The normalized distance (Vinnicombe distance) is defined by

$$\delta_\nu(G_1, G_2) = |\Psi(G_1, G_2)|_{\underset{\omega}{\max}} = \|\Psi(G_1, G_2)\|_\infty, \forall \quad \omega \in [0,\ \pi f_s] \tag{A.11}$$

One observes immediately from the structure of Ψ that

$$0 \le \delta_\nu(G_1, G_2) < 1. \tag{A.12}$$

If the condition of Eq. A.9 is not satisfied, by definition

$$\delta_\nu(G_1, G_2) = 1. \tag{A.13}$$

The Vinnicombe distance can be computed with the function *vgap.m* from the toolbox REDUC® [3].[3]

---

[2]The condition of Eq. A.9 is less restrictive than the condition used in Sect. 7.2.5 where two transfer functions with the same number of unstable poles and with the same number of encirclements of the origin have been considered.

[3]To be downloaded from the book website.

A.3 Robust Stability Condition

Using the *generalized stability margin* and the *Vinnicombe distance* between two transfer functions, one can express a robust stability condition (sufficient condition) for a controller K designed on the basis of the nominal model G_1 as follows. *Controller K which stabilizes model G_1 will also stabilize model G_2 if*

$$\delta_\nu(G_1, G_2) \leq b(K, G_1). \tag{A.14}$$

This condition can be replaced by a less restrictive condition, but which should be verified at all frequencies[4]:

$$|\Psi\,(G_1, G_2)| \leq \left|T(e^{-j\omega})\right|^{-1}, \;\forall\, \omega \in [0,\; \pi f_s]\,. \tag{A.15}$$

A.4 Notes and References

The original reference for *Vinnicombe distance* (ν-gap) and *generalized stability margin* is [4]. For a good pedagogical presentation, but with extensive use of the H_∞ norm, see [1, 2].

These concepts have been very useful for validation of reduced order controllers (see Chap. 9) and of the models identified in closed-loop operation (see Chap. 8).

References

1. Landau I, Zito G (2005) Digital control systems - design, identification and implementation. Springer, London
2. Zhou K, Doyle J (1998) Essentials of robust control. Prentice-Hall International, Upper Saddle River
3. Adaptech, Adaptech, 4 rue de la Tour de l'Eau, St. Martin dHères, France: REDUC® Controller order reduction by closed-loop identification (Toolbox for MATLAB®) (1999)
4. Vinnicombe G (1993) Frequency domain uncertainty and the graph topology. IEEE Trans Autom Control 38(9):1371–1383. doi:10.1109/9.237648

[4]This condition has to be compared with the conditions given in Sect. 7.2.5 (Eqs. (7.53)–(7.55)). Equation (A.15) can be interpreted as a generalization of these conditions.

Appendix B
Implementation of the Adaptation Gain Updating—The U-D Factorization

The adaptation gain equation is sensitive to round-off errors. This problem is comprehensively discussed in [1] where a U-D factorization has been developed to ensure the numerical robustness of the PAA. To this end, the adaptation gain matrix is rewritten as follows

$$F(t) = U(t)D(t)U^T(t) \tag{B.1}$$

where $U(t)$ is an upper triangular matrix with all diagonal elements equal to 1 and $D(t)$ is a diagonal matrix. This allows the adaptation gain matrix to remain positive definite so that the rounding errors do not affect the solution significantly.

Let

$$G(t) = D(t)V(t) \tag{B.2}$$
$$V(t) = U^T(t)\phi_f(t) \tag{B.3}$$
$$\beta(t) = 1 + V^T(t)G(t) \tag{B.4}$$
$$\delta(t) = \frac{\lambda_1(t)}{\lambda_2(t)} + V^T(t)G(t) \tag{B.5}$$

Define:

$$\Gamma(t) = \frac{U(t)G(t)}{\beta(t)} = \frac{F(t)\phi_f(t)}{1 + \phi_f^T(t)F(t)\phi_f(t)} \tag{B.6}$$

The U-D factorization algorithm of the parameter adaptation gain is given below.

Initialize U(0) and D(0) at time $t = 0$, this provides the initial value of the adaptation gain matrix $F(0) = U(0)D(0)U^T(0)$. At time $t + 1$, determine the adaptation gain $\Gamma(t)$ while updating $D(t + 1)$ and $U(t + 1)$ by performing the steps 1–6.

1. Compute $V(t) = U^T(t)\phi_f(t)$, $G(t) = D(t)V(t)$, $\beta_0 = 1$ and $\delta_0 = \frac{\lambda_1(t)}{\lambda_2(t)}$
2. For $j = 1$ to n_p (number of parameters) go through the steps 3–5
3. Compute

I.D. Landau et al., *Adaptive and Robust Active Vibration Control*,
Advances in Industrial Control, DOI 10.1007/978-3-319-41450-8

$$\beta_j(t) = \beta_{j-1}(t) + V_j(t)G_j(t)$$
$$\delta_j(t) = \delta_{j-1}(t) + V_j(t)G_j(t)$$
$$D_{jj}(t+1) = \frac{\delta_{j-1}(t)}{\delta_j(t)\lambda_1(t)} D_{jj}(t)$$
$$\Gamma_j(t) = G_j(t)$$
$$M_j(t) = -\frac{V_j(t)}{\delta_{j-1}(t)}$$

4. If $j = 1$ then go to step 6 else for $i = 1$ to $j - 1$ go through step 5
5. Compute

$$U_{ij}(t+1) = U_{ij}(t) + \Gamma_i(t)M_j(t)$$
$$\Gamma_i(t) = \Gamma_i(t) + U_{ij}(t)\Gamma_j(t)$$

6. For $i = 1$ to n_p do

$$\Gamma_i(t) = \frac{1}{\beta_{n_p}(t)}\Gamma_i(t)$$

A lower bound on the adaptation gain is simply obtained by maintaining the values of the elements of the diagonal matrix $D(t)$ above some specified threshold d_0 as follows:

$$d_i(t) = \begin{Bmatrix} d_0 \text{ or } d_i(t-1) \text{ if } d_i(t) \leq d_0 \\ d_i(t) \qquad \text{otherwise} \end{Bmatrix} \tag{B.7}$$

Notice that the implementation of such an algorithm is indeed simple to legitimate its use.[5]

Reference

1. Bierman G (1977) Factorization methods for discrete sequential estimation. Academic Press, New York

[5]The function *udrls.m* (MATLAB) available from the book website implements this algorithm.

Appendix C
Interlaced Adaptive Regulation: Equations Development and Stability Analysis

C.1 Equations Development

From Eq. (13.61), rewritten here, the development of the equations in order to obtain Eq. (13.63) is shown next.

The *a priori* error is given by

$$\varepsilon^{\circ}(t+1) = \upsilon(t+1) + \frac{q^{-d}B^{*}H_{S_0}H_{R_0}}{P_0}\left[\frac{B_Q}{A_Q} - \frac{\hat{B}_Q(t)}{\hat{A}_Q(t)}\right]w(t) \tag{C.1}$$

For constant $\hat{B}_Q(t)$ and $\hat{A}_Q(t)$ or neglecting the non-commutativity of time-varying operators, (C.1) can be written[6]

$$\varepsilon^{\circ}(t+1) = \upsilon(t+1) + \frac{q^{-d}BH_{S_0}H_{R_0}}{P_0}\left[\frac{B_Q}{A_Q} - \frac{\hat{B}_Q(t)}{\hat{A}_Q(t)}\right]w(t+1), \tag{C.2}$$

observing that

$$u_Q(t+1) = \frac{B_Q}{A_Q}w(t+1) \tag{C.3}$$

$$= B_Q w(t+1) - A_Q^{*}u_Q(t) \tag{C.4}$$

$$= B_Q w(t+1) - A_Q^{*}\hat{u}_Q(t) - A_Q^{*}\left(u_Q(t) - \hat{u}_Q(t)\right) \tag{C.5}$$

[6]Taking advantage of the notation $B = q^{-1}B^{*}$ one can conveniently use the relation $Bw(t+1) = B^{*}w(t)$.

I.D. Landau et al., *Adaptive and Robust Active Vibration Control*, Advances in Industrial Control, DOI 10.1007/978-3-319-41450-8 and also

$$\hat{u}_Q(t+1) = \hat{B}_Q(t)w(t+1) - \hat{A}_Q(t)\hat{u}_Q(t), \tag{C.6}$$

then (C.1) becomes

$$\begin{aligned}\varepsilon^\circ(t+1) = \upsilon(t+1) + \frac{q^{-d}BH_{S_0}H_{R_0}}{P_0}[\left(B_Q - \hat{B}_Q(t)\right)w(t+1)- \\ - \left(A_Q^* - \hat{A}_Q^*(t)\right)\hat{u}_Q(t) - A_Q^*\left(u_Q(t) - \hat{u}_Q(t)\right)].\end{aligned} \tag{C.7}$$

One can then define the *a posteriori* error as:

$$\begin{aligned}\varepsilon(t+1) = \upsilon(t+1) + \frac{q^{-d}BH_{S_0}H_{R_0}}{P_0}[\left(B_Q - \hat{B}_Q(t+1)\right)w(t+1)- \\ - \left(A_Q^* - \hat{A}_Q^*(t+1)\right)\hat{u}_Q(t) - A_Q^*\left(u_Q(t) - \hat{u}_Q(t)\right)].\end{aligned} \tag{C.8}$$

It is necessary to find an expression relating the difference $u_Q(t) - \hat{u}_Q(t)$ to the *a posteriori* error $\varepsilon(t+1)$. The measured output of the system, $y(t)$, is given by

$$y(t) = \hat{y}_1(t) + p(t), \tag{C.9}$$

where $\hat{y}_1(t)$ is the process output with the adaptive YKIIR compensator and $p(t)$ is the effect of the disturbance. Under the assumption that the ideal YKIIR compensator $\frac{B_Q(q^{-1})}{A_Q(q^{-1})}$ completely cancels out the disturbance $p(t)$, Eq. (C.9) becomes

$$y(t) = \hat{y}_1(t) - y_1(t), \tag{C.10}$$

where $y_1(t) = -p(t)$ is the process output with the ideal YKIIR compensator $\frac{B_Q}{A_Q}$. One can define

$$\hat{y}_1(t) = -\frac{q^{-d}B}{A}\frac{1}{S_0}\left[R_0 y(t) + H_{S_0}H_{R_0}\hat{u}_Q(t)\right] \tag{C.11}$$

as the plant output with the estimated YKIIR compensator and

$$y_1(t) = -\frac{q^{-d}B}{A}\frac{1}{S_0}\left[0 + H_{S_0}H_{R_0}u_Q(t)\right] \tag{C.12}$$

as the plant output with the ideal YKIIR compensator ($y(t)$ is zero in this case). Introducing these equations in (C.10), one obtains

$$y(t) = -\frac{q^{-d}B}{A}\frac{R_0}{S_0}y(t) + \frac{q^{-d}B}{A}\frac{H_{S_0}H_{R_0}}{S_0}\left(u_Q(t) - \hat{u}_Q(t)\right) \tag{C.13}$$

$$\left[1+\frac{q^{-d}B}{A}\frac{R_0}{S_0}\right]y(t)=\frac{q^{-d}BH_{S_0}H_{R_0}}{AS_0}\left(u_Q(t)-\hat{u}_Q(t)\right). \tag{C.14}$$

Therefore, since $P_0 = AS_0 + q^{-d}BR_0$, one gets

$$\varepsilon(t)=y(t)=\frac{q^{-d}BH_{S_0}H_{R_0}}{P_0}(u_Q(t)-\hat{u}_Q(t)) \tag{C.15}$$

noting that $\varepsilon(t) = y(t)$ if $\hat{B}_Q(t)$ and $\hat{A}^*_Q(t)$ are used. One introduces also the notation

$$\hat{u}^f_Q(t)=\frac{q^{-d}BH_{S_0}H_{R_0}}{P_0}\hat{u}_Q(t). \tag{C.16}$$

Turning back to (13.61) and using (C.15) and (C.16) as well as (13.59) one obtains:

$$\begin{aligned}\varepsilon^\circ(t+1)&=\upsilon(t+1)+\left(B_Q-\hat{B}_Q(t)\right)w^f(t)-\\&-\left(A^*_Q-\hat{A}^*_Q(t)\right)\hat{u}^f_Q(t)-A^*_Q\varepsilon(t).\end{aligned} \tag{C.17}$$

The *a posteriori* error equation becomes

$$\begin{aligned}\varepsilon(t+1)&=\upsilon(t+1)+\left(B_Q-\hat{B}_Q(t+1)\right)w^f(t)-\\&-\left(A^*_Q-\hat{A}^*_Q(t+1)\right)\hat{u}^f_Q(t)-A^*_Q\varepsilon(t).\end{aligned} \tag{C.18}$$

The above equation can be rewritten as

$$\varepsilon(t+1)=\frac{1}{A_Q}\left[\theta_1^T-\hat{\theta}_1^T(t+1)\right]\phi_1(t)+\upsilon^f(t+1)+\upsilon_1(t+1), \tag{C.19}$$

where $\upsilon^f(t+1)$ and $\upsilon_1(t+1)=-\left(A^*_Q-\hat{A}^*_Q(t+1)\right)\hat{u}^f_Q(t)$ are vanishing signals because $\upsilon^f(t+1)$ is the output of an asymptotically stable filter whose input is a Dirac pulse and $\hat{A}^*_Q(t+1)\rightarrow A^*_Q$ as shown next.

C.2 Stability Analysis of Interlaced Scheme (Sketch)

C.2.1 Estimation of $\hat{A}_Q$

Taking into account the structure of the Eq. (13.46) and the results of Chap. 4 and [1], one can immediately conclude that

$$\lim_{t\to\infty} \varepsilon_{D_p}(t) = 0. \tag{C.20}$$

and

$$\lim_{t\to\infty} \tilde{\theta}^T_{D_p}(t+1)\phi_{D_p}(t) = 0. \tag{C.21}$$

where $\tilde{\theta}^T_{D_p}(t+1) = \hat{\theta}^T_{D_p}(t+1) - \theta^T_{D_p}$.

From (C.21) one gets:

$$\begin{aligned}\tilde{\theta}^T_{D_p}(t)\phi_{D_p}(t-1) &= \sum_{i=1}^{n-1}\left(\hat{p}(t-i) + \hat{p}(t-2n+i)\right)\tilde{\alpha}_i(t) + \hat{p}(t-n)\tilde{\alpha}_n(t)\\ &= \left(\sum_{i=1}^{n-1}\left(z^{-i} + z^{-2n+i}\right)\tilde{\alpha}_i(t) + z^{-n}\tilde{\alpha}_n(t)\right)\hat{p}(t)\\ &\to 0 \quad \text{as} \quad t\to\infty\end{aligned} \tag{C.22}$$

where $\{\tilde{\alpha}_i\}_1^n = \left\{\hat{\alpha}_i(t) - \alpha_i\right\}_1^n$.

Based on the assumption that $\hat{p}(t)$ has n independent frequency components, the *Frequency Richness Condition for Parameter Convergence* holds. Therefore, the only solution to the above equation is $\lim_{t\to\infty}\tilde{\alpha}_i(t) = 0$, i.e., the parameters converge to their true values.

Since $A_Q(z^{-1}) = D_p(\rho z^{-1})$, then $\hat{A}_Q(z^{-1}) = \hat{D}_p(\rho z^{-1})$, one concludes that:

$$\lim_{t\to\infty} \hat{A}_Q(z^{-1}) = A_Q(z^{-1}) \tag{C.23}$$

C.2.2 Estimation of $B_Q(z^{-1})$

In all the cases the equation for the *a posteriori* adaptation error takes the form

$$\nu(t+1) = H(q^{-1})[\theta_1 - \hat{\theta}_1(t+1)]\Phi_1(t) \tag{C.24}$$

which allows to use straightforwardly for stability analysis the results of Chap. 4, Sect. 4.4.2 and [1].

For each choice of a the regressor and of the adaptation error a different positive real condition has to be satisfied for assuring asymptotic stability. The various options and the stability conditions are summarized in Table 13.2.

Reference

1. Landau ID, Lozano R, M'Saad M, Karimi A (2011) Adaptive control, 2nd edn. Springer, London

Appendix D
Error Equations for Adaptive Feedforward Compensation

D.1 Derivation of Eq. (15.27)—Chap. 15

Under Assumption H2 (perfect matching condition), the output of the primary path can be expressed as:

$$x(t+1) = -z(t+1) = -G(q^{-1})u(t+1) \tag{D.1}$$

where $u(t+1)$ is a dummy variable given by:

$$u(t+1) = -\, S^*(q^{-1})u(t) + R(q^{-1})y(t+1) = \theta^T\varphi(t) = [\theta_S^T, \theta_R^T]\begin{bmatrix}\varphi_y(t)\\ \varphi_u(t)\end{bmatrix} \tag{D.2}$$

where

$$\begin{aligned}\varphi^T(t) &= [-u(t), \ldots - u(t-n_S+1), y(t+1), \ldots y(t-n_R+1)]\\ &= [\varphi_u^T(t), \varphi_y^T(t)]\end{aligned} \tag{D.3}$$

and $y(t+1)$ is given by

$$y(t+1) = w(t+1) + \frac{B_M^*(q^{-1})}{A_M(q^{-1})}u(t) \tag{D.4}$$

This is illustrated in Fig. D.1.

For a fixed value of the parameter vector $\hat{\theta}$ characterizing the estimated filter $\hat{N}(q^{-1})$ of same dimension as the optimal filter $N(q^{-1})$, the output of the secondary path can be expressed by (in this case $\hat{z}(t) = \hat{z}^\circ(t)$ and $\hat{u}(t) = \hat{u}^\circ(t)$):

$$\hat{z}(t) = G(q^{-1})\hat{u}(t) \tag{D.5}$$

I.D. Landau et al., *Adaptive and Robust Active Vibration Control*,
Advances in Industrial Control, DOI 10.1007/978-3-319-41450-8

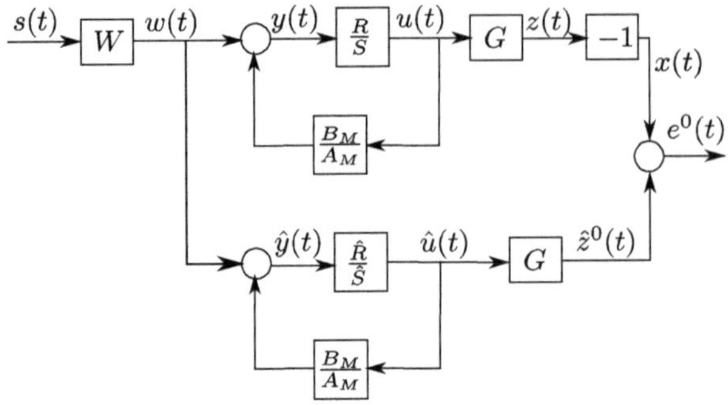

Fig. D.1 Equivalent representation of the system under the perfect matching hypothesis

where

$$\hat{u}(t+1) = \hat{\theta}^T \phi(t). \tag{D.6}$$

The key observation is that the dummy variable $u(t+1)$ can be expressed as:

$$\begin{aligned} u(t+1) &= \theta^T \phi(t) + \theta^T [\varphi(t) - \phi(t)] \\ &= \theta^T \phi(t) + \theta_S^T [\varphi_u - \phi_{\hat{u}}] + \theta_R^T [\varphi_y - \phi_{\hat{y}}] \end{aligned} \tag{D.7}$$

Define the dummy error (for a fixed vector $\hat{\theta}$)

$$\varepsilon(t+1) = u(t+1) - \hat{u}(t+1) \tag{D.8}$$

and the adaptation error

$$\nu(t+1) = -e(t+1) = z(t) - \hat{z}(t) = G(q^{-1})\varepsilon(t+1) \tag{D.9}$$

It results from (D.7) that:

$$u(t+1) = \theta^T \phi(t) - S^*(q^{-1})\varepsilon(t) + R(q^{-1})[y(t+1) - \hat{y}(t+1)] \tag{D.10}$$

But taking into account the expressions of $y(t)$ and $\hat{y}(t)$ given by (D.4) and (15.19), respectively, one gets:

$$u(t+1) = \theta^T \phi(t) - \left(S^*(q^{-1}) - \frac{R(q^{-1}) B_M^*(q^{-1})}{A_M(q^{-1})} \right) \varepsilon(t) \tag{D.11}$$

and therefore:

$$\varepsilon(t+1) = [\theta - \hat{\theta}]^T \phi(t) - \left(S^*(q^{-1}) - \frac{R(q^{-1}) B_M^*(q^{-1})}{A_M(q^{-1})} \right) \varepsilon(t) \tag{D.12}$$

This gives:

$$\frac{A_M S - B_M R}{A_M}\varepsilon(t+1) = [\theta - \hat{\theta}]^T \phi(t) \tag{D.13}$$

which can be rewritten as:

$$\varepsilon(t+1) = \frac{A_M(q^{-1})}{P(q^{-1})}[\theta - \hat{\theta}]^T \phi(t) \tag{D.14}$$

Taking now into account (D.9) one gets (15.27).

D.2 Adaptation Errors for Algorithm II—Chap. 15

For Algorithm II, the equation for the *a posteriori* error (15.34) becomes

$$\nu(t+1) = \frac{A_M G}{P\hat{G}}[\theta - \hat{\theta}(t+1)]^T \phi_f(t) \tag{D.15}$$

$$= \frac{A_M B_G^* \hat{A}_G}{P\hat{B}_G^* A_G}[\theta - \hat{\theta}(t+1)]^T \phi_f(t) \tag{D.16}$$

$$= \frac{b_1^G}{\hat{b}_1^G}\frac{A_M\left(B_G^*/b_1^G\right)\hat{A}_G}{P\left(\hat{B}_G^*/\hat{b}_1^G\right)A_G}[\theta - \hat{\theta}(t+1)]^T \phi_f(t) \tag{D.17}$$

One can now apply the result given in Chap. 4, Eqs. (4.125)–(4.131), with monic polynomials

$$H_1 = A_M\left(B_G^*/b_1^G\right)\hat{A}_G, \quad H_2 = P\left(\hat{B}_G^*/\hat{b}_1^G\right)A_G, \tag{D.18}$$

to obtain

$$\nu(t+1) = \frac{b_1^G}{\hat{b}_1^G}\left[\theta - \hat{\theta}(t+1)\right]^T \phi_f(t) + H_1^*(q^{-1})\left[\theta - \hat{\theta}(t)\right]^T \phi_f(t-1) - H_2^*(q^{-1})\nu(t). \tag{D.19}$$

and respectively

$$\nu^\circ(t+1) = \frac{b_1^G}{\hat{b}_1^G}\left[\theta - \hat{\theta}(t)\right]^T \phi_f(t) + H_1^*(q^{-1})\left[\theta - \hat{\theta}(t)\right]^T \phi_f(t-1) - H_2^*(q^{-1})\nu(t). \tag{D.20}$$

Therefore Eq. (15.36) will be exact if $b_1^G = \hat{b}_1^G$. This implies in practice that b_1^G and $\hat{b}_1^G$ should have the same sign and one needs to assume that their values are very close (which means that a good identification of G has been done). Same situation occurs for Algorithm III since one uses $\hat{G}$ instead of G.

D.3 Derivation of Eq. (15.86)—Chap. 15

For a fixed value of the parameter vector $\hat{\theta}$ characterizing the estimated filter $\hat{N}(q^{-1})$ of same dimension as the optimal filter $N(q^{-1})$, the output of the secondary path can be expressed by (in this case $\hat{z}(t) = \hat{z}^\circ(t)$, $\hat{u}(t) = \hat{u}^\circ(t)$ and $e(t) = e^\circ(t)$):

$$\hat{z}(t) = G\hat{u}(t) \tag{D.21}$$

with

$$\hat{u}(t) = \hat{u}_1(t) - \frac{B_K}{A_K}e(t) = \hat{u}_1(t) + \frac{B_K}{A_K}\nu(t), \tag{D.22}$$

where

$$\hat{u}_1(t+1) = \hat{\theta}^T\phi(t). \tag{D.23}$$

The key observation is that using [1, Eqs. (63)–(67)], the dummy variable $u(t+1)$ can be expressed as:

$$u(t+1) = \theta^T\phi(t) - S^*[u(t) - \hat{u}_1(t)] + R[y_1(t+1) - \hat{y}_1(t+1)]. \tag{D.24}$$

Define the dummy error (for a fixed vector $\hat{\theta}$)

$$\varepsilon(t+1) = u(t+1) - \hat{u}_1(t+1) - KG\varepsilon(t+1) \tag{D.25}$$

and the adaptation error becomes:

$$\nu(t+1) = -e(t+1) = -x(t+1) - \hat{z}(t+1) = G\varepsilon(t+1). \tag{D.26}$$

Taking into account the (D.22) and (D.26), $u(t+1)$ becomes:

$$u(t+1) = \theta^T\phi(t) - S^*[u(t) - \hat{u}(t) + \frac{B_K B_G}{A_K A_G}\varepsilon(t)] \\ + R[y_1(t+1) - \hat{y}_1(t+1)]. \tag{D.27}$$

It results from (D.27) by taking into account the expressions of $u_1(t)$ and $\hat{u}_1(t)$ given by (67) of [1] and (15.31) that:

$$u(t+1) = \theta^T \phi(t) - \left[S^*(1 + \frac{B_K B_G}{A_K A_G}) - \frac{R(q^{-1}) B_M^*}{A_M} \right] \varepsilon(t). \tag{D.28}$$

Using Eqs. (D.22) and (D.25), one gets (after passing all terms in ε on the left hand side):

$$\varepsilon(t+1) = \frac{A_M A_G A_K}{P_{fb-ff}} [\theta - \hat{\theta}]^T \phi(t). \tag{D.29}$$

Taking now into account Eq. (D.26) one obtains Eq. (15.86).

D.4 Derivation of Eq. (16.16)—Chap. 16

Using hypothesis H2″ (Sect. 16.3), one can construct an equivalent closed-loop system for the primary path as in Fig. D.2.

Considering a $Q(q^{-1})$ filter as in (16.2), the polynomial $S(q^{-1})$ given in (16.1) can be rewritten as

$$S(q^{-1}) = 1 + q^{-1} S^* = 1 + q^{-1}((A_Q S_0)^* - B_Q B_M^*). \tag{D.30}$$

Under hypothesis H2″ (perfect matching condition), the output of the primary path can be expressed as

$$x(t) = -z(t) = -G(q^{-1}) u(t) \tag{D.31}$$

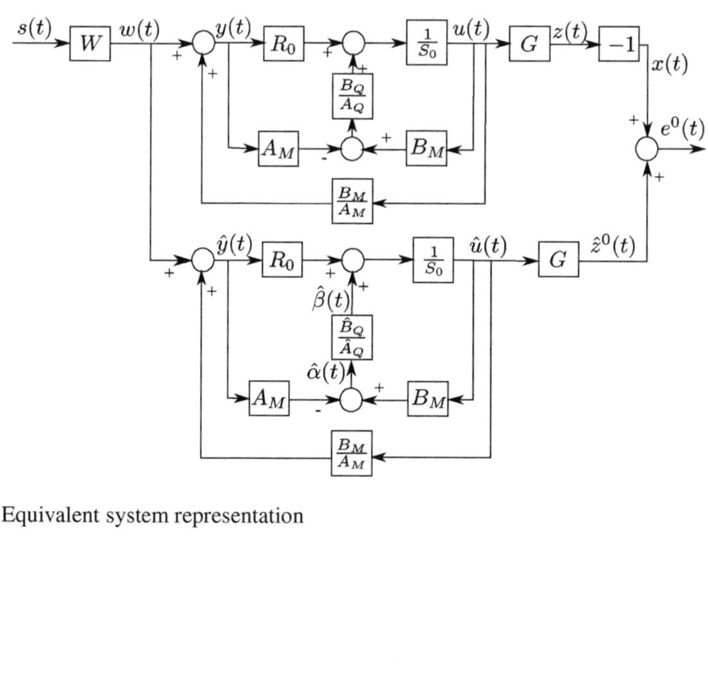

Fig. D.2 Equivalent system representation

and the input to the Youla–Kučera compensator as

$$y(t+1) = w(t+1) + \frac{B_M}{A_M} u(t+1) \tag{D.32}$$

where $u(t)$ is a dummy variable given by

$$\begin{aligned} u(t+1) &= -\ S^* u(t) + R y(t+1) \\ &= -\ ((A_Q S_0)^* - B_Q B_M^*) u(t) + (A_Q R_0 - B_Q A_M) y(t+1) \\ &= -\ (A_Q S_0)^* u(t) + A_Q R_0 y(t+1) + B_Q \left(B_M^* u(t) - A_M y(t+1) \right). \end{aligned} \tag{D.33}$$

Similarly, the output of the adaptive feedforward filter (for a fixed $\hat{Q}$) is given by

$$\hat{u}(t+1) = -\ (\hat{A}_Q S_0)^* \hat{u}(t) + \hat{A}_Q R_0 \hat{y}(t+1) + \hat{B}_Q \left(B_M^* \hat{u}(t) - A_M \hat{y}(t+1) \right). \tag{D.34}$$

The output of the secondary path is

$$\hat{z}(t) = G(q^{-1}) \hat{u}(t). \tag{D.35}$$

Define the dummy error (for a fixed estimated set of parameters)

$$\varepsilon(t) = -u(t) + \hat{u}(t) \tag{D.36}$$

and the adaptation error

$$\nu(t) = -e(t) = -(-z(t) + \hat{z}(t)) = -G(q^{-1})\varepsilon(t)). \tag{D.37}$$

Equation (D.33) can be rewritten as

$$\begin{aligned} u(t+1) = -\ & (A_Q S_0)^* \hat{u}(t) + A_Q R_0 \hat{y}(t+1) \\ & + B_Q (B_M^* \hat{u}(t) - A_M \hat{y}(t+1)) - (A_Q S_0)^* (u(t) - \hat{u}(t)) \\ & + A_Q R_0 (y(t+1) - \hat{y}(t+1)) \\ & + B_Q [B_M^* (u(t) - \hat{u}(t)) - A_M (y(t+1) - \hat{y}(t+1))]. \end{aligned} \tag{D.38}$$

Taking into consideration Eqs. (16.8), (D.32)

$$\begin{aligned} & B_Q [B_M^* (u(t) - \hat{u}(t)) - A_M (y(t+1) - \hat{y}(t+1))] = \\ & \qquad = B_Q \left[B_M^* \varepsilon(t) - A_M \frac{B_M^*}{A_M} \varepsilon(t) \right] = 0 \end{aligned} \tag{D.39}$$

and subtracting (D.34) from (D.38) one obtains

$$\begin{aligned}\varepsilon(t+1) = &-((-A_Q+\hat{A}_Q)S_0)^*\hat{u}(t)+(-A_Q+\hat{A}_Q)R_0\hat{y}(t+1) \\ &+(-B_Q+\hat{B}_Q)[B_M^*\hat{u}(t)-A_M\hat{y}(t+1)] \\ &-(A_QS_0)^*\varepsilon(t)+A_QR_0\frac{B_M^*}{A_M}\varepsilon(t).\end{aligned} \tag{D.40}$$

Passing the terms in $\varepsilon(t)$ on the left hand side, one gets:

$$\begin{aligned}\left[1+q^{-1}\left(\frac{A_M(A_QS_0)^*-A_QR_0B_M^*}{A_M}\right)\right]\varepsilon(t+1) &= \frac{A_QP_0}{A_M}\varepsilon(t+1) \\ &=(-A_Q^*+\hat{A}_Q^*)[-S_0\hat{u}(t)+R_0\hat{y}(t)] \\ &\quad+(-B_Q+\hat{B}_Q)[B_M\hat{u}(t+1)-A_M\hat{y}(t+1)]\end{aligned} \tag{D.41}$$

Using Eqs. (D.37) and (16.18) one gets Eq. (16.16).

Reference

1. Landau I, Alma M, Airimitoaie T (2011) Adaptive feedforward compensation algorithms for active vibration control with mechanical coupling. Automatica 47(10):2185–2196. doi:10.1016/j.automatica.2011.08.015

Appendix E
"Integral + Proportional" Parameter Adaptation Algorithm

"Integral + Proportional" parameter adaptation algorithms (IP-PAA) [1–3] should be considered in the context of AVC for two reasons:

- It allows to remove or to relax the positive real conditions for stability.
- It may accelerate the adaptation transients.

E.1 The Algorithms

The equations for the development of the *Integral + proportional* adaptation for adaptive feedforward compensation are identical to those given in Chap. 15 up to Eq. (15.35).

The specificity of the IP-PAA is that the estimated parameter vector $\hat{\theta}(t)$ is at each instant the sum of two components

$$\hat{\theta}(t) = \hat{\theta}_I(t) + \hat{\theta}_P(t), \tag{E.1}$$

where $\hat{\theta}_I(t)$ is the *integral* component generated through the type of algorithm introduced in Chap. 4 (these algorithms have memory) and a *proportional* component $\hat{\theta}_P(t)$ generated by an adaptation algorithm without memory.

The following IP-PAA is proposed:

$$\hat{\theta}_I(t+1) = \hat{\theta}_I(t) + \xi(t)F_I(t)\Phi(t)\nu(t+1), \tag{E.2a}$$

$$\hat{\theta}_P(t+1) = F_P(t)\Phi(t)\nu(t+1), \tag{E.2b}$$

$$\nu(t+1) = \frac{\nu^{\circ}(t+1)}{1 + \Phi^T(t)(\xi(t)F_I(t) + F_P(t))\Phi(t)}, \tag{E.2c}$$

I.D. Landau et al., *Adaptive and Robust Active Vibration Control*,
Advances in Industrial Control, DOI 10.1007/978-3-319-41450-8

$$F_I(t+1) = \frac{1}{\lambda_1(t)}\left[F_I(t) - \frac{F_I(t)\Phi(t)\Phi^T(t)F_I(t)}{\frac{\lambda_1(t)}{\lambda_2(t)} + \Phi^T(t)F_I(t)\Phi(t)}\right], \tag{E.2d}$$

$$F_P(t) = \alpha(t)F_I(t);\ \alpha(t) > -0.5, \tag{E.2e}$$

$$F(t) = \xi(t)F_I(t) + F_P(t), \tag{E.2f}$$

$$\xi(t) = 1 + \frac{\lambda_2(t)}{\lambda_1(t)}\Phi^T(t)F_P(t)\Phi(t), \tag{E.2g}$$

$$\hat{\theta}(t+1) = \hat{\theta}_I(t+1) + \hat{\theta}_P(t+1), \tag{E.2h}$$

$$0 < \lambda_1(t) \le 1,\ 0 \le \lambda_2(t) < 2,\ F_I(0) > 0, \tag{E.2i}$$

$$\Phi(t) = \phi_f(t), \tag{E.2j}$$

where $\nu(t+1)$ is the (filtered) adaptation error, $\lambda_1(t)$ and $\lambda_2(t)$ allow to obtain various profiles for the matrix adaptation gain $F_I(t)$ (see Sect. 4.3.4 and [1] for more details). For $\alpha(t) \equiv 0$, one obtains the algorithm with integral adaptation gain introduced in Sect. 4.3.3 (see also [4]). A detailed stability analysis can be found in [2].

The sufficient positive real conditions given in Chap. 15 for the *integral* type adaptation can be relaxed when using *integral + proportional* adaptation.

E.2 Relaxing the Positive Real Condition

One has the following result [2]:

Theorem E.1 *The adaptive system described by Eqs. (15.34), 15.44) and (E.2) for $\lambda_2(t) \equiv 0$ and $\lambda_1(t) \equiv 1$ is asymptotically stable provided that:*

(T1) It exists a gain K such that $\frac{H}{1+KH}$ is SPR,
(T2) The adaptation gains F_I and $F_P(t)$ and the observation vector $\Phi(t)$ satisfy

$$\sum_{t=0}^{t_1}\left[\Phi^T(t-1)\left(\frac{1}{2}F_I + F_P(t-1)\right)\Phi(t-1) - K\right]\nu^2(t) \ge 0 \tag{E.3}$$

for all $t_1 \ge 0$ or

$$\Phi^T(t)\left(\frac{1}{2}F_I + F_P(t)\right)\Phi(t) > K > 0, \tag{E.4}$$

for all $t \ge 0$.

The proof is given in [2]. The condition T1 is the consequence of the following result [2]:

Given the discrete transfer function

$$H(z^{-1}) = \frac{B(z^{-1})}{A(z^{-1})} = \frac{b_0 + b_1 z^{-1} + \cdots + b_{n_B} z^{-n_B}}{1 + a_1 z^{-1} + \cdots + a_{n_A} z^{-n_A}}, \tag{E.5}$$

under the hypotheses:

(H1) $H(z^{-1})$ has all its zeros inside the unit circle,
(H2) $b_0 \neq 0$,

there exists a positive scalar gain K such that $\frac{H}{1+KH}$ is SPR.

It is interesting to note that condition (E.3) implies that the regressor vector has the property

$$\sum_{t=0}^{t_1} \left[\Phi^T(t-1)\Phi(t-1)\right] > \varepsilon > 0, \tag{E.6}$$

which means that the trace of the covariance matrix of the regressor vector is positive, i.e., that the energy of the signal is greater than zero. The magnitude of the proportional gain will depend on how far the transfer function is from a SPR transfer function (level of K) and what is the energy of the regressor (which depends upon the disturbance).

E.3 Experimental Results

The AVC system considered in Chap. 15, has been used to carry on the experiment (see also Sect. 2.3). The adaptive feedforward compensator structure for the experiments has been $n_R = 3$, $n_S = 4$. A PRBS excitation on the global primary path will be considered as the disturbance. For the *adaptive* operation, algorithm FUPLR has been used with scalar adaptation gain ($\lambda_1(t) = 1$, $\lambda_2(t) = 0$). A variable $\alpha(t)$ in the IP-PAA has been chosen, starting with an initial value of 200 and linearly decreasing to 100 (over a horizon of 25 s). Time domain results obtained on the AVC system are shown in Fig. E.1. The advantage of using an IP-PAA is an overall improvement of the transient behaviour despite that the SPR condition on

$$H(q^{-1}) = \frac{A_M G}{P\hat{G}}$$

is not satisfied (the SPR condition is not satisfied around 83 Hz and around 116 Hz as shown in Fig. E.2). The improvement of performance can be explained by the relaxation of the SPR condition when using IP adaptation.

Figure E.3 shows the comparison between "Integral" and "Integral + Proportional" adaptation over an horizon of 1500 s (Fig. E.1 is a zoom of Fig. E.3 covering only the first 30 s after the introduction of the adaptive feedforward compensator). One can see that the various "spikes" which are obtained when using "Integral"

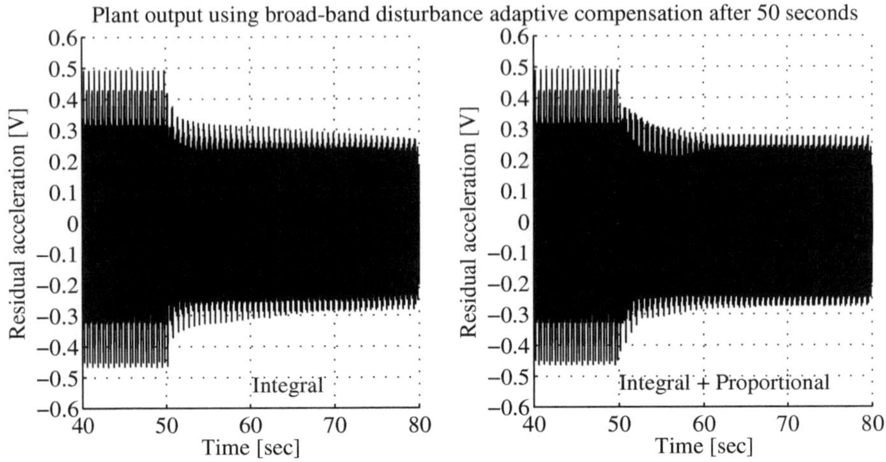

Fig. E.1 Real time results obtained with algorithm FUPLR using "Integral" scalar adaptation gain (*left*) and "Integral + Proportional" scalar adaptation gain (*right*)

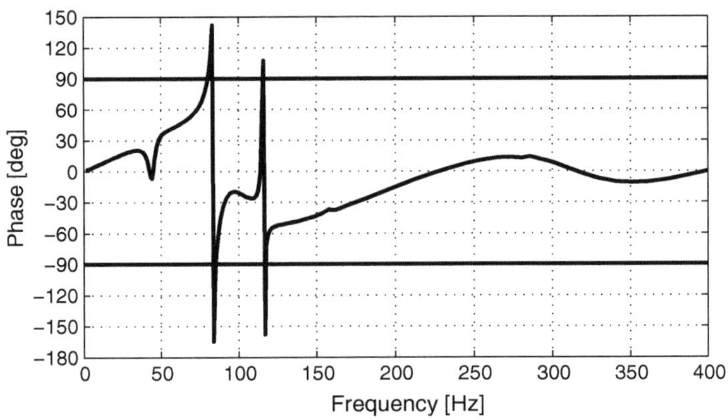

Fig. E.2 Phase of estimated $H(z^{-1})$ for FUPLR

adaptation and certainly caused by the violation of the SPR condition at some frequencies, are strongly attenuated when using "Integral + Proportional" adaptation. The attenuation obtained for the IP adaptation over the last 10 s shown in Fig. E.3 is of 13.45 dB, while for the I adaptation one has 12.99 dB. It is clear that IP adaptation gives better results even on a long run. For other experimental results see [5].

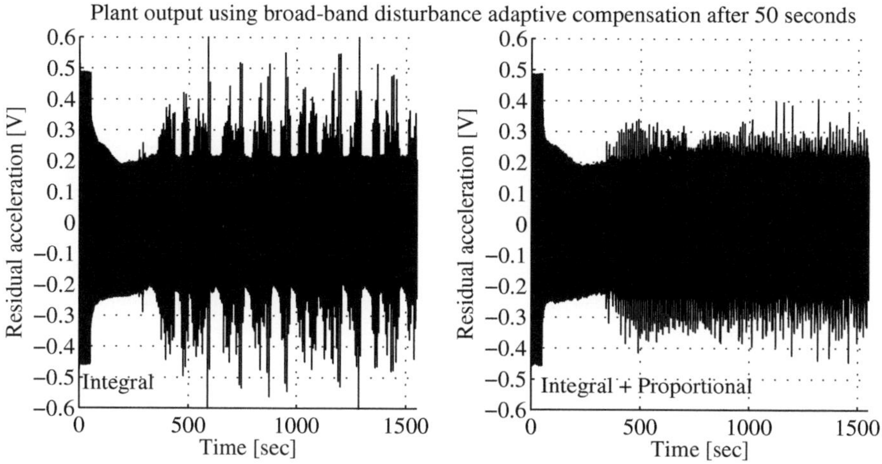

Fig. E.3 Real time results obtained with FUPLR using "Integral" scalar adaptation gain (*left*) and "Integral + Proportional" scalar adaptation gain (*right*) over 1500 s

References

1. Landau ID, Lozano R, M'Saad M, Karimi A (2011) Adaptive control, 2nd edn. Springer, London
2. Airimitoaie TB, Landau ID (2013) Improving adaptive feedforward vibration compensation by using integral+proportional adaptation. Automatica 49(5):1501–1505. doi:10.1016/j.automatica.2013.01.025
3. Tomizuka M (1982) Parallel MRAS without compensation block. IEEE Trans Autom Control 27(2):505–506. doi:10.1109/TAC.1982.1102907
4. Landau I, Alma M, Airimitoaie T (2011) Adaptive feedforward compensation algorithms for active vibration control with mechanical coupling. Automatica 47(10):2185–2196. doi:10.1016/j.automatica.2011.08.015
5. Airimitoaie TB (2012) Robust design and tuning of active vibration control systems. Ph.D. thesis, University of Grenoble, France, and University "Politehnica" of Bucharest, Romania

Index

I.D. Landau et al., *Adaptive and Robust Active Vibration Control*,
Advances in Industrial Control, DOI 10.1007/978-3-319-41450-8

MIX
Papier aus verantwortungsvollen Quellen
Paper from responsible sources
FSC® C105338

If you have any concerns about our products,
you can contact us on
ProductSafety@springernature.com

In case Publisher is established outside the EU,
the EU authorized representative is:
Springer Nature Customer Service Center GmbH
Europaplatz 3, 69115 Heidelberg, Germany

Printed by Libri Plureos GmbH
in Hamburg, Germany